# Springer Tracts in Modern Physics 110

W0079347

Editor: G. Höhler
Associate Editor: E. A. Niekisch

# Springer Tracts in Modern Physics

85 **Solid Surface Physics** With contributions by J. Hölzl, F. K. Schulte, H. Wagner

86 **Electron-Positron Interactions** By B. H. Wiik, G. Wolf

87 **Point Defects in Metals II:** Dynamical Properties and Diffusion Controlled Reactions With contributions by P. H. Dederichs, K. Schroeder, R. Zeller

88 **Excitation of Plasmons and Interband Transitions by Electrons** By H. Raether

89 Giant Resonance Phenomena in **Intermediate-Energy Nuclear Reactions** By F. Cannata, H. Überall

90* **Jets of Hadrons** By W. Hofmann

91 **Structural Studies of Surfaces** With contributions by K. Heinz, K. Müller, T. Engel, and K. H. Rieder

92 **Single-Particle Rotations in Molecular Crystals** By W. Press

93 **Coherent Inelastic Neutron Scattering in Lattice Dynamics** By B. Dorner

94 **Exciton Dynamics in Molecular Crystals and Aggregates** With contributions by V. M. Kenkre and P. Reineker

95 **Projection Operator Techniques in Nonequilibrium Statistical Mechanics** By H. Grabert

96 **Hyperfine Structure in 4d- and 5d-Shell Atoms** By S. Büttgenbach

97 **Elements of Flow and Diffusion Processes in Separation Nozzles** By W. Ehrfeld

98 **Narrow-Gap Semiconductors** With contributions by R. Dornhaus, G. Nimtz, and B. Schlicht

99 **Dynamical Properties of IV–VI Compounds** With contributions by H. Bilz, A. Bussmann-Holder, W. Jantsch, and P. Vogl

100* **Quarks and Nuclear Forces** Edited by D. C. Fries and B. Zeitnitz

101 **Neutron Scattering and Muon Spin Rotation** With contributions by R. E. Lechner, D. Richter, and C. Riekel

102 **Theory of Jets in Electron-Positron Annihilation** By G. Kramer

103 **Rare Gas Solids** With contributions by H. Coufal, E. Lüscher, H. Micklitz, and R. E. Norberg

104 **Surface Enhanced Raman Vibrational Studies at Solid/Gas Interfaces** By I. Pockrand

105 **Two-Photon Physics at $e^+e^-$ Storage Rings** By H. Kolanoski

106 **Polarized Electrons at Surfaces** By J. Kirschner

107 **Electronic Excitations in Condensed Rare Gases** By N. Schwentner, E.-E. Koch, and J. Jortner

108 **Particles and Detectors** Festschrift for Jack Steinberger Edited by K. Kleinknecht and T. D. Lee

109 **Metal Optics Near the Plasma Frequency** By F. Forstmann and R. R. Gerhardts

110* **Electrodynamics of the Semiconductor Band Edge** By A. Stahl and I. Balslev

111 **Surface Plasmons on Smooth and Rough Surfaces and on Gratings** By H. Raether

112 **Recent Experimental Tests of the Standard Theory of Electroweak Interactions** By C. Kiesling

* denotes a volume which contains a Classified Index starting from Volume 36

A. Stahl I. Balslev

# Electrodynamics of the Semiconductor Band Edge

With 42 Figures

Springer-Verlag Berlin
Heidelberg GmbH

Professor Dr. Arne Stahl

Institut für Theoretische Physik, Rheinisch-Westfälische Technische Hochschule
D-5100 Aachen, Fed. Rep. of Germany

Professor Dr. Ivar Balslev

Fysisk Institut, Odense Universitet, DK-5230 Odense M., Denmark

---

*Manuscripts for publication should be addressed to:*

Gerhard Höhler

Institut für Theoretische Kernphysik der Universität Karlsruhe
Postfach 6380, D-7500 Karlsruhe 1, Fed. Rep. of Germany

*Proofs and all correspondence concerning papers in the process of publication should be addressed to:*

Ernst A. Niekisch

Haubourdinstrasse 6, D-5170 Jülich 1, Fed. Rep. of Germany

---

Library of Congress Cataloging-in-Publication Data. Stahl, A. (Arne), 1931– Electrodynamics of the semiconductor band edge. (Springer tracts in modern physics; 110) Includes index. 1. Energy-band theory of solids. 2. Semiconductors. 3. Exciton theory. 4. Polaritons. 5. Electrodynamics. I. Balslev, I. II. Title. III. Title: Band edge. IV. Series. QC1.S797 vol. 110 539.7 s 86-24874 [QC176.8.E4]



ISBN 978-3-662-15170-9 ISBN 978-3-540-47178-3 (eBook)
DOI 10.1007/978-3-540-47178-3



Originally published by Springer-Verlag Berlin Heidelberg New York in 1987.
Softcover reprint of the hardcover 1st edition 1987



2153/3150-543210

# Preface

The last decade of semiconductor research has been characterized by a very strong interaction with the development of optoelectronic devices, and there are no signs of saturation in this trend. The continued interest in electrical and optical phenomena is not only confined to intraband transport processes and optical interband transitions but also extends to combined excitations as relevant for nonlinear optical response, optical response of photoexcited systems and injection lasers.

So far, the theoretical basis of this field has been marked by a predominance of microscopic methods making ample use of Green's-function techniques. In contrast to this, the present book emphasizes a new approach based on macroscopic constitutive equations. This approach contains essential information on electronic coherence and is thereby well suited for studying processes in which different interacting wave-like entities have comparable length scales. For typical semiconductors the length scale of 50–1000 Å is common to the exciton Bohr radius, the wavelengths of electrons and holes, and the wavelength of light near excitonic resonances. A full understanding of phenomena with this characteristic length scale is particularly important in view of the rapid development of materials research based on superlattices, quantum wells and space-charge layers near junctions.

The prerequisites necessary for handling the tools presented in this book are relatively modest in comparison with the above-mentioned microscopic methods. We hope that this will be appreciated especially by experimentally oriented physicists in this field. Although the book appears in a review series we have occasionally abandoned the review style in order to make the presentation as self-contained as possible. Thus some of the sections in the main text and especially the numerous appendices are added mainly for pedagogical reasons. We feel that these textbook-like additions are necessary in view of the fact that a substantial part of the material presented in the book has not been published elsewhere.

The authors would like to thank the University of Odense and the Rheinisch-Westfälische Technische Hochschule for hospitality and support during the authors' long mutual visits. The typing of the original manuscript was kindly and patiently done by Birgit Sørensen (OU), Söss Terkelsen (OU) and Josefine

Elbert (RWTH). Also acknowledged is the skilful art work by Tove Nyberg (OU). Finally the authors are indebted to Bob Collins for his gentle linguistic assistance.

Aachen and Odense,
January 1987

*Arne Stahl*
*Ivar Balslev*

# Contents

# 1. Introduction

The topic of the present book is the electrodynamics in semiconductors with focus on electronic transitions at the band edge. In a simple two-band model of a semiconductor we classify the transitions as interband or intraband processes (Fig. 1.1). The electrodynamics to be discussed is therefore concerned with three main topics:

a) Intraband dynamics covering the field of transport processes, i.e., the response of mobile carriers.
b) Interband dynamics involving the creation or annihilation of electron-hole pairs considered as virtual or real excitations.
c) Mixed type processes involving coupled interband and intraband degrees of freedom as encountered in nonlinear optics.

All three categories are well suited for a description with much emphasis on macroscopic coherence.

In the present book we shall make extensive use of the recently developed coherent wave theory (Stahl, 1979, 1984; Huhn and Stahl, 1984). Until now this approach has given new insight in important areas such as nonlinear response and half-space dynamics of excitonic and plasmonic excitations. The coherent

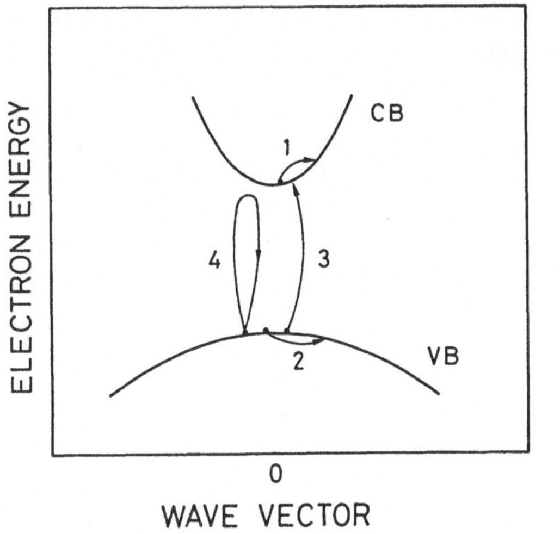


**Fig. 1.1.** Two band model of a semiconductor; electronic transitions in the model are classified as intraband processes (*1,2*) or as interband processes (*3,4*). In the context of a perturbative treatment one often further distinguishes between real (*3*) and virtual (*4*) interband transitions

wave theory is still in its developing stage, and so it is appropriate to give a broad and systematic review covering not only the approach itself and previous applications, but also potential areas of future applications.

Before we characterize the structure of the coherent edge dynamics let us first discuss electrodynamics in general. Macroscopic electrodynamics of condensed matter is based on Maxwell's field equations supplemented by relations describing the material response. These equations will be considered as constitutive equations in the sense of macroscopic electrodynamics, despite the fact that they sometimes reproduce also certain features commonly considered as "microscopic", as e.g. excitonic resonances. The field equations and the material equations forming the basis of electrodynamics are of very different nature. The field equations being founded on invariance principles such as gauge invariance are exact to any known degree of accuracy. The material equations are derived from approximate models of the underlying many-particle system and are always open to improvement.

## 1.1 Field Equations

The field equations giving the sources and vortices of the two fields $\boldsymbol{E}$ and $\boldsymbol{B}$ are

$$\begin{aligned} &\varepsilon_0 \nabla \cdot \boldsymbol{E} = \varrho_{tot} && \nabla \cdot \boldsymbol{B} = 0 \\ &\nabla \times \boldsymbol{E} = -\dot{\boldsymbol{B}} && \mu_0^{-1} \nabla \times \boldsymbol{B} = \varepsilon_0 \dot{\boldsymbol{E}} + \boldsymbol{J}_{tot} \ . \end{aligned} \tag{1.1}$$

It is often convenient to decompose the total charge $\varrho_{tot}$ and the total current $\boldsymbol{J}_{tot}$ into specified contributions:

$$\varrho_{tot} = \varrho_m + \varrho_p \tag{1.2a}$$

$$\boldsymbol{J}_{tot} = \boldsymbol{J}_m + \boldsymbol{J}_p + \boldsymbol{J}_M + \boldsymbol{J}_s \ . \tag{1.2b}$$

$\varrho_m$ is the density of monopolar charges, especially it comprises the charge of mobile charge carriers and ionized impurities. $\varrho_p$ is the polarization charge density related to electric multipoles. Thus,

$$\varrho_p = -\nabla \cdot \boldsymbol{P} \tag{1.3}$$

where $\boldsymbol{P}$ is the polarization density. The conduction current $\boldsymbol{J}_m$ is the current of mobile monopoles and is related to $\varrho_m$ by a conservation law

$$\dot{\varrho}_m + \nabla \cdot \boldsymbol{J}_m = 0 \ . \tag{1.4}$$

The polarization current

$$\boldsymbol{J}_p = \dot{\boldsymbol{P}} \tag{1.5}$$

satisfies the continuity equation

$$\dot{\varrho}_p + \nabla \cdot \boldsymbol{J}_p = 0 \ . \tag{1.6}$$

The further terms in (1.2b), viz. the magnetization current $\boldsymbol{J}_M$ and the supercurrent $\boldsymbol{J}_s$, have only been mentioned for the sake of completeness. They will not be further discussed in the present book.

The decomposition (1.2) only makes sense in the context of a specific model of matter. In our case this will be a two-band semiconductor. The distinction between mobile charges contributing to $\varrho_m, \boldsymbol{J}_m$ and bound charges contributing to $\varrho_p, \boldsymbol{J}_p$ coincides in this model with the distinction between intraband and interband processes.

## 1.2 Intraband Dynamics

The simplest form of a material equation is Ohm's law: $\boldsymbol{J}_m = \sigma \boldsymbol{E}$ where $\sigma$ denotes the steady state conductivity. The dynamic properties are accounted for by Drude's equation (Drude, 1900; Grosse, 1979)

$$\tau\left(\frac{d}{dt}+\frac{1}{\tau}\right)\boldsymbol{J}_m = \sigma \boldsymbol{E} \tag{1.7}$$

where $\tau$ is a relaxation time. A next step in the refinement is the "hydrodynamic" plasma theory with the constitutive equation (c.f. Sect. 6.5)

$$\tau\left(\frac{d}{dt}+\frac{1}{\tau}\right)\boldsymbol{J}_m + \beta\nabla\varrho_m = \sigma \boldsymbol{E} \ . \tag{1.8}$$

Here $\beta$ is a dispersion constant. In the form (1.8) the constitutive equation accounts for the dispersion of plasma waves.

A remarkable further refinement is achieved by going over to the kinetic level. The classical version of kinetic theory is based on a phase-space distribution function $g(\boldsymbol{r}, \boldsymbol{v}, t)$ with $\boldsymbol{r}$ and $\boldsymbol{v}$ being position and velocity. Then

$$\boldsymbol{J}_m = \int g(\boldsymbol{r}, \boldsymbol{v}, t)\boldsymbol{v}\, d^3\boldsymbol{v} \ ; \quad \varrho_m = \int g(\boldsymbol{r}, \boldsymbol{v}, t) d^3\boldsymbol{v} \ . \tag{1.9}$$

The simplest model equation for $g$ is of the form

$$\dot{g} + \boldsymbol{v}\nabla_{\boldsymbol{r}} g - \frac{e}{m_e}\boldsymbol{E}\nabla_{\boldsymbol{v}} g = -\tau^{-1}(g - g_0) \ . \tag{1.10}$$

$g_0$ is the equilibrium distribution, $e$ is the elementary charge and $m_e$ is the effective mass. Considering (1.10) as an extended form of a constitutive equation it is appropriate to interpret $e/m_e$ as a phenomenological parameter.

A further generalization is necessary when quantum coherence is involved. The distribution function $g$ must then be considered as the Wigner transform of the density matrix $C$ (Sect. 6.1). In the real space representation $C(\boldsymbol{r}_1, \boldsymbol{r}_2)$ is a function of two positions. Whenever $C(\boldsymbol{r}_1, \boldsymbol{r}_2) \neq 0$ there exists coherence in quantum mechanical sense between the two points $\boldsymbol{r}_1, \boldsymbol{r}_2$. Appearing in the context of macrophysics, the coherence property of $C$ is related to macroscopic quantum effects. This idea is known from the theory of superfluidity where off-diagonal long range order (ODLRO) in the density matrix characterizes the superfluid state (Penrose and Onsager, 1956; Yang, 1962). The transport equation for the density matrix generalizing (1.10) is of the form (Sect. 3.2, 7)

$$\dot{C} + i\left[\frac{\hbar}{2m_e}(\nabla_1^2 - \nabla_2^2) + \frac{e}{\hbar}\Big(\phi(\boldsymbol{r}_1) - \phi(\boldsymbol{r}_2)\Big)\right]C = (\dot{C})_{irr} \tag{1.11}$$

where $\boldsymbol{E} = -\nabla\phi$; $(\dot{C})_{irr}$ denoting the change rate of $C$ due to irreversible processes, replaces the relaxation term in (1.10).

As will be shown in Chaps. 2 and 3, the quantum kinetic theory provides an appropriate interface for merging interband and intraband dynamics of the semiconductor band edge.

## 1.3 Interband Dynamics

Turning to a survey of dielectric constitutive equations related to interband processes we begin with the simple form $\boldsymbol{P} = \varepsilon_0\chi\boldsymbol{E}$ with a constant susceptibility corresponding to the static limit. The dynamic case is most conveniently described in a Fourier representation such that

$$\boldsymbol{P}^\omega = \varepsilon_0\chi(\omega)\boldsymbol{E}^\omega \tag{1.12}$$

describes the response at frequency $\omega$. A typical frequency dependence of $\chi$ consists of a series of excitonic resonances followed by an absorption continuum above the gap frequency $\omega_g$ (see Fig. 4.1). A function with these properties is

$$\chi(\omega) = \sum_n \frac{f_n}{\omega_n^2 - \omega^2 - i\omega\Gamma_n} + \int_{\omega_g}^{\infty} \frac{a(\omega')d\omega'}{(\omega')^2 - \omega^2} . \tag{1.13}$$

In an equation-of-motion approach the discrete part of the spectrum is generated from a series of oscillators (see Sect. 4.8)

$$\boldsymbol{P} = \sum \boldsymbol{P}_n \tag{1.14a}$$

$$\ddot{\boldsymbol{P}}_n + \omega_n^2\boldsymbol{P}_n + \Gamma_n\dot{\boldsymbol{P}}_n = \varepsilon_0 f_n\boldsymbol{E} . \tag{1.14b}$$

An extension to nonlocal response is easily obtained by adding on the l.h.s. of (1.14b) a dispersive term $\beta_n\nabla^2\boldsymbol{P}_n$ analogous to (1.8).

In order to generate an absorption continuum the field must couple to a wave propagation system equivalent to a continuum of oscillators. Such a mechanism is provided by the relative motion of the electron-hole pair. We therefore shall consider a macroscopic bilocal pair wave function $Y(\boldsymbol{r}_h, \boldsymbol{r}_e)$. As will be shown in Chaps. 2, 3 the linear response of $Y$ to the electric field $\boldsymbol{E}$ is described by a wave equation in electron-hole configuration space:

$$\dot{Y} + i\left(\omega_g - \frac{\hbar}{2m_h}\nabla_h^2 - \frac{\hbar}{2m_e}\nabla_e^2 + \hbar^{-1}V_{eh}(\boldsymbol{r}_e - \boldsymbol{r}_h)\right)Y$$

$$= i\hbar^{-1}M_0\boldsymbol{E}\delta(\boldsymbol{r}_e - \boldsymbol{r}_h) \ . \tag{1.15}$$

In the spirit of macroscopic electrodynamics the parameters in (1.15) are the phenomenological constants $\omega_g$, $\hbar m_h^{-1}$, $\hbar m_e^{-1}$, $M_0^2\hbar^{-1}$. The function $\hbar^{-1}V_{eh}$ describes the electron-hole interaction. The dielectric response is related to the real part of $Y$ on the diagonal ($\boldsymbol{r}_e = \boldsymbol{r}_h$):

$$P(\boldsymbol{r}) = 2M_0\,\mathrm{Re}\{Y(\boldsymbol{r},\boldsymbol{r})\} \ . \tag{1.16}$$

By solving (1.15) and applying (1.16) one finds the contribution of the band edge to real and imaginary part of $\chi(\omega)$. The structure of $\chi$ exhibits all the typical features as the excitonic resonances and the absorption continuum above $\omega_g$ (Sect. 4.5). Thus (1.15) can be considered as the interband constitutive equation of the band edge in a similar sense as the oscillator equation (1.14b) characterizes a single resonance.

The use of the bilocal amplitude $Y$ and the constitutive equation (1.15) is particularly useful when studying wave propagation perturbed by obstacles such as point charges, phonons, surfaces, space charge layers etc. because the interactions between these obstacles and the polariton waves are particularly easy to express in the electron-hole configuration space. The study of the wave propagation in the full $\boldsymbol{r}_h$, $\boldsymbol{r}_e$ configuration space makes it possible to derive exact polariton Greens functions (Sect. 7.2, Appendix E).

The coherent wave description of pair production via (1.15) is further distinguished from a conventional treatment by its ability of being merged with the quantum transport equation (1.11) to a closed nonlinear set of constitutive equations for the band edge. Since these equations relate the coherent pair amplitude to the intraband density matrices, they can express a transfer of coherence from interband to intraband dynamics and vice versa.

Regarding the relation to Ginzburg-Landau theory (Ginzburg and Landau, 1950) it might be helpful to compare the e-h pair amplitude $Y$ with the wave function $\psi$ considered in *GL* theory. Similar to $Y$, the function $\psi$ is a macroscopic wave function related to Fermion pairs (Cooper pairs). A difference between $\psi$ and $Y$ is that $\psi$ depends only on one spatial coordinate interpreted as the center of the Cooper pair. A further difference is that $\psi$ is related to supercurrent while $Y$ refers to a dielectric property.

## 1.4 Survey of the Book

The derivation of the edge equations is inspired by the Bloch equations for a two-level transition (Feynman et al., 1957). As a starting point we shall use in Chap. 2 a dynamical tight-binding model where the semiconductor is con-

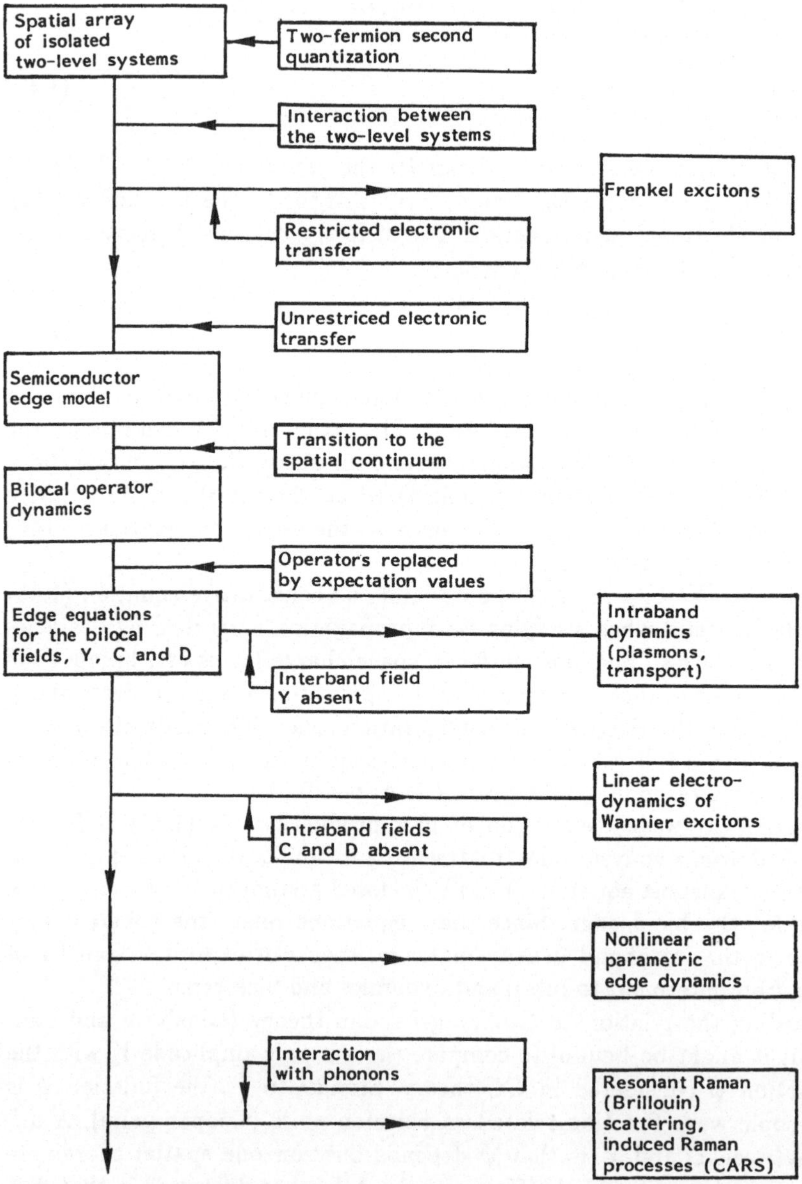


**Fig. 1.2.** Stages in the development of the dynamical band-edge model

structed from an array of two-level atoms with overlapping wave functions. This model will be generalized in different directions as shown in Fig. 1.2. Refinements transforming the model into a realistic form are discussed in Chap. 3.

In Chaps. 4, 5 we treat the linear dynamics of interband transitions. It is demonstrated how to find relevant solutions to (1.15) in various cases. While the unbounded semiconductor is treated in Chap. 4, we discuss in Chap. 5 the half-space geometry. Thus Chap. 5 is the place for a critical review of dead layer and boundary condition problems of exciton polaritons. Chapter 6 is devoted to intraband dynamics. Apart from the half space problem in Sect. 6.8 most of the results discussed in Chap. 6 have been obtained before by other methods. However, it is illustrative to consider the old results in the new context of coherent band edge dynamics. In Chap. 7 we discuss the application of the band edge model to nonlinear and parametric processes, such as coherent nonlinear response, Raman scattering, response of biexcitons and processes in a dense electron-hole plasma.

There are very important aspects of band edge dynamics which are not included in the book, namely

a) phenomena involving a static magnetic field,
b) phonon assisted transitions in indirect gap materials,
c) processes in a dense exciton gas,
d) photoconductivity,
e) carrier injection effect,
f) stimulated emission involving impurities, and
g) spontaneous excitonic recombination.

A further development of the coherent wave approach might well be helpful for the exploration of some of the above topics. However, this is not attempted here.

# 2. From the Two-Level System to the Two-Band System

## 2.1 Dynamics of a Two-Level Transition

One of the most basic models in quantum physics besides the harmonic oscillator is the two-level system. It gives insight into the nature of a quantum transition. The importance of the two-level system for the understanding of semiconductor dynamics is due to the fact that a semiconductor may be visualized as an array of interacting two-level systems. This aspect has been exploited previously, e.g. by Goll and Haken (1978) and by Gan and Yang (1982). This will also be the approach adopted in this book. Here we want to prepare the discussion of interacting two-level systems by a short tutorial introduction to the dynamics of noninteracting two level systems. In our presentation we do not quite follow the standard procedure because we want to prepare for a generalization describing the two band system. Thus we start from a Heisenberg picture instead of the more common density matrix representation (Feynman, Vernon and Hellwarth, 1957), and we introduce positions for the two level atoms from the very beginning in order to describe the dynamics in solids.

Let us consider a collection of two-level atoms with level spacing $E_g = \hbar\omega_g$. The atoms are counted by an index $j$ which later will be related to their positions. The upper state is assumed as a nondegenerate $s$ state (spin degeneracy will be neglected). We denote the upper state of atom $j$ by the symbol $|u, j\rangle$. The lower state is assumed as a threefold degenerate $p$ state; the corresponding symbol is $|\ell, \lambda, j\rangle$, $\lambda$ counting the sublevels; for spinless $p$ states $\lambda$ corresponds to the direction of polarization. For each atom we introduce a set of Fermion operators:

$c_j^\dagger$ creates an electron in the state $|u, j\rangle$;
$c_j$ annihilates an electron in the state $|u, j\rangle$;
$d_{\lambda j}$ creates an electron in the state $|\ell, \lambda, j\rangle$;
$d_{\lambda j}^\dagger$ annihilates an electron in the state $|\ell, \lambda, j\rangle$.

Note that our notation as far as the lower state is concerned actually is a *defect particle* representation, or as one would say in semiconductor physics:

$d^\dagger_{\lambda j}$ creates a hole in the state $|\ell, \lambda, j\rangle$;
$d_{\lambda j}$ annihilates a hole in the state $|\ell, \lambda, j\rangle$.

The Fermi operators obey the usual anticommutation relations

$$[c_j, c^\dagger_k]_+ \equiv c_j c^\dagger_k + c^\dagger_k c_j = \delta_{jk} \; ; \quad [d_{\nu j}, d^\dagger_{\lambda k}]_+ = \delta_{\nu\lambda}\delta_{jk} \tag{2.1}$$

all others are anticommuting.

If we attribute an energy $\hbar\omega_g/2$ to the level $|u, j\rangle$ and $-\hbar\omega_g/2$ to the level $|\ell, \lambda, j\rangle$ the energy operator of the undisturbed atomic system becomes

$$H_A = \tfrac{1}{2}\hbar\omega_g \sum_j \left( c^\dagger_j c_j - \sum_\lambda d_{\lambda j} d^\dagger_{\lambda j} \right) . \tag{2.2}$$

With our assumption about the states there will be no electric dipole moment in the energy states, but in the transition a dipole is produced. The component $\lambda$ of the dipole moment of atom $j$ is represented by the operator

$$\hat{\pi}_{\lambda j} = M_0 (d_{\lambda j} c_j + c^\dagger_j d^\dagger_{\lambda j}) \tag{2.3}$$

$M_0$ being the dipole matrix element.

Assume that the atoms are exposed to the electromagnetic field of a light wave. The electric field component $\lambda$ at position $j$ is denoted by $E_{\lambda j}$. In the dipole approximation the electromagnetic interaction energy is then given by

$$H_{EM} = -\sum_{\lambda j} \hat{\pi}_{\lambda j} E_{\lambda j} \quad . \tag{2.4}$$

Higher multipoles are discussed in Appendix B.

A convenient set of variables which may be used in studying the dynamics of the atomic system under the influence of the field consists of the following four quantities:

$\hat{s}^\dagger_{\lambda j} = c^\dagger_j d^\dagger_{\lambda j}$ creation operator for an electron-hole pair at site $j$;
$\hat{s}_{\lambda j} = d_{\lambda j} c_j$ annihilation operator for a pair at site $j$;
$\hat{n}_j = c^\dagger_j c_j$ number of electrons in the state $|u, j\rangle$;
$\hat{p}_{\lambda\nu, j} = d^\dagger_{\lambda j} d_{\nu j}$ tensorial "number" of holes in the subspace spanned by the states $|\ell, \lambda, j\rangle$.

Note that the number operators for the different states of the hole multiplet are the diagonal elements of the tensor $\hat{p}_{\lambda\nu, j}$. If one thinks of electrons instead of holes in the lower state, then the pair operators must be interpreted as describing upward and downward transitions, respectively.

The next step will be to set up Heisenberg equations of motion for the four operators $\hat{n}_j$, $\hat{p}_{\lambda\nu,j}$, $\hat{s}^{\dagger}_{\lambda j}$, $\hat{s}_{\lambda j}$ with the Hamiltonian being the sum of (2.2,4)

$$H_{TL} = H_A + H_{EM} \quad . \tag{2.5}$$

One arrives after some straightforward algebra at the following set of equations

$$\dot{\hat{s}}_{\lambda j} + i\omega_g \hat{s}_{\lambda j} = \frac{iM_0}{\hbar} \sum_\nu E_{\nu j}[\delta_{\nu\lambda}(1-\hat{n}_j) - \hat{p}_{\nu\lambda j}] \tag{2.6a}$$

$$\dot{\hat{s}}^{\dagger}_{\lambda j} = i\omega_g \hat{s}^{\dagger}_{\lambda j} = -\frac{iM_0}{\hbar} \sum_\nu E_{\nu j}[\delta_{\nu\lambda}(1-\hat{n}_j) - \hat{p}_{\lambda\nu j}] \tag{2.6b}$$

$$\dot{\hat{n}}_j = -\frac{iM_0}{\hbar} \sum_\lambda (\hat{s}_{\lambda j} - \hat{s}^{\dagger}_{\lambda j}) E_{\lambda j} \tag{2.6c}$$

$$\dot{\hat{p}}_{\lambda\nu j} = -\frac{iM_0}{\hbar}(E_{\lambda j}\hat{s}_{\nu j} - \hat{s}^{\dagger}_{\lambda j} E_{\nu j}) \quad . \tag{2.6d}$$

Let us now assume that the atoms are placed on a lattice with elementary volume $\Omega$, and that the field is slowly varying on the spatial scale defined by $\Omega$. Then many atoms will be exposed to the same conditions, and the law of large numbers tells us that the expectation values of atomic variables behave like deterministic variables. On the other hand, it can be seen by taking the expectation of (2.6) that the expectation values of the four quantities $\hat{n}_j$, $\hat{p}_{\lambda\nu,j}$, $\hat{s}_{\lambda j}$, $\hat{s}^{\dagger}_{\lambda j}$ formally obey the same equations as the corresponding operators. We therefore introduce a set of macrovariables related to these expectation values. In doing so we take the opportunity to go over from pure number quantities to densities by applying a factor $\Omega^{-1}$, and we replace the discrete subscript $j$ counting the atoms by a continuous macrocoordinate $\boldsymbol{r}$ interpolating the lattice sites (Appendix A). Together with this step we make the following change of notation

| | | |
|---|---|---|
| $\Omega^{-1}\langle\hat{s}_{\lambda j}\rangle \rightarrow s_\lambda(\boldsymbol{r})$ | downward transition amplitude | |
| $\Omega^{-1}\langle\hat{s}^{\dagger}_{\lambda j}\rangle \rightarrow s^*_\lambda(\boldsymbol{r})$ | upward transition amplitude | (2.7) |
| $\Omega^{-1}\langle\hat{n}_j\rangle \rightarrow n(\boldsymbol{r})$ | electron density in upper state | |
| $\Omega^{-1}\langle p_{\lambda\nu j}\rangle \rightarrow p_{\lambda\nu}(\boldsymbol{r})$ | hole density in lower state. | |

Besides the spatial dependence the above defined quantities will in general also possess a time dependence. This time dependence develops according to a set of equations similar to (2.6)

$$\dot{s}_\lambda + i\omega_g s_\lambda = \frac{iM_0}{\hbar} \sum_\nu E_\nu[\delta_{\nu\lambda}(N-n) - p_{\nu\lambda}] \tag{2.8a}$$

$$\dot{s}_\lambda^* - i\omega_g s_\lambda^* = -\frac{iM_0}{\hbar}\sum_\nu E_\nu[\delta_{\nu\lambda}(N-n) - p_{\lambda\nu}] \tag{2.8b}$$

$$\dot{n} = -\frac{iM_0}{\hbar}\sum_\lambda (s_\lambda - s_\lambda^*)E_\lambda \tag{2.8c}$$

$$\dot{p}_{\lambda\nu} = -\frac{iM_0}{\hbar}(E_\lambda s_\nu - s_\lambda^* E_\nu) \ . \tag{2.8d}$$

$N = \Omega^{-1}$ is the reference density of all two-level atoms present.

Instead of components we may of course use a vector-tensor notation; then the set of equations (2.8) takes on the following form

$$\dot{\boldsymbol{s}} + i\omega_g \boldsymbol{s} = \frac{iM_0}{\hbar}\boldsymbol{E}[(N-n)\underset{\sim}{1} - \underset{\sim}{p}] \tag{2.9a}$$

$$\dot{\boldsymbol{s}}^* - i\omega_g \boldsymbol{s}^* = -\frac{iM_0}{\hbar}[(N-n)\underset{\sim}{1} - \underset{\sim}{p}]\boldsymbol{E} \tag{2.9b}$$

$$\dot{n} = -\frac{iM_0}{\hbar}(\boldsymbol{s} - \boldsymbol{s}^*)\boldsymbol{E} \tag{2.9c}$$

$$\dot{\underset{\sim}{p}} = -\frac{iM_0}{\hbar}(\boldsymbol{E}\otimes\boldsymbol{s} - \boldsymbol{s}^*\otimes\boldsymbol{E}) \tag{2.9d}$$

where $\otimes$ denotes a dyadic product.

These equations describe the dynamics of an $s-p$ two-level transition with an arbitrary state of polarization. In most presentations the tensorial properties of (2.9) are suppressed by the assumption that $\boldsymbol{E}$ is a polarized field, say in direction $\lambda$. Then of the vector $\boldsymbol{s}$ only the component $s_\lambda$ will be excited. The tensor $\underset{\sim}{p}$ then has $p_{\lambda\lambda}$ as the only nonvanishing element. This allows us to set up a quasiscalar version of (2.8) where only the nonvanishing components are retained. Dropping all subscripts one then obtains from (2.8):

$$\dot{s} + i\omega_g s = \frac{iM_0}{\hbar}(N-n-p) \tag{2.10a}$$

$$\dot{n} = -\frac{iM_0}{\hbar}(s-s^*)E \tag{2.10b}$$

$$\dot{p} = -\frac{iM_0}{\hbar}(s-s^*)E \ . \tag{2.10c}$$

(The equation corresponding to (2.8b) has been omitted because it has become redundant). Equations (2.10) are known as the atomic Bloch equations because in an appropriate notation they turn out to be isomorphic to Bloch's celebrate equations describing the dynamics of the magnetization vector in a time dependent magnetic field (Bloch, 1946).

To make contact with electrodynamics let us consider the dipole density conventionally termed the polarization. From (2.3) and (2.10) we find

$$P = M_0(s + s^*) = 2M_0 \operatorname{Re}\{s\} \ . \tag{2.11}$$

If then we take the real part of (2.10a) and multiply by $2M_0$ we find

$$\dot{P} = 2\omega_g M_0 \operatorname{Im}\{s\} \ . \tag{2.12}$$

On the other hand we know from electrodynamics that $\dot{P}$ is the polarization current $J_P$. We therefore have

$$J_P = -i\omega_g M_0(s - s^*) \ . \tag{2.13}$$

If we use (2.11,12) we can bring the dynamic equations of the two-level transition into the form most convenient for studying electrodynamics:

$$\ddot{P} + \omega_g^2 P = \frac{2M_0\omega_g}{\hbar}(N - n - p)E \tag{2.14a}$$

$$\dot{n} = \frac{1}{\hbar\omega_g} E J_P \tag{2.14b}$$

$$\dot{p} = \frac{1}{\hbar\omega_g} E J_P \ . \tag{2.14c}$$

In the linear response regime, i.e. for $n, p \ll N$, (2.14a) describes forced harmonic oscillation of the transition polarization. For higher excitation when the electron-hole densities $n, p$ become comparable to the saturation density $N$ the oscillator responds with an apparently reduced strength. The two remaining equations represent the creation of electron-hole pairs by the power density

$$\Pi = E J_P \tag{2.15}$$

fed into the system. Equations (2.14b,c) therefore express the energy balance.

For realistic fits to experimental data it is necessary to take into account the dissipative interaction with the environment. In the simplest possible approximation this is done by introducing two phenomenological relaxation times, $T_1$ and $T_2$; $T_1$ describes a relaxation of the densities $n$ and $p$ towards their thermal equilibrium values, while $T_2$ measures the time it takes for the free oscillation of the transition polarization to be removed by dephasing processes. When these phenomenological relaxation terms are added the constitutive equations of the two level material become

$$\ddot{P} + T_2^{-1}\dot{P} + \omega_g^2 P = \frac{2M_0^2\omega_g}{\hbar}(N - n - p)E \tag{2.16a}$$

$$\dot{n} = \frac{1}{\hbar\omega_g} E J_P - T_1^{-1}(n - n_0) \tag{2.16b}$$

$$\dot{p} = \frac{1}{\hbar\omega_g} E J_P - T_1^{-1}(p - p_0) \ . \tag{2.16c}$$

The constitutive equations of a two-level material have found many beautiful applications. They are the basis of a theory of the laser effect (Haken, 1970) and they can be used to describe a variety if optical pulse phenomena (McCall, Hahn, 1969; Lamb, 1971). Last but not least the linear approximation to (2.14a) makes it evident that a quantum transition with energetic distance $\hbar\omega_g$ in linear-response theory is equivalent to a harmonic oscillator with characteristic frequency $\omega_g$.

## 2.2 Classification of Exciton Models According to Mobility

In the preceding section we have studied a model consisting of isolated two-level atoms which could only interact with an external light wave. For convenience these atoms were placed on a lattice with elementary volume $\Omega$, but so far we have made no use of an important possibility inherent in the crystalline structure. If the atoms come close enough together, then the wave functions belonging to states on different sites begin to overlap, and electrons and holes become mobile. Models making some allowance for this type of mobility are named exciton models (Agranovich, Ginzburg, 1984).

Exciton models are classified according to the degree of mobility which is allowed. In so-called Wannier-type models the electron in the upper state and the hole in the lower state are given independent degrees of freedom. In order to describe all allowed configurations of the pair in a Wannier-type model one therefore needs a six-dimensional configuration space.

An exciton model with a reduced mobility is the so-called Frenkel model. In this model it is assumed that the excited electron-hole pair can move, but without becoming separated. The Frenkel model is therefore obtained from the Wannier model by imposing the constraint that only the subspace with vanishing relative distance is allowed.

Going on in this manner one could also classify the two-level model of Sect. 2.1 as a special exciton model, viz. the model excluding any kind of mobility by imposing a second constraint. On the other hand, starting from the Bloch-type theory, models with mobile carriers are obtained by a generalization.

In the Wannier case two additional features have to be considered. Firstly one must generalize the relevant dynamical quantities which were the four functions $s(\boldsymbol{r})$, $s^*(\boldsymbol{r})$, $n(\boldsymbol{r})$, $p(\boldsymbol{r})$ to the configuration space which is assumed to be accessible for electron-hole pairs. This amounts to the introduction of functions of two coordinates into the constitutive equations. In order to have a short name we shall refer to that procedure as a "bilocal" description. The second generalization that has to be added to Bloch's theory in order to make it a Wannier type model are extra terms in the constitutive equations to account for mobility. They can for example be derived from a mobility part added to the Hamiltonian (2.5). In order to cope with the continuum description adopted in (2.7) it will be particularly convenient to treat mobility within an effective mass approximation.

In the Frenkel case the bilocal generalization is not needed because the space spanned by the coordinates of the Frenkel pair is precisely that in which the electric field is prescribed. Adding mobility is therefore the only generalization which must be applied when going over from a Bloch-type model to the Frenkel case. This will be done in the next section.

## 2.3 Frenkel Excitons

As mentioned in the last section a Frenkel model can be set up using the same set of variables as in atomic Bloch theory. In order to keep the equations as simple as possible, we assume that we have only one state of polarization and apply the quasiscalar notation as introduced in Sect. 2.1. Because of the slightly different notation we list once more the relevant set of variables (in operator form):

$\hat{s}_j^\dagger = c_j^\dagger d_j^\dagger$ creator of a Frenkel pair at site $j$;
$\hat{s}_j = d_j c_j$ annihilator of a Frenkel pair at site $j$;
$\hat{n}_j = c_j^\dagger c_j$ electron number in state $|u, j\rangle$
$\hat{p}_j = d_j^\dagger d_j$ hole number in state $|\ell, j\rangle$.

The model becomes a Frenkel-type model if we retain the Hamiltonian (2.5) but add a term accounting for mobility in the restricted sense explained in the last section. Let us therefore write

$$H_{Fr} = H_A + H_{EM} + H_{MFr} \ . \tag{2.17}$$

The Frenkel-type mobility part $H_{MFr}$ must have the property that it transfers a Frenkel pair from one site to another. Evidently this is achieved by the ansatz

$$H_{MFr} = \sum_{ik} T_{ik} \hat{s}_i^\dagger \hat{s}_k \ . \tag{2.18}$$

Hermiticity requires that the "hopping matrix" $T_{ik}$ is symmetric

$$T_{ik} = T_{ki} \ . \tag{2.19}$$

If we are not interested in saturation properties we can restrict the discussion to the linearized form of the equation for the pair operator $\hat{s}_j$. If we calculate

$$\dot{\hat{s}}_j = \frac{i}{\hbar}[H_{Fr}, \hat{s}_j] \tag{2.20}$$

and neglect terms involving $n$ or $p$, we arrive at

$$\dot{\hat{s}}_j + i\omega_g \hat{s}_j + \frac{i}{\hbar}\sum_k T_{jk}\hat{s}_k = \frac{iM_0}{\hbar}E \ . \tag{2.21}$$

Let us go over to expectation values and introduce a continuum description in terms of densities as explained in Sect. 2.1 (compare also Appendix A). Then (2.21) takes on the form

$$\dot{s} + i\omega_g s + \frac{i}{\hbar}\int T(\boldsymbol{r}')s(\boldsymbol{r}-\boldsymbol{r}')d^3\boldsymbol{r}' = \frac{iM_0}{\hbar}NE \ . \tag{2.22}$$

Since $T(\boldsymbol{r})$ will be of short range we may expand under the integral

$$\begin{aligned}\int T(\boldsymbol{r}')s(\boldsymbol{r}-\boldsymbol{r}')d^3\boldsymbol{r}' \approx \int T(\boldsymbol{r}')[&s(\boldsymbol{r}) - (\boldsymbol{r}'\nabla)s(\boldsymbol{r}) \\ &+ \tfrac{1}{2}(\boldsymbol{r}'\nabla)^2 s(\boldsymbol{r}) + \ldots]d^3\boldsymbol{r}' \ .\end{aligned} \tag{2.23}$$

In the simplest case of an isotropic material (2.23) reduces to

$$\int T(\boldsymbol{r}')s(\boldsymbol{r}-\boldsymbol{r}')d^3\boldsymbol{r}' \approx s(\boldsymbol{r})\int T(\boldsymbol{r}')d^3\boldsymbol{r}' + \tfrac{1}{6}\nabla^2 s(\boldsymbol{r})\int (\boldsymbol{r}')^2 T(\boldsymbol{r}')d^3\boldsymbol{r}' \ . \tag{2.24}$$

When inserted into (2.22) the term with the zero-order moment of $T$ can be absorbed into a renormalized frequency $\omega_g$. The second-order term may be written in a notation stressing similarity to the kinetic energy in an effective mass approximation. The final form of (2.22) is then

$$\dot{s} + i\left(\omega_g - \frac{\hbar}{2m^*}\nabla^2\right)s = \frac{iM_0}{\hbar}NE \tag{2.25}$$

which is a Schrödinger-like equation with a source term coupling to the electric field. It is the constitutive equation of a medium with Frenkel-type excitations. Despite of its Schrödinger-like appearance (2.25) should be considered as a material equation in the sense of macroscopic electrodynamics.

Some authors, among them Hopfield in his pioneer work (Hopfield, 1958, 1969), have used instead of (2.25) for the complex exciton amplitude $s$ an equation for the polarization $P$. Since $P$ is still related to the real part of $s$ by (2.11), an equation for $P$ can be found by eliminating $\mathrm{Im}\{s\}$ from (2.25). If this is done in the long wavelength limit (neglect of higher order derivatives) one arrives at the following equation

$$\ddot{P} + \omega_g^2 P - \frac{\hbar\omega_g}{m^*}\nabla^2 P = \frac{2M_0^2\omega_g}{\hbar} NE \ . \tag{2.26}$$

This is the standard model equation of a medium with a propagating oscillator-like resonance. A medium with such a resonance is termed spatially dispersive. If a spatially dispersive medium is excited by a plane wave electric field

$$E = E_0 \exp\left(i\boldsymbol{q}\cdot\boldsymbol{r} - i\omega t\right)$$

the response according to (2.26) becomes

$$P = \varepsilon_0 \chi E \tag{2.27}$$

with the dielectric susceptibility $\chi$ being a function of $\omega$ and $\boldsymbol{q}$:

$$\chi(\omega, \boldsymbol{q}) = \frac{2M_0^2 N\omega_g}{\hbar\varepsilon_0}\left(\omega_g^2 + \frac{\hbar\omega_g}{m^*}q^2 - \omega^2\right)^{-1} \ . \tag{2.28}$$

The topic of spatial dispersion has been treated in great detail, e.g. in the book by Agranovich and Ginzburg (1984). We postpone our discussion of the matter to Chap. 4 where we shall have at our disposal a Wannier-type model which is more closely related to the topic of this book than a Frenkel-type model.

## 2.4 A Tight-Binding Model of the Semiconductor Band Edge

The essential features of the electronic structure of a semiconductor are the existence of at least two bands separated by a gap with the Fermi level lying somewhere in the gap. The lower valence band is then almost filled and the upper conduction band almost empty. In the so called "tight-binding" approach this kind of structure is obtained if one places a collection of two-level atoms on a lattice and assumes that there is a sufficient overlap between atomic states on neighbouring sites in order to allow for mobility. This model will serve for deducing the principal structure of band-edge dynamics. Later on we shall generalize the model in such a way that it fits into the framework of a fully developed semiconductor band structure.

Our tight-binding two-band model uses as building blocks the two-level atoms as they were discussed in Sect. 2.1. Again, as in the Frenkel-type model of Sect. 2.3, we add to the two-level Hamiltonian (2.5) a mobility term $H_M$, which now has to allow for independent mobility of electrons and holes. In Sect. 2.2 we mentioned this unrestricted kind of mobility as characteristic for Wannier-type exciton models. In a notation similar to (2.18) we may write $H_M$ as follows

$$H_M = \sum_{ik} (T^c_{ik} c^\dagger_i c_k + T^v_{ik} d^\dagger_i d_k). \tag{2.29}$$

$T^c_{ik}$ transfers electrons in the upper level between different atoms; the superscript "$c$" refers to "conduction-band". $T^v_{ik}$ transfers holes in the lower level; the superscript "$v$" refers to "valence band".

Because now electrons and holes can move independently, the system can develop new excitations which were not present in the two-level system or in the Frenkel model. They correspond to a coherent excitation of an electron-hole pair where electron and hole are found at different sites. In order to take care of these excitations we generalize our former set of variables by introducing bilocal operators. For simplicity, at least at the beginning, the quasiscalar notation of (2.10) will be adopted (see also Fig. 2.1):

| | |
|---|---|
| $\hat{Y}_{ij} = d_i c_j$ | annihilates a hole at site $i$ and an electron at site $j$; |
| $\hat{Y}^\dagger_{ij} = c^\dagger_j d^\dagger_i$ | creates an electron at site $j$ and a hole at site $i$; |
| $\hat{C}_{ij} = c^\dagger_i c_j$ | transfers an electron from site $j$ to site $i$; |
| $\hat{D}_{ij} = d^\dagger_i d_j$ | transfers a hole from site $j$ to site $i$. (2.30) |

The operators used for the two level system are found on the diagonal of the bilocal operators

$$\hat{s}_j = \hat{Y}_{jj}; \quad \hat{s}^\dagger_j = \hat{Y}^\dagger_{jj}; \quad \hat{n}_j = \hat{C}_{jj}; \quad \hat{p}_j = \hat{D}_{jj} \ . \tag{2.31}$$

Next we have to find the equations of motion for the operators (2.30) if the Hamiltonian is the sum

$$H = H_{TL} + H_M$$

where $H_{TL}$ and $H_M$ are given by (2.5) and (2.29). We only give the results which can be calculated by some lengthy but straightforward algebra (see Appendix F):

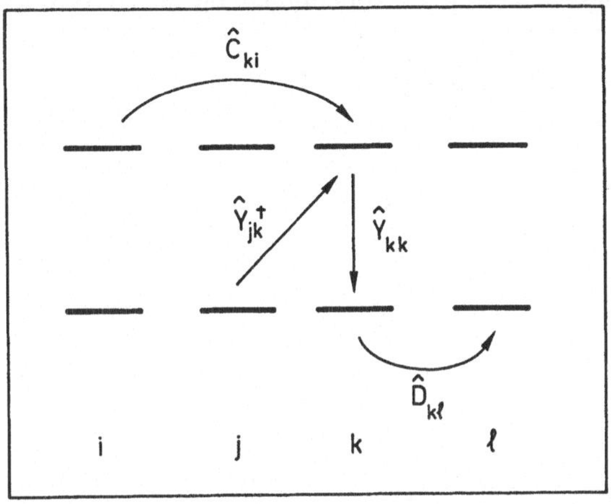


**Fig. 2.1.** Relevant variables in a tight-binding two-band model. The arrows refer to transfer of an electron. Note that the transfer of a hole in the valence band goes in opposite direction

$$\dot{\hat{Y}}_{ij} + i\left[\omega_g \hat{Y}_{ij} + \frac{1}{\hbar}\sum_{\ell}(T^c_{j\ell}\hat{Y}_{i\ell} + T^v_{i\ell}\hat{Y}_{\ell j})\right]$$

$$= \frac{iM_0}{\hbar}(\delta_{ij}E_j - E_i\hat{C}_{ij} - E_j\hat{D}_{ji}) \tag{2.32a}$$

$$\dot{\hat{C}}_{ij} + \frac{i}{\hbar}\sum_{\ell}(T^c_{j\ell}\hat{C}_{i\ell} - T^c_{\ell i}\hat{C}_{\ell j}) = -\frac{iM_0}{\hbar}(E_i\hat{Y}_{ij} - E_j\hat{Y}^{\dagger}_{ji}) \tag{2.32b}$$

$$\dot{\hat{D}}_{ij} + \frac{i}{\hbar}\sum_{\ell}(T^v_{j\ell}\hat{D}_{i\ell} - T^v_{\ell i}\hat{D}_{\ell j}) = -\frac{iM_0}{\hbar}(E_i\hat{Y}_{ji} - E_j\hat{Y}^{\dagger}_{ij}) \ . \tag{2.32c}$$

An equation for $\hat{Y}^{\dagger}_{ij}$ would be redundant being the adjoint of (2.32a).

Because equations similar to (2.32) will appear in many different forms we have decided to give them a name and propose to call them *band-edge equations*. The set of band-edge equations is subdivided into the interband part, represented by (2.32a) and its adjoint, and the intraband part, represented by the two equations (2.32b,c). Most remarkable in any version of band-edge equations is that one has a closed set of equations for the set of variables (2.30) which by this very property are identified as the relevant variables for describing the dynamics of a band edge.

In their form (2.32) as operator equations defined on lattice points, the band-edge equations are not well suited for practical calculations. It is therefore desirable again to apply the continuum approximation and to go over to expectation values. Either manipulation is closely connected with a well-established standard approximation: For the continuum limit it is the effective-mass approximation, for the use of expectation values it is the self-consistent field approximation. Let us begin with a discussion of the continuum approximation:

If we apply the band-limited sampling procedure as explained in Appendix A to the field operators we arrive at a set of operators defined on the continuum

$$c_B(\boldsymbol{r}) = \sum_j c_j \Delta(\boldsymbol{r} - \boldsymbol{R}_j) \qquad c^{\dagger}_B(\boldsymbol{r}) = \sum_j c^{\dagger}_j \Delta(\boldsymbol{r} - \boldsymbol{R}_j)$$

$$d_B(\boldsymbol{r}) = \sum_j d_j \Delta(\boldsymbol{r} - \boldsymbol{R}_j) \qquad d^{\dagger}_B(\boldsymbol{r}) = \sum_j d^{\dagger}_j \Delta(\boldsymbol{r} - \boldsymbol{R}_j) \ . \tag{2.33}$$

It would have been nice to call the band-limited operators "macroscopic", but because the distinction "microscopic-macroscopic" is already used in so many different senses, we do not want to add to this confusion. The operators (2.33) obey anticommutation relations, for example,

$$[c_B(\boldsymbol{r}), c^{\dagger}_B(\boldsymbol{r}')]_+ = \delta_B(\boldsymbol{r} - \boldsymbol{r}'); \qquad [d_B(\boldsymbol{r}), d^{\dagger}_B(\boldsymbol{r}')]_+ = \delta_B(\boldsymbol{r} - \boldsymbol{r}') \ . \tag{2.34}$$

The bilocal operators $\hat{Y}_{ij}$, $\hat{C}_{ij}$, $\hat{D}_{ij}$ too, are transformed to a band-limited form, e.g.

$$\hat{Y}_B(\boldsymbol{r}_1, \boldsymbol{r}_2) = \sum_{ij} \hat{Y}_{ij} \Delta(\boldsymbol{r}_1 - \boldsymbol{R}_i) \Delta(\boldsymbol{r}_2 - \boldsymbol{R}_j) \ . \tag{2.35}$$

Let us make the convention that we are allowed, in a context where mistakes are excluded, to replace the space-consuming notation $\hat{Y}_B(\boldsymbol{r}_1, \boldsymbol{r}_2)$ by the shorter symbol $\hat{Y}_{12}$. The convention means that number subscripts as "1" or "2" stand for continuous variables in contrast to letter subscripts "$i, j$" used in the notation of quantities referring to discrete points. The convention also includes that the subscript "$B$" referring to band limitation is dropped.

With the help of (2.35) we can transform the band-edge equations to a continuum form. As an example we consider the interband equation which becomes

$$\begin{aligned} &\dot{\hat{Y}}_{12} + i\omega_g \hat{Y}_{12} + \frac{i}{\hbar} \int (T^c_{23} \hat{Y}_{13} + T^v_{13} \hat{Y}_{32}) d^3 \boldsymbol{r}_3 \\ &= \frac{iM_0}{\hbar} (\delta_{12} E_2 - E_1 \hat{C}_{12} - E_2 D_{21}) \ . \end{aligned} \tag{2.36}$$

In going over from (2.32a) to (2.36) one makes use of the completeness and orthonormality properties of the sampling functions $\Delta(\boldsymbol{r} - \boldsymbol{R}_j)$ as given in the appendix. Note that the electric field, although appearing in (2.32a) in a discontinuous form, must not be treated by the interpolation procedure because being a macroscopic field it is already band-limited. In the sense of our general convention $\delta_{12}$ is the band-limited $\delta$ function.

To the continuous form we apply the effective mass approximation which means that we make a moment expansion of the transfer kernels $T^c$ and $T^v$ in a similar manner as in Sect. 2.3. For the sake of simplicity we assume that only isotropic second order moments contribute, but it should be stressed that, if there is need, more complicated situations are tractable. We then obtain

$$\begin{aligned} \int (T^c_{32} \hat{Y}_{13} + T^v_{13} \hat{Y}_{32}) d^3 \boldsymbol{r}_3 = \hat{Y}_{12} &\int (T^c(\boldsymbol{r}') + T^v(\boldsymbol{r}')) d^3 \boldsymbol{r}' \\ &+ \tfrac{1}{6} \int (\boldsymbol{r}')^2 T^c d^3 \boldsymbol{r}' \nabla_2^2 \hat{Y}_{12} \\ &+ \tfrac{1}{6} \int (\boldsymbol{r}')^2 T^v d^3 \boldsymbol{r}' \nabla_1^2 \hat{Y}_{12} \ . \end{aligned} \tag{2.37}$$

The first term goes into a renormalization of $\omega_g$ while the second moment terms will be interpreted as the effective mass kinetic energy contributions by the identification:

$$\frac{1}{6} \int (\boldsymbol{r}')^2 T^c d^3 \boldsymbol{r}' = -\frac{\hbar^2}{2m_e} \qquad \frac{1}{6} \int (\boldsymbol{r}')^2 T^v d^3 \boldsymbol{r}' = -\frac{\hbar^2}{2m_h} \tag{2.38}$$

where $m_e$, $m_h$ are the effective masses of electrons and holes, respectively.

This identification gets support from a consideration of the Hamiltonian (2.29) in band-limited continuum representation. Concentrating on the conduction band part we get from applying the transcription to the continuum

$$H_M = \int T^c(\boldsymbol{r}_1 - \boldsymbol{r}_2) c^\dagger(\boldsymbol{r}_1) c(\boldsymbol{r}_2) d^3\boldsymbol{r}_1 d^3\boldsymbol{r}_2 \ . \tag{2.39}$$

Transforming to

$$\boldsymbol{r} = (\boldsymbol{r}_1 + \boldsymbol{r}_2)/2 \qquad \boldsymbol{r}' = \boldsymbol{r}_1 - \boldsymbol{r}_2$$

and expanding to second order in $\boldsymbol{r}'$ one obtains

$$\begin{aligned} H_M^c = \int T^c(\boldsymbol{r}') \Big[ & c^\dagger(\boldsymbol{r}) c(\boldsymbol{r}) - \frac{(\boldsymbol{r}')^2}{4} (\nabla c^\dagger)(\nabla c) \\ & + \frac{(\boldsymbol{r}')^2}{8} ((\nabla^2 c^\dagger) c + c^\dagger (\nabla^2 c)) \Big] d^3\boldsymbol{r}\, d^3\boldsymbol{r}' \ . \end{aligned} \tag{2.40}$$

Applying integration by parts and the isotropy argument, one ends up with

$$\begin{aligned} H_M^c = & \int T^c(\boldsymbol{r}') d^3\boldsymbol{r}' \int c^\dagger(\boldsymbol{r}) c(\boldsymbol{r}) d^3\boldsymbol{r} \\ & - \tfrac{1}{6} \int (\boldsymbol{r}')^2 T^c(\boldsymbol{r}') d^3\boldsymbol{r}' \int \nabla c^\dagger \nabla c\, d^3\boldsymbol{r} \ . \end{aligned} \tag{2.41}$$

Comparing the second term in (2.41) with (2.38) it is found to be of the standard effective mass form

$$H_M^c = \frac{\hbar^2}{2m_e} \int \nabla c^\dagger \nabla c\, d^3\boldsymbol{r} + \text{const} \ . \tag{2.42}$$

Evidently (2.42) allows a spectral representation in $\boldsymbol{k}$-space in terms of the band structure $E_c(\boldsymbol{k})$ :

$$H_M^c = \int E_c(\boldsymbol{k}) c^\dagger(\boldsymbol{k}) c(\boldsymbol{k}) d^3\boldsymbol{k} + \text{const} \qquad E_c(\boldsymbol{k}) = \frac{\hbar^2 \boldsymbol{k}^2}{2m_e} \ . \tag{2.43}$$

All steps used in the derivation of (2.43) are easily generalized to more complicated band structures $E(\boldsymbol{k})$.

Given now (2.36) and its intraband counterparts in effective mass form, the final step is to go over to expectation values. As long as we consider $E$ as an external field only and do not take into account electron-hole interactions, this step is trivial because then the equations are linear in the operators. We therefore immediately go over to the expectation-value version of the equations by just dropping the "hat":

$$\dot{Y}_{12} + i\left(\omega_g - \frac{\hbar}{2m_h}\nabla_1^2 - \frac{\hbar}{2m_e}\nabla_2^2\right)Y_{12}$$
$$= \frac{iM_0}{\hbar}(\delta_{12}E_2 - E_1C_{12} - E_2D_{21}) \tag{2.44a}$$

$$\dot{C}_{12} + \frac{i\hbar}{2m_e}(\nabla_1^2 - \nabla_2^2)C_{12} = -\frac{iM_0}{\hbar}(E_1Y_{12} - E_2Y_{21}^*) \tag{2.44b}$$

$$\dot{D}_{12} + \frac{i\hbar}{2m_h}(\nabla_1^2 - \nabla_2^2)D_{12} = -\frac{iM_0}{\hbar}(Y_{21}E_1 - Y_{12}^*E_2) \ . \tag{2.44c}$$

Despite all extensions and refinements to be discussed in the next chapter, equations (2.44) form the core of the constitutive equations of band edge dynamics. We therefore shall discuss in the next section some properties referring to the equations themselves independently of special solutions.

## 2.5 Preliminary Discussion of Band-Edge Equations

### 2.5.1 Variables

The bilocal expectation values entering the constitutive equations (2.44) have the following meaning: $C(\mathbf{r}_1, \mathbf{r}_2) = \langle c^\dagger(\mathbf{r}_1)c(\mathbf{r}_2)\rangle \equiv C_{12}$ is the so called single-particle density matrix (Landau and Lifschitz, 1981) of conduction band electrons. Its diagonal part is the electron density

$$n(\mathbf{r}) = C(\mathbf{r}, \mathbf{r}) \ . \tag{2.45}$$

As will be discussed later in more detail (Chap. 6), $C$ is related to the phase-space distribution function $f_e(\mathbf{r}, \mathbf{k})$ which is the key quantity in the theory of transport processes. The relation between $C$ and $f_e$ is

$$f_e(\tfrac{1}{2}(\mathbf{r}_1 + \mathbf{r}_2), \mathbf{k}) = \int C(\mathbf{r}_1, \mathbf{r}_2)\exp\left[i\mathbf{k}(\mathbf{r}_1 - \mathbf{r}_2)\right]d^3(\mathbf{r}_1 - \mathbf{r}_2) \ . \tag{2.46}$$

From the equivalence between density matrix and distribution function it follows that $C$ must also contain the necessary information for the calcuation of transport currents. So, e.g., the electric current carried by the intraband motion of electrons turns out to be

$$J_e(\mathbf{r}) = \frac{e\hbar}{2im_e}(\nabla_1 - \nabla_2)C\bigg|_{\mathbf{r}_1 = \mathbf{r}_2 = \mathbf{r}} \ . \tag{2.47}$$

$D(\mathbf{r}_1, \mathbf{r}_2) = \langle d^\dagger(\mathbf{r}_1)d(\mathbf{r}_2)\rangle \equiv D_{12}$ is the density matrix of holes in the valence band. Its properties are analogous to the corresponding properties of the conduction-band density matrix $C$.

$Y(\boldsymbol{r}_1, \boldsymbol{r}_2) = \langle d(\boldsymbol{r}_1) c(\boldsymbol{r}_2) \rangle \equiv Y_{12}$ will be called the "electron-hole wave function" or "pair wave function". In the context of exciton theory also the name "coherent exciton wave function" might be appropriate. The specification of $Y$ as being a "coherent" wave function stresses the fact that the source term on the r.h.s. of (2.44a) leads to a well defined phase relation between the electric field $E$ and the wave function $Y$. The bilocality of $Y$ refers to electron-hole configuration space. The coherence property of $Y$ is therefore coherence in six-dimensional configuration space. As will become clear in later sections the production of a coherent wave function $Y$ involves collective excitation of many electron-hole pairs. A construct resembling the coherent exciton wave $Y$ in many respects is the Cooper-pair wave function $\psi$ used in the Ginzburg-Landau theory of superconductivity (Landau and Lifschitz, 1960).

On the diagonal $Y(\boldsymbol{r}_1, \boldsymbol{r}_2)$ contains information about the polarization due to interband processes

$$P(\boldsymbol{r}) = M_0 (Y(\boldsymbol{r}, \boldsymbol{r}) + Y^*(\boldsymbol{r}, \boldsymbol{r})) \ . \tag{2.48}$$

The use of a complex function $Y$ indicates that $Y$ is a composite of effectively two physical variables. This structure reminds of the quasispin property of a two-level transition. The real part of $Y$ determines the polarization (2.48), while the imaginary part of $Y$ determines the current density.

### 2.5.2 Gross Structure of Equations

Consider the intraband equation (2.44b); a close relationship between this equation and the Wlassow-Boltzmann equation of transport theory (compare Sect. 6.1) suggests an ordering of terms according to the following scheme:

$$\dot{C}_{12} = (\dot{C}_{12})_{\mathrm{drift}} + (\dot{C}_{12})_{\mathrm{source}} \qquad \text{with} \tag{2.49}$$

$$(\dot{C}_{12})_{\mathrm{drift}} = -\frac{i\hbar}{2m_e} (\nabla_1^2 - \nabla_2^2) C_{12} \tag{2.49a}$$

$$(\dot{C}_{12})_{\mathrm{source}} = -\frac{iM_0}{\hbar} (E_1 Y_{12} - E_2 Y_{21}^*) \ . \tag{2.49b}$$

The name "drift"-term for (2.49a) has been taken over from transport theory. In the special case which is considered here the "drift" is a free flight of the effective-mass electrons. In later generalizations the drift motion will also be considered under the influence of given (eventually self-consistent) potential fields.

The interpretation of (2.49b) as a source term is suggested if we think of it as a generalization of the power term in the energy balance (2.14). The new feature in comparison to the two-level system is that the source term involves off-diagonal contributions of $Y_{12}$.

The structure of the intra-valenceband equation (2.44c) is practically identical to (2.44b); we therefore write it

$$\dot{D}_{12} = (\dot{D}_{12})_{\text{drift}} + (\dot{D}_{12})_{\text{source}} \tag{2.50}$$

with an identical interpretation of respective terms. Coming to the interband equation (2.44a), the structure is at least superficially similar and we therefore apply again the decomposition

$$\dot{Y}_{12} = (\dot{Y}_{12})_{\text{drift}} + (\dot{Y}_{12})_{\text{source}} \tag{2.51}$$

$$(\dot{Y}_{12})_{\text{drift}} = -i\left(\omega_g - \frac{\hbar}{2m_h}\nabla_1^2 - \frac{\hbar}{2m_e}\nabla_2^2\right)Y_{12} \tag{2.51a}$$

$$(\dot{Y}_{12})_{\text{source}} = \frac{iM_0}{\hbar}(\delta_{12}E_2 - E_1C_{12} - E_2D_{21}) \ . \tag{2.51b}$$

The naming of (2.51a) as "drift" is more or less formal; it is based on the analogy with (2.49a) and ignores important differences between (2.51a) and (2.49a). So, e.g., (2.51a) is an elliptic differential operator while (2.49a) is hyperbolic. This difference implices that the intraband equation in the classical limit describes moving particles while the interband equation remains an equation for a quantum mechanical wave phenomenon.

Like the corresponding term in the constitutive equation (2.10) for the two level system, the source (2.51b) contains a term linear in the field and the saturating terms depending on $C$ and $D$. Again, as in the intraband sources, there is a typical dependence on the off-diagonal parts of $C$ and $D$ respectively.

As a shortband notation we shall frequently write the differential operators describing the drift motion as follows:

$$\Omega_{eh} = \omega_g - \frac{\hbar}{2m_h}\nabla_1^2 - \frac{\hbar}{2m_e}\nabla_2^2 \tag{2.52a}$$

$$\Omega_{ee} = \frac{\hbar}{2m_e}(\nabla_1^2 - \nabla_2^2) \qquad \Omega_{hh} = \frac{\hbar}{2m_h}(\nabla_1^2 - \nabla_2^2) \ . \tag{2.52b}$$

### 2.5.3 Bilocal Coherence

As to be seen in (2.45, 47, 48) the relevant electromagnetic quantities are found on the diagonal of the bilocal functions $Y$, $C$ and $D$. Then, what kind of information do provide the off-diagonal parts? The answer, supported by the discussion of special solutions in later chapters, is that the off-diagonal part of any of the three functions is related to quantum coherence. In realistic models this coherence will not extend to infinite distance and therefore the dependence of $Y$, $C$ and $D$ on $|\boldsymbol{r}_1 - \boldsymbol{r}_2|$ will be such that a nonvanishing result is only obtained if

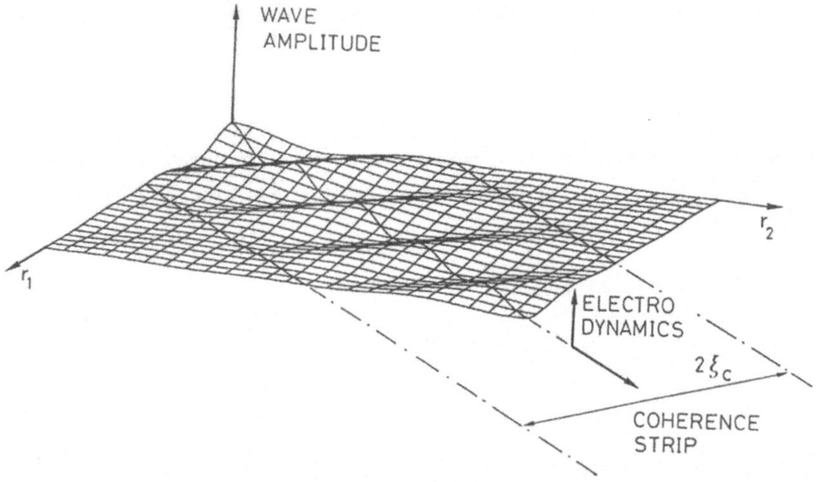


**Fig. 2.2.** Coherence region in a two dimensional model of configuration space. The line $r_1 = r_2$ contains information on electrodynamic properties

$$|r_1 - r_2| \lesssim \xi_c$$

where the typical length $\xi_c$ may be named the coherence length. A two-dimensional illustration of this situation is shown in Fig. 2.2. In applications to exciton theory (Sect. 4.5) the excitonic Bohr-radius $a_B$ takes over the role of $\xi_c$ with respect to $Y$. Another example is the application of the intraband equation to surface quantization (Sect. 6.2). There the layer thickness is the relevant coherence length.

## 2.6 Bose-Type Coherent States

In the preceding sections was developed a wave picture of electron-hole pair excitations in a semiconductor. In this and the following section we shall discuss the question of how this wave picture is related to the particle picture. A further aspect of this discussion will be to give some insight into a Schrödinger type approach to band edge dynamics, as opposed to the Heisenberg picture which has been adopted in the preceding sections.

The common notion that at not too high densities, pairs of Fermions behave like Bosons, gets a quantitative meaning if we look at the commutation relations for the electron-hole pair operators $\hat{Y}$ and $\hat{Y}^\dagger$ (see Appendix F):

$$[\hat{Y}_{ij}, \hat{Y}^\dagger_{k\ell}] = \delta_{ik}\delta_{j\ell} - \delta_{ik}C_{\ell j} - \delta_{j\ell}D_{ki} \ . \tag{2.53}$$

The first term which is unity in a generalized sense reflects the approximate Boson property of Fermion pairs. When comparing (2.53) with the band edge equations (2.23a) one can see that the Boson approximation goes together with a linearization of the dynamics of the transition, because the saturation terms in (2.32a) come directly from the non-Bose terms in (2.53). Furthermore one knows, e.g. by looking at (2.14a), that a linearized transition dynamically resembles forced harmonic oscillation. It has therefore become usual to study pair excitations in linear approximation by Bose-type exciton oscillator models (Hopfield, 1958). Other applications of the Bose oscillator model are lattice vibrations and the field oscillators of the quantized electromagnetic field.

Our primary interest is in the *driven* motion of the exciton oscillator because this is the type of motion which comes nearest to what we have been studying in the previous sections. In suitably normalized form an oscillator driven by a force $F(t)$ is represented by a Hamiltonian

$$H = \frac{\omega_0}{2}(\hat{q}^2 + \hat{p}^2) - \hat{q}F = H_0 - \hat{q}F \ . \tag{2.54}$$

The canonically conjugate variables $\hat{q}$, $\hat{p}$ have to obey the commutation relation

$$[\hat{p}, \hat{q}] = \frac{\hbar}{i} \ . \tag{2.55}$$

The canonical equations of motion which at the same time are Heisenberg's equations of motion become

$$\dot{\hat{q}} = \omega_0 \hat{p}; \quad \dot{\hat{p}} = -\omega_0 q + F \ . \tag{2.56}$$

In terms of initial operators at $t = 0$ the solutions to (2.56) are

$$\hat{q}(t) = \hat{q}(0)\cos\omega_0 t + \hat{p}(0)\sin\omega_0 t + \int_0^t F(t')\sin(\omega_0(t-t'))dt'$$

$$\hat{p}(t) = -\hat{q}(0)\sin\omega_0 t + \hat{p}(0)\cos\omega_0 t + \int_0^t F(t')\cos(\omega_0(t-t'))dt' \ . \tag{2.57}$$

So far there has been no mention of states. When thinking about states in connection with the harmonic oscillator, in the first place one remembers the eigenstates of the unperturbed Hamiltonian $H_0$, the so called number states $|n\rangle$. If the oscillator is an exciton mode, the number state $|n\rangle$ means that the mode is occupied by $n$ excitons. We quote some results related to these states $|n\rangle$ which can be found in any textbook on quantum theory (Merzbacher, 1970). The eigenvalue equation is

$$H_0|n\rangle = E_n|n\rangle; \quad E_n = \hbar\omega_0(n + \tfrac{1}{2}) \ . \tag{2.58}$$

In order to derive the spectrum of $H_0$ and to find explicit representations for the states $|n\rangle$ it is useful to consider exciton creation and annihilation operators

$$b^\dagger = (2\hbar)^{-1/2}(\hat{q} - i\hat{p}) \qquad b = (2\hbar)^{-1/2}(\hat{q} + i\hat{p}) \tag{2.59}$$

with the properties

$$bb^\dagger - b^\dagger b = 1 \tag{2.60}$$

$$b|n\rangle = n^{1/2}|n-1\rangle \qquad b^\dagger|n\rangle = (n+1)^{1/2}|n+1\rangle \ . \tag{2.61}$$

With the help of (2.61) a general number state $|n\rangle$ can be created from the ground state $|0\rangle$ by repeated application of $b^\dagger$

$$|n\rangle = (n!)^{-1/2}(b^\dagger)^n|0\rangle \ . \tag{2.62}$$

Number states form a convenient basis whenever one is discussing counting experiments with a small number of particles involved. It is therefore quite natural that number states are dominant in high energy physics. In solid state physics the situation is different because a typical probe in a solid state experiment is rather often a coherent wave formed by phase sensitive superposition of many number states. Within the framework of quantum theory this situation is described by the concept of coherent states (Glauber, 1963). A textbook introduction to coherent states is given by Merzbacher (1970).

A suitable characterization of coherent states is provided by their property that they are produced when a quantum system is excited from its ground state by means of an external force. This force has to be represented by a function of time which is "classical" in the sense that it is not an operator. For our example of a Bose-type exciton mode this means that we should find a coherent state when we transform the Heisenberg picture solution (2.57) to the Schrödinger picture.

In the Schrödinger picture the evolution of the system is represented by a state vector $|\psi(t)\rangle$ obeying the equation

$$(H_0 - \hat{q}F(t))|\psi\rangle = i\hbar\frac{\partial}{\partial t}|\psi\rangle \ . \tag{2.63}$$

Instead of directly attacking the problem of solving the Schrödinger equation with an explicitly time-dependent Hamiltonian, one can gain insight into the nature of the solution using the equivalence between the Schrödinger and the Heisenberg method. Assuming that at $t = 0$ the exciton mode is in its ground state

$$|\psi(0)\rangle = |0\rangle, \tag{2.64}$$

the calculation of the time dependent expectation values of $\hat{q}$ and $\hat{p}$ in the Heisenberg picture gives

$$\bar{q} \equiv \langle 0|\hat{q}(t)|0\rangle = \int_0^t F(t')\sin(\omega_0(t-t'))dt' \tag{2.65a}$$

$$\bar{p} \equiv \langle 0|\hat{p}(t)|0\rangle = \int_0^t F(t')\cos(\omega_0(t-t'))dt' . \tag{2.65b}$$

These values must also be found in the Schrödinger picture, hence one has

$$\langle\psi|\hat{q}(0)|\psi\rangle = \bar{q}; \quad \langle\psi|\hat{p}(0)|\psi\rangle = \bar{p} . \tag{2.66}$$

Using (2.59) we calculate expectation values for the nonhermitian operators $b$, $b^\dagger$

$$\beta = \langle\psi|b|\psi\rangle = (2\hbar)^{-1/2}(\bar{q} + i\bar{p}); \tag{2.67a}$$

$$\beta^* = \langle\psi|b^\dagger|\psi\rangle = (2\hbar)^{-1/2}(\bar{q} - i\bar{p}) . \tag{2.67b}$$

Let us further consider the m.s. fluctuation around these expectation values

$$\overline{\Delta\beta^2} \doteqdot \langle\psi|(b^\dagger - \beta^*)(b-\beta)|\psi\rangle = \psi|b^\dagger b|\psi\rangle - |\beta|^2 . \tag{2.68}$$

Exploiting the equivalence between Schrödinger- and Heisenberg pictures we can calculate $\langle\psi|b^\dagger b|\psi\rangle$ from

$$\langle\psi|b^\dagger b|\psi\rangle = (2\hbar)^{-1}\langle 0|(\hat{q}(t) - i\hat{p}(t))(\hat{q}(t) + i\hat{p}(t))|0\rangle . \tag{2.69}$$

The result is

$$\langle\psi|b^\dagger b|\psi\rangle = |\beta|^2 . \tag{2.70}$$

Comparing this with (2.68) we arrive at the conclusion that

$$\overline{\Delta\beta^2} = 0 . \tag{2.71}$$

This means that $|\psi\rangle$ must have been an eigenstate of $b$ with eigenvalue $\beta$. Since $|\psi\rangle$ is a coherent state the result can be stated as a theorem on coherent states:

Coherent states of the exciton mode fulfil an eigenvalue equation

$$b|\beta\rangle = \beta|\beta\rangle . \tag{2.72}$$

In (2.72) we have switched over to the standard Dirac notation with the eigenvalue as descriptor of the ket. Equation (2.72) can even be considered as defining the class of coherent states because any complex number $\beta$ can be reached by an appropriate choice of $F(t)$.

The complex eigenvalue $\beta$ of the non-Hermitian operator $b$ is called the complex mode amplitude of the exciton mode. While the complex amplitude in the state of driven oscillation is sharp, the real amplitude-operators $\hat{p}$ and $\hat{q}$ do not attain sharp values. But at least their fluctuation takes on the minimum which is necessary to comply with the uncertainty principle, i.e.

$$\langle\beta|(\hat{q}-\overline{q})^2|\beta\rangle\langle\beta|(\hat{p}-\overline{p})^2|\beta\rangle = \hbar^2/4 \tag{2.73}$$

for any coherent state $|\beta\rangle$.

The expansion of a coherent state in terms of quantum number states is

$$\begin{aligned}|\beta\rangle &= \exp\left(-\tfrac{1}{2}|\beta|^2\right)\sum(n!)^{-1}(\beta b^\dagger)^n|0\rangle \\ &= \exp\left(\beta b^\dagger - \tfrac{1}{2}|\beta|^2\right)|0\rangle\end{aligned} \tag{2.74}$$

as can be easily verified by inserting (2.74) into (2.72) and making use of (2.61). From this expansion it follows that in a state $|\beta\rangle$ the number of particles fluctuates according to a Poisson distribution. The probability to find $n$ excitons in the coherently excited mode is

$$w_n(\beta) = e^{-\overline{n}}\overline{n}^n/n!; \quad \overline{n} = |\beta|^2 \ . \tag{2.75}$$

A general property of the Poisson distribution is that the relative fluctuation goes like the inverse square root of the mean value:

$$(\overline{\Delta n^2})^{1/2}/\overline{n} = |\beta|^{-1} \ . \tag{2.76}$$

In a coherent state with macroscopic amplitude, i.e., $|\beta|^2 \gg 1$, one therefore has not only sharp values for the field amplitudes but also for the energy. Such a macroscopic coherent state is therefore to be considered as the quantum theoretical description of a classical field mode.

We close with a few remarks on the possible generalization to a multimode excitation. The multimode Hamiltonian is a sum of terms resembling (2.54)

$$H = \tfrac{1}{2}\sum_k [\omega_k(q_k^2 + p_k^2) - q_k F_k(t)] \quad . \tag{2.77}$$

Then multimode coherent states can be introduced as direct products of single mode states. Using (2.74) they are written as follows

$$|\{\beta\}\rangle = \exp\left[\sum_k(\beta_k b^\dagger - \tfrac{1}{2}|\beta_k|^2)\right]|0\rangle \ . \tag{2.78}$$

"$|0\rangle$" then is to be interpreted as the multimode ground state where all exciton oscillators are in the unoccupied state.

## 2.7 Coherent Pair States

In the last section an exciton mode was treated as if it were a true Bose oscillator. But we know that this is only an approximation because the exciton is a pair of Fermions which obey the exclusion principle. Therefore of the whole series of number states $|n\rangle$ only the states $|0\rangle$ = "unoccupied" and $|1\rangle$ = "occupied" are admitted. As a consequence the nice picture of reaching such a high excitation that the mode becomes a classical field oscillator is only applicable to a true Bose oscillator but not to an exciton mode. The exciton mode is saturated long before it can become classical.

Fortunately this does not mean that the idea of forming a macroscopic coherent state with a classical amplitude is completely wrong for Fermion systems. This can for example be learned from the BCS theory of superconductivity (Bardeen, Cooper, Schrieffer, 1957). In this theory it is shown that a macroscopic coherent state can be formed from paired Fermion states if one considers a multimode system. The paired Fermions in BCS theory are electrons of a single band, but with different spin. They are known as Copper pairs. It will turn out that the state of a coherently excited semiconductor has a very similar structure to that of the BCS state. The difference will be that the pairing in the semiconductor refers to electron-hole pairs instead of Cooper pairs. This result points to a close relationship between superconductivity and quantum optics which has inspired some interesting work in either field. Di Rienzo and coworkers (Di Rienzo et al., 1978) have applied ideas from quantum optics to the theory of superconductivity, Comte and Nozières (1982) have used ideas taken from BCS theory to study highly excited semiconductors. The difference between the work by Comte and Nozières and the problem we are confronted with in this section is that Comte and Nozière have studied stationary states as they occur in a plasma which has undergone intraband relaxation whereas we are aiming at the dynamics of coherent excitation before scattering takes over. In this respect our point of view is more like that adopted in the paper by Arecchi and coworkers (Arecchi et al., 1972) on coherence in two-level systems.

What we want to demonstrate in this section is the following: If one studies the dynamics of the semiconductor model of Sect. 2.4 in Schrödinger representation the state of the system turns out to be a coherent pair state with a structure known from BCS theory.

We shall use the discrete version of the semiconductor model. The necessary multimode structure in this model is introduced via the manifold of atoms which were counted by their site-index $j$. Electron-hole pairs are produced and annihilated by pair operators

$$\hat{Y}_{ij}^{\dagger} = c_j^{\dagger} d_i^{\dagger} \qquad \hat{Y}_{ij} = d_i c_j. \tag{2.79}$$

The commutation relations for these operators have been given in (2.53). In an approximate sense $\hat{Y}$ and $\hat{Y}^{\dagger}$ are Bose operators.

Let us introduce the operator $\hat{Y}^\dagger\{\eta\}$ creating a multimode pair state with amplitudes $\eta_{ij}$

$$\hat{Y}^\dagger\{\eta\} = \sum_{ij} \eta_{ij} \hat{Y}^\dagger_{ij} \ . \tag{2.80}$$

$\hat{Y}^\dagger\{\eta\}$ can be considered a generalization of the product $\beta b^\dagger$ appearing in (2.74). This serves as a motivation for introducing coherent pair states as those states which can be represented in a form analogous to (2.74)

$$|\{\eta\}\rangle = N \exp\left[\sum_{ij} \eta_{ij} \hat{Y}^\dagger_{ij}\right] |0\rangle \ . \tag{2.81}$$

Since the normalization used in (2.74) does not hold any longer we have introduced the new normalization factor $N$. If $|\{\eta\}\rangle$ is factorized into a product of exponentials and each factor is expanded we arrive at

$$|\{\eta\}\rangle = N \prod_{ij} (1 + \eta_{ij} \hat{Y}^\dagger_{ij}) |0\rangle \ . \tag{2.82}$$

Higher order terms give no contribution because of the exclusion principle. It proves convenient to have at our disposal a representation of coherent pair states where $\eta_{ij}$ is diagonalized. Since we have assumed no symmetry properties of $\eta_{ij}$ we cannot expect this diagonalization to be achieved by a unitary transformation. But at least it is always possible to find a pair of regular matrixes $\underset{\sim}{\boldsymbol{R}}$ and $\underset{\sim}{\boldsymbol{S}}$ which allow to set up the diagonal representation

$$\eta_{ij} = \sum_q R_{iq} \eta_q S_{qj} \ . \tag{2.83}$$

The new diagonal form of the state $|\{\eta\}\rangle$ then becomes

$$|\{\eta\}\rangle = N \prod_q (1 + \eta_q \hat{Y}^\dagger_q) |0\rangle \tag{2.84}$$

$$\hat{Y}^\dagger_q = c^\dagger_q d^\dagger_q \quad \text{with} \tag{2.84a}$$

$$d^\dagger_q = \sum d^\dagger_i R_{iq} \qquad c^\dagger_q = \sum S_{qj} c^\dagger_j \ . \tag{2.85}$$

In order to preserve the Fermi-Dirac commutation rules the annihilation operators have to be defined by means of the inverse transformations

$$d_q = \sum_i (\underset{\sim}{\boldsymbol{R}}^{-1})_{qi} d_i \qquad c_q = \sum_j c_j (\underset{\sim}{\boldsymbol{S}}^{-1})_{jq} \ . \tag{2.86}$$

In diagonal representation the coherent pair state looks most similar to the BCS-type state. This similarity becomes even more obvious if the normalization constant $N$ is distributed among the factors under the product in such a way that each factor becomes normalized. Then the coherent pair state takes on the BCS-like form

$$|\{\eta\}\rangle = \prod(u_q + v_q c_q^\dagger d_q^\dagger)|0\rangle \tag{2.87}$$

$$\prod u_q = N; \quad u_q\eta_q = v_q; \quad |u_q|^2 + |v_q|^2 = 1. \tag{2.87a}$$

The form (2.87) of the coherent pair state is distinguished by the property that the amplitudes are simply related to expectation values of the corresponding pair annihilation operators

$$u_q^* v_q = \langle\{\eta\}|\hat{Y}_q|\{\eta\}\rangle \ . \tag{2.88}$$

So far coherent pair states were introduced in a formal manner by the guess that (2.81) might provide a relevant generalization to (2.74). We still have to show that these states are also coherent in the sense of the last section. There we have classified a state as coherent if it can be excited from the ground state by means of a "classical" force.

In Appendix J a proof is given that within the semiconductor model of Sect. 2.4 this concept of "coherence by excitation" applies to the class of coherent pair states. The theorem which can be proved states that the model semiconductor evolves through time dependent pair states $|\{\eta(t)\}\rangle$ if it is excited from the ground state $|0\rangle$ by an arbitrary time dependent classical field $E(t)$.

In the case of true Boson systems, coherent states are distinguished by their property that in such a state the annihilation operator is replaced by a complex number function. For a coherent pair state this is in general not true unless the state was diagonalized in the sense of (2.87). But, as can be learned from the theory of superconductivity (De Gennes, 1966) and also from the work on coherence in two-level systems (Arecchi et al., 1972), such a replacement is allowed at least in an approximate sense if a many Fermion system is excited to a coherent state far from saturation, i.e., $|v| \ll 1$. We shall make use of this possibility in Sect. 3.2 where products of operators are approximated by products of number functions.

# 3. Refinements of the Band-Edge Model

## 3.1 Survey of Desirable Refinements

In the last chapter the essential structure of the constitutive equations of a band-edge model was developed. But this model has still to undergo a number of refinements in order to apply it to realistic experimental situations. This chapter will be devoted to the development of these refinements. But before this is attempted we shall, in this introductory section, give some comments on the things to be done in this chapter.

### 3.1.1 Coulomb Force and Induced Fields

The electric field introduced in Chap. 2 was always imagined as the transverse field of a light wave driving the transition dipoles. But in a band-edge model with independently mobile carriers this is not the only electromagnetic interaction that can occur. If an electron-hole pair is separated it becomes a pair of monopoles, and these monopoles then are influenced by longitudinal fields in a number of ways:

i) there may be external longitudinal fields exerting a potential $\Phi^{ex}$ that is felt by the monopoles;
ii) the electron-hole densities themselves act as sources of longitudinal fields described by an induced potential $\Phi^{in}$; $\Phi^{ex}$ and $\Phi^{in}$ form together what is known as the "self-consistent field" $\Phi$;
iii) the Coulomb-type interaction between those pairs which are tied together in the coherent wave $Y$, will as an internal property of the pair not be mediated by self-consistent fields. Instead it will give rise to an extra term in the constitutive equations. This term will be responsible for the formation of the bound state resonances known as Wannier excitons.
iv) The excitonic polarization acts as an additional source of longitudinal fields if the exciton is of the longitudinal type.

A self-consistency requirement also applies to the transverse part of the electromagnetic field. It means that the polarization identified as the diagonal part of the coherent electron-hole wave $Y$ will act as a source emitting induced electromagnetic radiation. This has to be accounted for by adding to the band-edge equations the electromagnetic wave equation with a source term derived from $Y$.

All these refinements will be discussed in Sect. 3.2.

### 3.1.2 Electron-Phonon Interaction

The system of electrons and holes does not form a closed system but is interacting with a number of other dynamic structures present in a semiconductor. Of particular importance is the electron-phonon interaction whose impact on band-edge dynamics will be explored in Sect. 3.3. We shall take into account two types of interactions, viz. the so-called Fröhlich interaction being mediated by long range electric fields and the deformation potential interaction which is a direct short range interaction.

### 3.1.3 Band Structure

The naïve tight binding model of Sect. 2.4 has to be replaced by a realistic band structure. In the first place this means that the atomic states forming the microscopic basis of the approach in Sect. 2.4 will be replaced by Wannier states taken from band-structure calculations. The use of realistic Wannier functions entails a number of complications that were ignored in the simple approach of Chap. 2. This will confront us with the following problems:

i) to think of the influence of crystal symmetry;
ii) to take care of possible degeneracies (including spin);
iii) to face the fact that Wannier functions are not as localized as atomic orbitals.

Our strategy, when dealing with these complications, will be to avoid the construction of a "perfect" set of band edge equations where all these complications are taken care of. In order not to obscure the important conceptual points we shall stick to the quasiscalar simplification of the last chapter wherever possible and only mark the entry points of the different refinements in such a way that they can be added at need.

### 3.1.4 Irreversible Processes

The band-edge equations as they were derived in Sect. 2.4 are reversible because they come directly from Hamiltonian mechanics without any manipulations like projection or phase randomization. But there is some need also to consider irreversible influences on band edge dynamics. This is most clearly seen for the intraband part. As already mentioned in Sect. 2.5, the intraband part is a kinetic equation with additional sources provided by interband processes. Then in order to make the kinetic equations complete there should also be some

Boltzman-type collision terms added. These terms are of vital importance if one wants to study how the electron-hole plasma produced by coherent excitation reaches a state of intraband equilibrium. That such a state as a long lived intermediate state occurs before recombination takes over is well known from studies on highly excited semiconductors.

Other important irreversible processes influencing the intraband part of band-edge dynamics are generation and recombination processes of various kinds. In principle there should also be incorporated the processes describing exchange of carriers between the band state system and the dopants.

As far as our present understanding goes, the irreversible contributions to the interband part of band-edge dynamics are not as rich as in the intraband case. In most applications it seems to be sufficient to introduce the $T_2$-type relaxation known already from the two-level system [compare (2.16)].

Irreversible contributions to the equations will be discussed in Sect. 3.7.

## 3.2 Coulomb Interaction and Self-Consistent Fields

As announced in the last section the bare edge equation (2.44) must be extended in order to take care of different Coulomb-type interactions in the system. Using the compact notation of (2.52), but with a more general meaning of the $\Omega$ operators, the extended edge equations are written as

$$\dot{Y}_{12} + i\Omega_{eh}Y_{12} = iM_0\hbar^{-1}(\delta_{12}E_2 - E_1C_{12} - E_2D_{21}) + X_Y \tag{3.1a}$$

$$\dot{C}_{12} + i\Omega_{ee}C_{12} = -iM_0\hbar^{-1}(E_1Y_{12} - E_2Y_{21}^*) + X_C \tag{3.1b}$$

$$\dot{D}_{12} + i\Omega_{hh}D_{12} = -iM_0\hbar^{-1}(Y_{21}E_1 - Y_{12}^*E_2) + X_D \; ; \tag{3.1c}$$

$$\Omega_{eh} = \omega_g - \frac{\hbar}{2m_h}\nabla_1^2 - \frac{\hbar}{2m_e}\nabla_2^2 + \hbar^{-1}[e(\Phi_1 - \Phi_2) + V_{eh}] \tag{3.2a}$$

$$\Omega_{ee} = \frac{\hbar}{2m_e}(\nabla_1^2 - \nabla_2^2) + e\hbar^{-1}(\Phi_1 - \Phi_2) \tag{3.2b}$$

$$\Omega_{hh} = \frac{\hbar}{2m_h}(\nabla_1^2 - \nabla_2^2) - e\hbar^{-1}(\Phi_1 - \Phi_2) \; . \tag{3.2c}$$

$V_{eh}$ is the electron-hole interaction potential. In the original Wannier model which is well confirmed in the low excitation regime, $V_{eh}$ is assumed to be a dielectrically screened Coulomb potential (Knox, 1963):

$$V_{eh} = -\frac{e^2}{4\pi\varepsilon_0\varepsilon|\boldsymbol{r}_1 - \boldsymbol{r}_2|} \; . \tag{3.3}$$

The choice of the screening constant $\varepsilon$ in (3.3) requires a decision depending on the situation to be treated. We shall discuss this problem separately in Sect. 3.4.

Except where otherwise stated, $\varepsilon$ will denote the static dielectric constant of the undoped semiconductor.

The terms $X_Y$, $X_C$, $X_D$ are exchange terms relevant at high excitations. These will be discussed at the end of this section.

In general the electric field $\boldsymbol{E}$ and the scalar potential $\Phi$ are each composed of two parts

$$\boldsymbol{E} = \boldsymbol{E}^{ex} + \boldsymbol{E}^{in} \tag{3.4}$$

$$\Phi = \Phi^{ex} + \Phi^{in} \tag{3.5}$$

where $\boldsymbol{E}^{ex}$ and $\Phi^{ex}$ are external fields while $\boldsymbol{E}^{in}$ and $\Phi^{in}$ are self-consistent induced fields. By "external" we mean that the sources of $\boldsymbol{E}^{ex}$ and $\Phi^{ex}$ are not derived from $Y$, $C$ or $D$.

The induced potential $\Phi^{in}$ obeys the Poisson equation

$$\varepsilon_0 \varepsilon \nabla^2 \Phi^{in} = -\varrho^{in} \tag{3.6}$$

where the induced charge density $\varrho^{in}$ consists of monopolar contributions $\varrho_m^{in}$ from $C$ and $D$, and a dipolar contribution $\varrho_d^{in}$ from the pair wavefunction $Y$ :

$$\varrho^{in}(\boldsymbol{r}) = \varrho_m^{in}(\boldsymbol{r}) + \varrho_d^{in}(\boldsymbol{r}) = -e[C(\boldsymbol{r},\boldsymbol{r}) - D(\boldsymbol{r},\boldsymbol{r})] - \nabla \boldsymbol{P}(\boldsymbol{r}) \ . \tag{3.7}$$

The interband polarization $\boldsymbol{P}$ is derived from $Y$ as in (2.48)

$$\boldsymbol{P}(\boldsymbol{r}) = 2\boldsymbol{M}_0 \operatorname{Re}\{Y(\boldsymbol{r},\boldsymbol{r})\} \ . \tag{3.8}$$

The induced field $\boldsymbol{E}^{in}$ in the most general case may consist of an irrotational (longitudinal) part $\boldsymbol{E}_\ell^{in}$ and a solenoidal (transverse) part $\boldsymbol{E}_t^{in}$. The longitudinal field is derived from the potential $\Phi^{in}$

$$\boldsymbol{E}_\ell^{in} = -\nabla \Phi^{in} \ . \tag{3.9}$$

Therefore the possible sources of $\boldsymbol{E}_\ell^{in}$ are the charge densities given by (3.7).

The transverse part $\boldsymbol{E}_t^{in}$ alone, is derived from a wave equation where the corresponding polarization

$$\boldsymbol{P}_t(\boldsymbol{r}) = [2\boldsymbol{M}_0 \operatorname{Re}\{Y(\boldsymbol{r},\boldsymbol{r})\}]_t \tag{3.10}$$

acts as a source

$$c^2 \varepsilon_0 \nabla^2 \boldsymbol{E}_t^{in} - \varepsilon_0 \tilde{\varepsilon} \ddot{\boldsymbol{E}}_t^{in} = \ddot{\boldsymbol{P}}_t \ . \tag{3.11}$$

The quantity $\tilde{\varepsilon}$ will be called the residual dielectric constant. $\tilde{\varepsilon}$ is related, but rarely equal to the residual dielectric constant used when treating single excitonic resonances (Hopfield and Thomas, 1963). The proper choice of $\tilde{\varepsilon}$ will be discussed in Sect. 3.4 together with the already mentioned problem of choosing the screening constant $\varepsilon$.

When treating exciton polaritons it is convenient to generalize (3.11) in such a way that it accounts for longitudinal as well as for transverse fields related to the interband polarization. The generalization amounts to replacing (3.11) by the equation

$$-c^2\varepsilon_0\nabla\times\nabla\times\boldsymbol{E}^{in} - \varepsilon_0\tilde{\varepsilon}\boldsymbol{E}^{in} = 2M_0\,\mathrm{Re}\{\ddot{Y}(\boldsymbol{r},\boldsymbol{r})\} \; . \tag{3.11a}$$

Note that the possibility to treat transverse and longitudinal fields on an equal footing as in (3.11a), is related to the use of a local gauge (Appendix B). The typical property of this gauge is that the dipolar coupling to the electric field has the same form irrespective of the longitudinal or transverse character of the field. As a consequence one can also forget about this distinction in the source terms on the r.h.s. of (3.1).

The intuitive reasoning in the above presentation of electromagnetic interactions in the edge equations is greatly helped by the possibility of maintaining the traditional distinction between bound and mobile charges and their respective currents. The "bound charges" are treated in terms of $Y$ and the mobile charges are derived from $C$ and $D$. Thus we can keep this useful distinction although it is understood to have only an operational sense. In a more fundamental understanding the same carriers are participating in "bound" $Y$-type and in mobile $C$-type or $D$-type mechanisms.

The structure laid down in equations (3.1–11) gets additional support from a derivation based on the standard Hamiltonian of a two band semiconductor with electromagnetic interactions (Huhn and Stahl, 1984). Without repeating the lengthy calculation let us briefly discuss a few aspects of this.

The starting point is the Hamiltonian of Sect. 2.4, extended by two additional terms

$$H = H_{TL} + H_M + H_{ex} + H_c \; . \tag{3.12}$$

$H_{TL}$ is the two-level Hamiltonian in (2.5), $H_M$ is the mobility term (2.29), and $H_{ex}$ given by

$$H_{ex} = -e\sum_j(\hat{C}_{jj} - \hat{D}_{jj})\Phi^{ex}(\boldsymbol{R}_j) \tag{3.12a}$$

describes monopole interaction with an external field [the dipolar part in our gauge appears in $H_{EM}$; compare (B.23–28) and (2.5)]. $H_c$ is the Coulomb interaction

$$H_c = \frac{1}{2}\int \psi^\dagger(\boldsymbol{x}_1)\psi^\dagger(\boldsymbol{x}_2)V(\boldsymbol{x}_1 - \boldsymbol{x}_2)\psi(\boldsymbol{x}_2)\psi(\boldsymbol{x}_1)d^3\boldsymbol{x}_1 d^3\boldsymbol{x}_2 \; . \tag{3.12b}$$

$V(\boldsymbol{x}_1 - \boldsymbol{x}_2)$ is the properly screened two point-charge interaction, e.g. the dielectrically screened potential (3.3) (compare also Sect. 3.4 and Appendix H).

Note that $H_c$ is formulated on the microscopic scale, denoted by $\boldsymbol{x}$ in contrast to the interpolating macroscale $\boldsymbol{r}$ (Appendix A). In a two-band tight-binding model the field operators $\psi, \psi^\dagger$ are expanded in terms of tight binding orbitals $t_c, t_v$

$$\psi(\boldsymbol{x}) = \sum [c_j t_c(\boldsymbol{x} - \boldsymbol{R}_j) + d_j^\dagger t_v(\boldsymbol{x} - \boldsymbol{R}_j)]$$

$$\psi^\dagger(\boldsymbol{x}) = \sum [c_j^\dagger t_c(\boldsymbol{x} - \boldsymbol{R}_j) + d_j t_v(\boldsymbol{x} - \boldsymbol{R}_j)] \ . \tag{3.13}$$

The orbital functions are assumed to obey the approximate selection rules:

$$\int t_c(\boldsymbol{x} - \boldsymbol{R}_i) t_c(\boldsymbol{x} - \boldsymbol{R}_j) d^3\boldsymbol{x} = \int t_v(\boldsymbol{x} - \boldsymbol{R}_i) t_v(\boldsymbol{x} - \boldsymbol{R}_j) d^3\boldsymbol{x} = \delta_{ij}$$

$$\int t_c(\boldsymbol{x} - \boldsymbol{R}_i) t_v(\boldsymbol{x} - \boldsymbol{R}_j) d^3\boldsymbol{x} = 0$$

$$e \int t_c(\boldsymbol{x} - \boldsymbol{R}_i) \boldsymbol{x} t_v(\boldsymbol{x} - \boldsymbol{R}_j) d^3\boldsymbol{x} = M_0 \delta_{ij} \ . \tag{3.14}$$

With the help of (3.14), $H_c$ can be transformed into a multipole expansion. In this way it is seen that $H_c$ describes the interaction of an assembly of monopoles and dipoles localized on lattice sites. The monopoles are related to the dynamical variables $\hat{C}_{jj}$ and $\hat{D}_{jj}$ whereas the dipoles belong to $(\hat{Y}_{jj} + \hat{Y}_{jj}^\dagger)$.

Considering the contributions of the extra terms $H_{ex} + H_c$ in (3.12) to the equations of motion, it is seen that $H_{ex}$ poses no serious problems because the contributions resulting from $H_{ex}$ are at most linear in the dynamical variables $\hat{Y}$, $\hat{C}$ or $\hat{D}$ respectively. The structure of the terms derived from $H_{ex}$ can be shown to coincide exactly with the potential field terms in (3.1) if the fields are identified with $\Phi^{ex}$ and $-\nabla\Phi^{ex}$, respectively.

Special consideration must be given to the contributions from $H_c$ to the edge equations because the commutators

$$[\hat{Y}_{ij}, H_c], \quad [\hat{C}_{ij}, H_c], \quad [\hat{D}_{ij}, H_c]$$

in general produce terms which are bilinear in the dynamical variables. They are treated according to the following rules:

1. The terms are rearanged in order to identify self-consistent fields;
2. Wherever possible a normal order is established with creation operators standing on the left side of annihilation operators;
3. The sources of self-consistent fields are replaced by their expectation values, and the exchange terms $X_Y$, $X_C$, $X_D$ are added according to a time dependent Hartree-Fock approximation.

In order to illustrate the procedure let us consider as an example the monopole-monopole term from the commutator $[\hat{Y}_{ij}, H_c]$. This term is found to be

$$[\hat{Y}_{ij}, H_c]_{mm} = \sum_k e^2(V_{jk} - V_{ik}) d_i (\hat{C}_{kk} - \hat{D}_{kk}) c_j \tag{3.15}$$

$$V_{jk} = (4\pi\varepsilon_0\varepsilon|\boldsymbol{R}_j - \boldsymbol{R}_k|)^{-1} \ . \tag{3.15a}$$

The first rule, viz. identification of self-consistent fields, is easily met. Evidently the combination

$$\hat{\Phi}_m(\boldsymbol{R}_i) = -e \sum_k V_{ik}(\hat{C}_{kk} - \hat{D}_{kk}) \tag{3.16}$$

is the operator-valued potential field produced by the monopolar charges associated with $\hat{C}$ and $\hat{D}$.

The second rule was to establish a normal order. In our example (3.15), this amounts to placing $d_i$ behind the bracket $(\hat{C}_{kk} - \hat{D}_{kk})$. This commutation transforms (3.15) into

$$[\hat{Y}, H_c] = e[\hat{\Phi}_m(\boldsymbol{R}_i) - \hat{\Phi}_m(\boldsymbol{R}_j)]\hat{Y}_{ij} - e^2(V_{ij} - V_{ii})\hat{Y}_{ij} \ . \tag{3.17}$$

In the first term of (3.17) the operator-valued fields $\hat{\Phi}_m$ are now replaced by their expectation values in the spirit of the so-called self-consistent field approximation (Ehrenreich and Cohen, 1959). In this and the closely related random phase approximation (RPA) it is assumed that (3.16) and similar expressions are dominated in their expectation value by a contribution having its origin in collective motion (Bohm and Pines, 1953; Nozières and Pines, 1958; Suhl and Werthammer, 1961; Egri, 1985).

If one wants to go beyond the Hartree-type self-consistent field approximation one can add the corresponding exchange terms in the spirit of a Hartree-Fock approximation. In case of (3.17) this amounts to the followig replacements

$$\langle\hat{\Phi}_m(\boldsymbol{R}_i)\hat{Y}_{ij}\rangle = -e \sum_k V_{ik}(\langle c_k^\dagger c_k d_i c_j\rangle - \langle d_k^\dagger d_k d_i c_j\rangle)$$

is replaced by

$$\begin{aligned}\langle\hat{\Phi}_m(\boldsymbol{R}_i)\hat{Y}_{ij}\rangle \approx & -e \sum_k V_{ik}(\langle c_k^\dagger c_k\rangle\langle d_i c_j\rangle - \langle c_k^\dagger c_j\rangle\langle d_i c_k\rangle \\ & - \langle d_k^\dagger d_k\rangle\langle d_i c_j\rangle + \langle d_k^\dagger d_i\rangle\langle d_k c_j\rangle) \\ & = \Phi_m(\boldsymbol{R}_i)Y_{ij} - e\sum_k V_{ik}(C_{kj}Y_{ik} - D_{ki}Y_{kj})\end{aligned} \tag{3.18}$$

and analogously

$$\langle\hat{\Phi}(\boldsymbol{R}_j)\hat{Y}_{ij}\rangle \approx \Phi_m(\boldsymbol{R}_j)Y_{ij} - e\sum_k V_{jk}(C_{kj}Y_{ik} - D_{ki}Y_{kj}) \ . \tag{3.19}$$

The general rule for generating all relevant terms in the Hartree-Fock approximation is to consider all products of expectation values that can be formed with factors being either $Y$, $C$ or $D$.

While the terms in (3.18, 19) are nonlinear in the dynamic variables, the last term in (3.17) is a linear one. It describes the Coulombic interaction of the two point charges associated with the position vectors $\boldsymbol{r}_1$ and $\boldsymbol{r}_2$ in the wave function $Y(\boldsymbol{r}_1, \boldsymbol{r}_2)$. The nonlinear character of the contributions (3.18, 19) means that their relevance is restricted to highly excited semiconductors. In addition the exchange terms in (3.18, 19) exhibit a nonlocal structure. In the continuum limit they will become integrals. The edge equations with exchange included therefore become integro-differential equations. The nonlocality is related to quantum-coherence because it is based on a comparison between the range of the interaction $V$ and a relevant coherence length. If the relevant coherence length of either $C$, $D$ or $Y$ in (3.18, 19) is small compared to the range of $V$, then the nonlocality can be expressed by differential operators by means of a moment expansion similar to the procedure described in Sect. 2.4.

When the rules demonstrated in the above example are applied to the complete set of equations of motion generated from the Hamiltonian (3.12) one finds the structure of edge dynamics laid down in (3.1–11). The exchange terms which are not listed explicitely are easily generated from the corresponding Hartree-terms as shown in (3.18, 19). In discussing solutions to the edge equations in Sect. 7.7 the leading exchange term will be denoted as $\Omega_x Y_{ij}$ and interpreted as a nonlocal screening of the potential $V_{eh}$. Other exchange terms might be more naturally considered as additional source terms or as dynamic contributions to band renormalization. So, e.g., in a Green's function approach to band edge dynamics (Haug and Schmitt-Rink, 1984) part of (3.18, 19) appears as a contribution to the self-energy.

The use of the RPA in the derivation of the edge equations (3.1) indicates that corrections might be necessary. As can be learned from the extensive work on the validity of the RPA (Nozieres and Pines, 1958; Bohm and Pines, 1953) necessary corrections refer to residual short range interactions. In our approach these will be accounted for by irreversible terms (Sect. 3.7).

A problem that needs further discussion is the screening of the two point charge interaction $V_{eh}$, and the choice of the residual dielectric constant $\tilde{\varepsilon}$. These topics will be discussed in Sect. 3.4 and in Appendix H.

## 3.3 Electron-Phonon Interaction

Although the name "phonon" appeals to the particle picture it is quite common to treat lattice vibrations as wavelike objects. In the long wavelength limit the displacement characteristic for the mode under consideration is described by a wavefunction $\xi(\boldsymbol{r}, t)$. The multi-purpose symbol $\xi$ will be used in the same manner as by Cardona (1982). The spatial dependence is to be understood in the sense of a band-limited interpolation (Appendix A).

It is known that the influence of phonons on electronic processes near the semiconductor band edge is mainly exerted via two types of interactions (Mahan, 1981; Zawadzki, 1982; Bir and Pikus, 1972). One is called the Fröhlich interaction (Fröhlich, Pelzer, and Zienau, 1950). It is mediated by long-range potential fields. A typical source of such potential fields is a longitudinal optical (LO) phonon. In piezoelectric crystals also acoustic phonons can participate in a Fröhlich-type interaction. The second relevant kind of electron-phonon interaction is the so-called deformation-potential interaction. This interaction is of short range and therefore there is no necessity to introduce a mediating potential field. The deformation-potential interaction is considered as a direct local interaction and is expressed as a shift in the local band edge proportional to the phonon amplitude $\xi$.

The above mentioned classification of possible electron-phonon interactions offers an attractive structure for phenomenological reasoning. This structure will be handled in a similar way as it was done in the preceding sections with the results of band-structure theory. In actually following this plan one meets the problem that the project tends to become subdivided into a large number of different cases, as can be seen from the (yet incomplete) listing of a number of relevant examples:

| Phonon | Electronic mechanism | Interaction |
|---|---|---|
| LA | interband | def.pot.cond.band |
| LA | interband | def.pot.val.band |
| LA | intraband | def.pot.cond.band |
| LA | interband | piezoel.Fröhlich |
| TO | interband | def.pot. |
| LO | interband | Fröhlich |
| LO | intraband | Fröhlich |

From the diversity of the subject it is clear that somehow we must rationalize the presentation. We therefore go on using the multi-purpose phonon amplitude $\xi(\boldsymbol{r})$. In applications to acoustical phonons $\xi$ replaces in a symbolic sense the elastic strain tensor $\underset{\sim}{\boldsymbol{\varepsilon}}$. Then $\xi$ may have up to six independent components. In the long-wavelength limit the relevant dynamics of $\xi$ is elastomechanics of crystals (Nye, 1972). In dealing with optical modes $\xi$ will usually be identified with the charge displacement vector $\boldsymbol{w}$ introduced by Huang (1951) in his phenomenological theory of optical lattice modes. For nonpolar optical phonons were Huang's theory does not apply, $\xi$ may be identified with the mode amplitude. In discussing the pertinent contributions of $\xi$ to the electronic band-edge equations we have to distinguish between the Fröhlich type and the deformation-potential type:

### 3.3.1 Fröhlich-Interaction

Fröhlich interaction takes place if the lattice mode is accompanied by a dielectric polarization. In a symbolically abbreviated form the polarization related to $\xi$ is written as

$$P_\xi = \underset{\sim}{Q}\xi \ . \tag{3.29}$$

In case of an optical phonon described by Huang's displacement vector $\boldsymbol{w}$, $\underset{\sim}{Q}$ is proportional to the effective charge. Another example is the application of (3.29) to the piezoelectric polarization produced by an acoustic strain; then $\underset{\sim}{Q}$ stands for the piezoelectric tensor.

Given $P_\xi$, one can calculate the contribution of $\xi$ to the electric potential field $\Phi$ entering the electronic band-edge equations. This is done by solving Poisson's equation:

$$\varepsilon_0\varepsilon\nabla^2\Phi_\xi = \nabla\cdot P_\xi \ . \tag{3.30}$$

Again, as in Sect. 3.2, the screening constant $\varepsilon$ has to be adjusted to the model in the sense that it accounts for those modes which, if present, are kept outside the dynamic equations (see also Sect. 3.4). The solution of (3.30) then has to be inserted into the drift operators $\Omega_{eh}$, $\Omega_{ee}$, $\Omega_{hh}$. If one assumes that besides $\Phi_\xi$ other contributions to $\Phi$ are absent or negligible, then, e.g., $\Omega_{eh}$ (3.2a) becomes

$$\Omega_{eh}(\xi) = \omega_g - \frac{\hbar}{2m_h}\nabla_1^2 - \frac{\hbar}{2m_e}\nabla_2^2 + \frac{1}{\hbar}[e\Phi_\xi(\boldsymbol{r}_1) - e\Phi_\xi(\boldsymbol{r}_2) + V_{eh}] \ . \tag{3.31}$$

Analogous expressions are found for $\Omega_{ee}$ and $\Omega_{hh}$.

Sometimes one is only interested in the influence of the phonon on the electronic system while the reaction force on the phonons is neglected. An example where this is the case is spontaneous Raman scattering (Sect. 7.1). But in general the backward influence of the electronic system on $\xi$ is of equal interest.

An elegant method to derive this backward influence is furnished by the use of a Lagrangian (Appendix C). Let us assume that the bare phonon (without electron-phonon interaction) obeys an oscillator equation

$$\mu(\ddot{\xi} + \omega_0^2\xi) = 0 \ . \tag{3.32}$$

This equation is derived from an action principle with the Lagrange density (Kittel, 1963)

$$\Lambda_\xi = \tfrac{1}{2}\mu\dot{\xi}^2 - \tfrac{1}{2}\mu\omega_0^2\xi^2 \ . \tag{3.33}$$

The integrated Lagrangian of the phonon system then is

$$L_\xi = \int(\tfrac{1}{2}\mu\dot{\xi}^2 - \tfrac{1}{2}\mu\omega_0^2\xi^2)d^3\boldsymbol{r} \ . \tag{3.34}$$

In Appendix C it is shown that the interaction term in (3.31) is derived from an interaction Lagrangian

$$L_{int} = -e\int|Y(\boldsymbol{r}_1,\boldsymbol{r}_2)|^2[\Phi_\xi(\boldsymbol{r}_1) - \Phi_\xi(\boldsymbol{r}_2)]d^3\boldsymbol{r}_1d^3\boldsymbol{r}_2 \ . \tag{3.35}$$

By solving (3.30) one can replace $\Phi_\xi$ in (3.35) by a convolution of $\xi$ with the Coulomb propagator

$$\Phi_\xi(\boldsymbol{r}_i) = -\frac{1}{4\pi\varepsilon_0\varepsilon}\int\frac{\nabla[Q\xi(\boldsymbol{r})]}{|\boldsymbol{r}-\boldsymbol{r}_i|}d^3\boldsymbol{r} \tag{3.36}$$

yielding

$$\begin{aligned}L_{int} = \frac{e}{4\pi\varepsilon_0\varepsilon}&\int\left(\frac{1}{|\boldsymbol{r}-\boldsymbol{r}_1|}-\frac{1}{|\boldsymbol{r}-\boldsymbol{r}_2|}\right)\nabla[Q\xi(\boldsymbol{r})]\\ &\times|Y(\boldsymbol{r}_1,\boldsymbol{r}_2)|^2 d^3\boldsymbol{r}\, d^3\boldsymbol{r}_1 d^3\boldsymbol{r}_2 \ .\end{aligned} \tag{3.37}$$

It is now easy to derive the equation of motion for $\xi$ under the influence of $Y$ from the action principle

$$\delta\int(L_\xi + L_{int})dt = 0 \ . \tag{3.38}$$

Since we only ask for the phonon equation, the variation is only performed with respect to $\xi$. Applying the standard Euler-Lagrange procedure under the assumption that $\xi$ is the displacement vector $\boldsymbol{w}$ of Huang's model one finds the equation of motion:

$$\begin{aligned}\mu(\ddot{\boldsymbol{w}}+\omega_0^2\boldsymbol{w}) = -\frac{eQ}{4\pi\varepsilon_0\varepsilon}\nabla&\int\left(\frac{1}{|\boldsymbol{r}-\boldsymbol{r}_1|}-\frac{1}{|\boldsymbol{r}-\boldsymbol{r}_2|}\right)\\ &\times|Y(\boldsymbol{r}_1,\boldsymbol{r}_2)|^2 d^3\boldsymbol{r}_1\, d^3\boldsymbol{r}_2 \ .\end{aligned} \tag{3.39}$$

In this case $Q = \tilde{q}/\Omega^{1/2}$, $\tilde{q}$ being the effective charge. The factor $\Omega^{-1/2}$ accounts for the fact that $\boldsymbol{w}(\boldsymbol{r})$ is an interpolated function according to the rules in Appendix A. A treatment of the exciton-phonon interaction within the band-edge model is then based on (3.39) together with an interband equation for $Y$ where the drift operator (3.31) is used.

The Fröhlich-type phonon influence on intraband dynamics is easily obtained by inserting $\Phi_\xi$ from (3.36) into the intraband drift operators (3.2b,c):

$$\Omega_{ee}(\xi) = \frac{\hbar}{2m_e}(\nabla_1^2-\nabla_2^2)+\frac{e}{\hbar}[\Phi_\xi(\boldsymbol{r}_1)-\Phi_\xi(\boldsymbol{r}_2)] \ , \tag{3.40a}$$

$$\Omega_{hh}(\xi) = \frac{\hbar}{2m_h}(\nabla_1^2-\nabla_2^2)-\frac{e}{\hbar}[\Phi_\xi(\boldsymbol{r}_1)-\Phi_\xi(\boldsymbol{r}_2)] \ . \tag{3.40b}$$

The problem of finding the reaction force from intraband motion on the phonon oscillator is somewhat more complicated than in the interband case. The reason is that $C$ and $D$ have to obey constraints. Therefore the simple Lagrange procedure used in the interband case does not apply. We postpone the problem to Sect. 6.7 where it will be discussed in connection with the polaron model.

### 3.3.2 Deformation Potential

The short range electron-phonon interaction can be treated formally along very similar lines as the long-range Fröhlich interaction if one replaces the Coulomb potential (3.36) by a contact potential

$$\Delta_\nu(\boldsymbol{r}) = d_\nu \xi(\boldsymbol{r}) \ ; \quad \nu = c, v \ . \tag{3.41}$$

$\Delta_\nu$ describes the local variation of the respective band edge with the phonon coordinate $\xi$. In order to keep the sign convention adopted in the Fröhlich case it is convenient to express $\Delta_\nu$ in terms of electron energies also for the valence band. The deformation potential constant $d_\nu$ will in general be a multicomponent parameter, and then (3.41) includes a summation. An example is the deformation potential of acoustical phonons. Denoting the strain tensor by $\underset{\sim}{\boldsymbol{\varepsilon}}$ the most general form would be

$$\Delta_\nu = \sum_{\kappa\lambda} d_\nu^{\kappa\lambda} \varepsilon_{\kappa\lambda} \tag{3.42}$$

but in most cases it is sufficient to consider the approximation

$$\Delta_\nu = d_\nu \Theta \ ; \quad \Theta = \sum_\lambda \varepsilon_{\lambda\lambda} \ . \tag{3.43}$$

When the deformation potentials are inserted into the drift operators of the electronic system one obtains

$$\Omega_{eh}(\xi) = \omega_g - \frac{\hbar}{2m_h}\nabla_1^2 - \frac{\hbar}{2m_e}\nabla_2^2 + \frac{1}{\hbar}[V_{eh} - d_v\xi(\boldsymbol{r}_1) + d_c\xi(\boldsymbol{r}_2)] \tag{3.44a}$$

$$\Omega_{ee}(\xi) = \frac{\hbar}{2m_e}(\nabla_1^2 - \nabla_2^2) - \frac{dc}{\hbar}[\xi(\boldsymbol{r}_1) - \xi(\boldsymbol{r}_2)] \tag{3.44b}$$

$$\Omega_{hh}(\xi) = \frac{\hbar}{2m_h}(\nabla_1^2 - \nabla_2^2) + \frac{dv}{\hbar}[\xi(\boldsymbol{r}_1) - \xi(\boldsymbol{r}_2)] \ . \tag{3.44c}$$

Note that, for simplicity, potential terms other than the deformation potential have not been considered in (3.44); in general $\Phi$-type contributions of different origin must be expected to occur together with the deformation potentials $\Delta_\nu$. In particular it is possible that $\Phi$ comprises a Fröhlich potential arising from the same phonon as the deformation potential considered. In this case of competitive interaction, in general the long-range Fröhlich mechanism turns out to be much stronger, and one may neglect therefore the corresponding contribution from $\Delta$.

The reaction force exerted by the interband excitation $Y$ on the phonon via the deformation potential again is found by invoking a Lagrange formulation. Since the reasoning is analogous to the Fröhlich case we immediately give the resulting phonon equation:

$$\mu(\ddot{\xi} + \omega_0^2\xi) = \int |Y(\mathbf{r}_1, \mathbf{r}_2)|^2 [d_v\delta(\mathbf{r} - \mathbf{r}_1) - d_c\delta(\mathbf{r} - \mathbf{r}_2)] d^3\mathbf{r}_1\, d^3\mathbf{r}_2 \ . \tag{3.45}$$

The electron-phonon coupling gives rise to nonlinear interactions. An interesting application of this nonlinearity are so called induced Raman processes. We shall discuss these processes in Sect. 7.3.

## 3.4 On the Choice of Epsilon

In the preceding two sections the concept of an environmental dielectric constant was used. For example, in the wave equation (3.12) we introduced the concept of a residual dielectric constant $\tilde{\varepsilon}$, and in (3.30) for the Fröhlich potential $\Omega_\xi$ a suitable $\varepsilon$ was to be used. In addition, the electron-hole interaction is always screened dielectrically, and so an appropriate dielectric constant should be assigned to $V_{eh}$ in the drift operator $\Omega_{eh}$. In this section we shall discuss in some detail the meaning of these dielectric parameters. In particular we want to stress that the proper choice of the $\varepsilon$ parameter depends on the context and the model in which it is going to be used. To give an example: In Hopfield's single resonance polariton model the higher resonances and the interband continuum contribute to the "background", whereas in full band-edge dynamics the total response of the lowest edge is treated as a part of the system. As a result the "background" to be used with the coherent wave approach is the response of higher band edges.

In the discussion of choosing $\varepsilon$ one must treat two cases separately: (a) The Coulomb interaction $V_{eh}$, the closely related induced potentials $\Phi^{in}$ and the longitudinal fields derived from these potentials; (b) induced displacement currents appearing in the dynamical field equations. It is convenient in both cases to split up the charge and current densities into the following contributions:

$$\varrho = \varrho^{ex} - e[C(\mathbf{r},\mathbf{r}) - D(\mathbf{r},\mathbf{r})] - \nabla \mathbf{P}_{ion} - 2M_0\nabla\,\mathrm{Re}\{Y(\mathbf{r},\mathbf{r})\} - \nabla\tilde{\mathbf{P}} \tag{3.46a}$$

$$\mathbf{J} = \mathbf{J}^{ex} + \mathbf{J}_m + \frac{\partial}{\partial t}[\mathbf{P}_{ion} + 2M_0\,\mathrm{Re}\,\{Y(\mathbf{r},\mathbf{r})\} + \tilde{\mathbf{P}}] \ . \tag{3.46b}$$

$\mathbf{J}_m$ is the intraband current density (from $C$ and $D$), $\mathbf{J}^{ex}$ is the external current density (usually zero), $\varrho^{ex}$ is the external charge density such as that of ionized impurities, $\mathbf{P}_{ion}$ is the ionic polarization, and $\tilde{\mathbf{P}}$ is the residual interband polarization not included in the actual edge dynamics.

### 3.4.1 Longitudinal Self-Consistent Fields

As described in Sect. 3.2 the induced longitudinal fields have their origin in the two point-charge interaction $V_{eh}$. So the use of a screening constant $\varepsilon$ in $V_{eh}$ entails the use of the same constant in the Poisson equation (3.6). If the potential describing the screening by a dense plasma is used (see Appendix H), then Poisson's equations should be replaced by a more general equation. For example the Debye-Hückel interaction (H.13) goes with the following equation for the potential:

$$(\nabla^2 - \kappa_0^2)\Phi^{in} = -\varepsilon_0 \varrho^{in} \ . \tag{3.47}$$

In the choice of the two point-charge interaction it has proved successful to consider the static response of those charges which are not accounted for explicitely by $\varrho^{in}$. The static approximation is justified by the fact that $\Phi^{in}$ usually is slowly varying at least compared to the typical electronic resonance frequency $\omega_g$. In the absence of a plasma then it is convenient to let the total interband contribution [the last two terms in (3.46)] be described by a constant interband susceptibility $\chi_{inter}$. If further the time variation is slower than dipole-allowed lattice resonances, then also $P_{ion}$ is a simple constant $\varepsilon_0 \chi_{ion}$ times the induced electric field. In this simple case the relevant $\varepsilon$ (3.47) is

$$\varepsilon_s = 1 + \chi_{inter} + \chi_{ion} \ . \tag{3.48}$$

This value of $\varepsilon$ is the static dielectric constant of the undoped (and unexcited) semiconductor. In a theory refined to include dipolar lattice vibrations as a part of the dynamical system, $\chi_{ion}$ should be dropped and the relevant $\varepsilon$ is the "high frequency dielectric constant" $\varepsilon_\infty = 1 + \chi_{inter}$ (see Fig. 3.1).

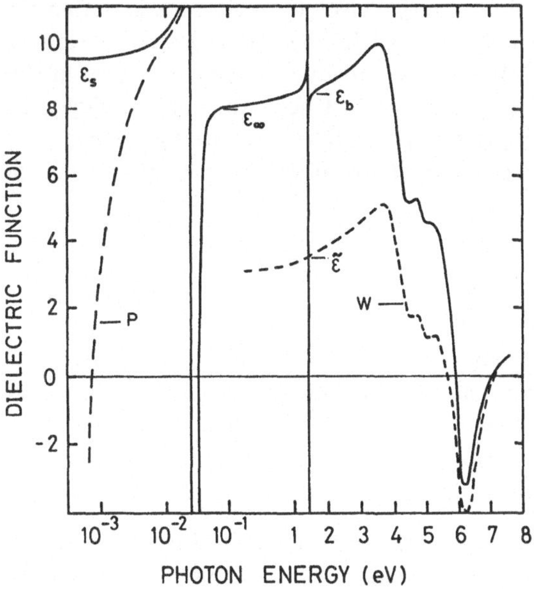


**Fig. 3.1.** Dielectric function of a typical semiconductor. The *dashed curve* marked $P$ is the dielectric function in the presence of a plasma. The curve marked $W$ is the response in the absence of the lowest band edge. $\varepsilon_s$, $\varepsilon_\infty$, $\varepsilon_b$ and $\tilde{\varepsilon}$ are discussed in the text

In the electron-hole interaction

$$V_{eh} = -\frac{e^2}{4\pi\varepsilon_0\varepsilon}\,\frac{1}{|\boldsymbol{r}_1 - \boldsymbol{r}_2|} \tag{3.48a}$$

appearing in the interband operator $\Omega_{eh}$ the use of a static screening-constant $\varepsilon$ becomes questionable when the frequency $|\omega - \omega_g|$ of the electron-hole motion is comparable or higher than the lattice frequencies responsible for $\chi_{\text{ion}}$ in (3.48) (Knox, 1963).

Another important deviation from the simple form (3.51) occurs in photoexcited and heavily doped semiconductors. The intraband part of $\varrho^{in}$ [see (3.46)] then gives rise to plasma type screening in addition to the dielectrical screening. The static limit of the Lindhard screening theory is discussed in Appendix H. More elaborate calculations based on many-body Green's function techniques have been done by Rice (1974), Zimmermann (1976), Haug and Schmitt-Rink (1984).

### 3.4.2 $\tilde{\varepsilon}$ in the Dynamical Field Equations

The dynamical field equation is

$$-c^2\varepsilon_0\nabla\times\nabla\times\boldsymbol{E} - \varepsilon_0\ddot{\boldsymbol{E}} = \dot{\boldsymbol{J}} \ . \tag{3.49}$$

In most applications $\boldsymbol{J}^{ex}$ is zero. For time variations much faster than the ionic resonances we may neglect $\boldsymbol{P}_{ion}$ in (3.46). Then (3.49) can be written as

$$-c^2\varepsilon_0\nabla\times\nabla\times\boldsymbol{E} - \varepsilon_0\tilde{\varepsilon}\ddot{\boldsymbol{E}} = 2M_0\,\mathrm{Re}\{\ddot{Y}(\boldsymbol{r},\boldsymbol{r})\} + \dot{\boldsymbol{J}}_m \ . \tag{3.50}$$

$\tilde{\varepsilon} \equiv 1 + \tilde{P}/\varepsilon_0 E^{in}$ is the residual (interband) dielectric constant accounting for the higher interband transitions not included in the dynamics of the lowest edge. As shown in Sect. 4.5, all excitonic resonances and the continuum of the lowest edge are included in $\mathrm{Re}\{Y(\boldsymbol{r},\boldsymbol{r})\}$. A typical behaviour of $\tilde{\varepsilon}$ is shown in Fig. 3.1. $\tilde{\varepsilon}$ is generally different from Hopfield and Thomas' "background dielectric constant" $\varepsilon_b$ applied when describing the response as a single resonance plus a background (see Sect. 4.8).

## 3.5 Band-Edge Complexities

Up to now we have considered a very simple band edge characterized by a pair of nondegenerate band extrema at $k = 0$. The bands were assumed to have a parabolic, isotropic behaviour in reciprocal space. In real semiconductors the band structure is much more complex in the following respects:

i) There are several bands, and each pair of one full and one empty band forms an edge in a general sense. The corresponding interband transitions are usually classified in terms of critical points with vanishing gradient in $\boldsymbol{k}$-space of the interband energy (Bassani and Pastori Parravicini, 1975).

ii) The band extrema in most semiconductors are degenerate. The sources of degeneracy are electron spin and crystal symmetry, and complex mixing effects occur when both sources contribute.

iii) Band extrema in semiconductors are occasionally characterized by noticeable nonparabolicity and anisotropy, the latter in form of ellipsoidal or warped energy surfaces in $\boldsymbol{k}$-space.

In the present section we shall briefly discuss the modifications of the band-edge equations caused by non-parabolicity, band anisotropy and degeneracies of band extrema. We retain the structure of a single edge considered to be the lowest one and structured as a band extremum at $\boldsymbol{k} = 0$. We shall return to the response of other types of critical points in Sect. 4.10. By analogy with the expansion (2.37) it is clear that higher orders in the wave vector $\boldsymbol{k}$ can be included in the band structure $E(\boldsymbol{k})$ of the actual band. The general operators appearing in $\Omega_{eh}$, $\Omega_{ee}$ and $\Omega_{hh}$ are $E_v(-i\nabla)$ and $E_c(-i\nabla)$ which can be expanded to higher orders in the gradient operator. Similarly if, for example, $E_c(\boldsymbol{k})$ is anisotropic, the reciprocal effective mass is a second rank tensor. This property is again described by replacing $\boldsymbol{k}$ by $(-i\nabla)$ in $E_c(\boldsymbol{k})$ and inserting the operator into $\Omega_{eh}$ and $\Omega_{ee}$.

We then consider the modifications of the band edge equations caused by band degeneracies. In this case one must treat $Y$, $C$, and $D$ as multicomponent functions. The number of components in $C$ and $D$ is the square of the degeneracy of the conduction band and the valence band, respectively, while $Y$ has a number of components equal to the products of the degeneracies.

Let us discuss as an example the 4×4 component function $D$ describing the density matrix of a $p_{3/2}$ valence band. For clarity we neglect coupling to interband transitions. A repeated derivation of the dynamics of holes in such a band structure yields

$$\dot{D}_{12}^{ab} + \frac{i}{\hbar}\left(\sum_c H_v^{bc}(-i\nabla_2)D_{12}^{ac} - \sum_d H_v^{da}(-i\nabla_1)D_{12}^{db}\right) = 0 \tag{3.55}$$

where $a, b, c, d$ count the 4 valence band sublevels. $H_v^{bc}(\boldsymbol{k})$ is the effective mass Hamiltonian matrix having as eigenvalues $E_v^n(\boldsymbol{k})$ $(n = 1, 2, 3, 4)$ the parabolic band structure of the valence band. (Due to time reversal the four eigenvalues are pairwise degenerate (Dresselhaus, Kip and Kittel, 1955; Kane, 1956). It is seen that the simple structure of $\Omega_{hh}$ used previously is replaced by a much more complicated operator.

Still neglecting interband-intraband coupling one obtains for the interband constitutive equation

$$\dot{Y}_{12}^{ai} + \frac{i}{\hbar}\left[\sum_j H_c^{ij}(-i\nabla_2)Y_{12}^{aj} + \sum_b H_v^{bi}(-i\nabla_1)Y_{12}^{bc}\right.$$

$$\left. + \hbar\omega_g Y_{12}^{ai} + V_{eh}Y_{12}^{ai}\right] = \frac{i}{\hbar}M_0^{ai}\delta_{12}E_1 \ . \tag{3.56}$$

Here $a, b$ and $i, j$ count sublevels of valence band and conduction band, respectively, and $H_c^{ij}(\boldsymbol{k})$ is the effective mass Hamiltonian matrix. For the common case of a $s_{1/2}$ conduction band $H_c^{ij}$ is a 2×2 diagonal matrix. In (3.56) the transition dipole is a two-indexed vector.

As seen from (3.55, 56) the $\Omega$ operators are described by matrices in case of band degeneracies. This complication tends to obscure the fundamental structure of the constitutive equations. In the present work we therefore generally retain the simple structure without indices, summations etc. For some purposes, however, we need to remind the reader that we are usually dealing with multicomponent functions $Y, C$, and $D$ and corresponding matrix structures of $\Omega_{eh}$, $\Omega_{ee}$, and $\Omega_{hh}$. In this case we use a greek letter (usually $\lambda$) as subscript for specifying the pair sublevel (not the band sublevel). In this notation (3.56) may be written as

$$\dot{Y}_\lambda(\boldsymbol{r}_1, \boldsymbol{r}_2) + i\Omega_{eh}^{\lambda\mu}Y_\mu(\boldsymbol{r}_1, \boldsymbol{r}_2) = \frac{i}{\hbar}\boldsymbol{M}_{0\lambda}\boldsymbol{E}(\boldsymbol{r}_2)\delta_B(\boldsymbol{r}_1 - \boldsymbol{r}_2) \tag{3.57}$$

with the summation convention applied to the repeated index $\mu$. Here the number subscripts are replaced by the conventional notation for spatial variables. For an $s_{1/2} - p_{3/2}$ edge, $\lambda$ and $\mu$ run from 1 to 8, and the 8×8 matrix $\Omega_{eh}^{\lambda\mu}$ is built up from elements of the lower order matrices $H_c^{ij}$ and $H_v^{ab}$ mentioned previously.

The structure of the transition dipole matrix element $M_{0\lambda}$ for different sublevels is related to the vectorial nature of the dipole. We shall generally treat the transition dipoles as vectors $\boldsymbol{M}_{0\lambda}$ ($\lambda = 1, 2\ldots$) in three dimensional space.

## 3.6 The Real-Space Structure of the Transition Dipole

The coupling between the band edge and the electric field is described in the preceding sections by the dipole matrix element $\boldsymbol{M}_0$ characterizing the transitions between states localized on the same site. However, for nonvanishing overlap of the wave functions used, the radiative interaction also involves transitions between states at different sites. In the present section we shall calculate how the source terms on the right hand side of the edge equation (2.32 and 3.1) are modified by finite wave function overlap.

Let us consider the contribution from the electromagnetic field to the rate of change of $\hat{Y}_{ij}$ :

$$(\dot{\hat{Y}}_{ij})_{\text{source}} = \frac{i}{\hbar}[H_{EM}, \hat{Y}_{ij}] \qquad \text{where} \tag{3.58}$$

$$H_{EM} = \sum_{k\ell} H_{k\ell}^{EM} \ . \tag{3.59}$$

It is shown in Appendix B that the lowest order in the multipole expansion of $H_{k\ell}^{EM}$ is the electric dipole contribution given by

$$H_{k\ell}^{D} = -(d_k c_\ell + c_\ell^\dagger d_k^\dagger)\boldsymbol{m}_{k\ell}\boldsymbol{E}(\boldsymbol{R}_{k\ell}) \ , \tag{3.60}$$

where $\boldsymbol{E}$ is the electric field, $\boldsymbol{R}_{k\ell}$ is the coordinate of a point in the neighbourhood of $\boldsymbol{R}_k$ and $\boldsymbol{R}_\ell$ [e.g. $\boldsymbol{R}_{k\ell} = (\boldsymbol{R}_k + \boldsymbol{R}_\ell)/2$]; and

$$\boldsymbol{m}_{k\ell} = -e\int w_v^*(\boldsymbol{x} - \boldsymbol{R}_k)(\boldsymbol{x} - \boldsymbol{R}_{k\ell})w_c(\boldsymbol{x} - \boldsymbol{R}_\ell)d^3\boldsymbol{x} \ . \tag{3.61}$$

Using the commutator relation (F.3) in Appendix F one finds

$$(\dot{\hat{Y}})_{\text{source}} = -\boldsymbol{m}_{ij}\boldsymbol{E}(\boldsymbol{R}_{ij}) + \sum_\ell C_{\ell j}\boldsymbol{m}_{i\ell}\boldsymbol{E}(\boldsymbol{R}_{i\ell}) + \sum_k D_{ik}\boldsymbol{m}_{kj}\boldsymbol{E}(\boldsymbol{R}_{kj}) \ . \tag{3.62}$$

In (3.61) and in Appendix B, $w_v$ and $w_c$ are localized wave functions of the valence band and the conduction band, respectively. The most accurate wave functions to be used here are the Wannier functions obtained by an appropriate Fourier transformation of the Bloch functions $\varphi_v(\boldsymbol{k}, \boldsymbol{x})$ and $\varphi_c(\boldsymbol{k}, \boldsymbol{x})$ (Knox, 1963). By using this relation between Wannier and Bloch functions it is possible to express the transition dipole $\boldsymbol{m}_{ij}$ in terms of familiar characteristics of the Bloch representation such as band energies and interband momentum matrix elements. Details of this calculation are given in Appendix D. The result is

$$\boldsymbol{m}_{ij} = \frac{e\hbar\Omega}{im_0(2\pi)^3}\int_{BZ} \frac{\boldsymbol{p}(\boldsymbol{k})\exp\left[i\boldsymbol{k}(\boldsymbol{R}_j - \boldsymbol{R}_i)\right]}{E_c(\boldsymbol{k}) - E_v(\boldsymbol{k})} d^3\boldsymbol{k} \tag{3.63}$$

where

$$\boldsymbol{p}(\boldsymbol{k}) = \frac{\hbar}{i}\int \varphi_v^*(\boldsymbol{k}, \boldsymbol{x})\nabla\varphi_c(\boldsymbol{k}, \boldsymbol{x})d^3\boldsymbol{x} \ . \tag{3.64}$$

In the continuum limit we must perform a band limited interpolation of $\boldsymbol{m}_{ij}$. The resulting dipole density $\boldsymbol{M}(\boldsymbol{r})$ depends only on the relative coordinate $\boldsymbol{r} = \boldsymbol{R}_j - \boldsymbol{R}_i$. Hence

$$\boldsymbol{M}(\boldsymbol{r}) = \frac{e\hbar}{im_0(2\pi)^3}\int_{BZ} \frac{\boldsymbol{p}(\boldsymbol{k})\exp\left(i\boldsymbol{k}\cdot\boldsymbol{r}\right)}{E_c(\boldsymbol{k}) - E_v(\boldsymbol{k})} d^3\boldsymbol{k} \ . \tag{3.65}$$

Defining $\boldsymbol{M}_0$ as the integrated strength of $\boldsymbol{M}(\boldsymbol{r})$ we get

$$\boldsymbol{M}_0 = \int \boldsymbol{M}(\boldsymbol{r}) d^3\boldsymbol{r} = \frac{e\boldsymbol{p}(0)}{im_0\omega_g} \quad . \tag{3.66}$$

Before estimating the width of the smeared-out dipole density $\boldsymbol{M}(\boldsymbol{r})$ let us write down the modified equations for the interband dynamics and the polarization field [cf. (3.1a), (B.31) in Appendix B)]

$$\begin{aligned}\dot{Y}_{12} + i\Omega_{eh}Y_{12} = \frac{i}{\hbar}[&\boldsymbol{M}(\boldsymbol{r}_1 - \boldsymbol{r}_2)\boldsymbol{E}(\boldsymbol{r}_{12}) - \int C_{02}\boldsymbol{M}(\boldsymbol{r}_1 - \boldsymbol{r}_0)\boldsymbol{E}(\boldsymbol{r}_{01}) d^3\boldsymbol{r}_0 \\ &- \int D_{10}\boldsymbol{M}(\boldsymbol{r}_0 - \boldsymbol{r}_2)\boldsymbol{E}(\boldsymbol{r}_{02}) d^3\boldsymbol{r}_0\end{aligned} \tag{3.67a}$$

$$\boldsymbol{P}(\boldsymbol{r}_{12}) = \int \boldsymbol{M}(\boldsymbol{r}_1 - \boldsymbol{r}_2)(Y_{12} + Y^*_{12}) d^3(\boldsymbol{r}_1 - \boldsymbol{r}_2) \tag{3.67b}$$

where $\boldsymbol{r}_{12}$ is the coordinate of the reference point between $\boldsymbol{r}_1$ and $\boldsymbol{r}_2$ (midpoint or center of mass). The finite width of $\boldsymbol{M}(\boldsymbol{r})$ is usually unimportant for the last two terms on the right hand side of (3.67a) and for the source terms in the intraband edge equations (3.1b–c). The interband equation then becomes:

$$\begin{aligned}\dot{Y}_{12} + i\Omega_{eh}Y_{12} = \frac{i}{\hbar}[&\boldsymbol{M}(\boldsymbol{r}_1 - \boldsymbol{r}_2)\boldsymbol{E}(\boldsymbol{r}_{12}) \\ &- \boldsymbol{M}_0 C_{12}\boldsymbol{E}(\boldsymbol{r}_1) - \boldsymbol{M}_0 D_{21}\boldsymbol{E}(\boldsymbol{r}_2)] \quad .\end{aligned} \tag{3.68}$$

If the $\boldsymbol{k}$ dependence of $\boldsymbol{p}(\boldsymbol{k})$ and the interband energy $E_c(\boldsymbol{k}) - E_v(\boldsymbol{k})$ is neglected in the integral (3.65), then $\boldsymbol{M}(\boldsymbol{r})$ is equal to $\boldsymbol{M}_0\delta_B(\boldsymbol{r})$ where $\delta_B$ is the band limited delta function. A rough estimate of how much wider is $\boldsymbol{M}(\boldsymbol{r})$ than $\delta_B(\boldsymbol{r})$ can be obtained as follows: We assume that $\boldsymbol{p}(\boldsymbol{k}) \approx \boldsymbol{p}(0)$ in the Brillouin zone, and that the interband energy is parabolic in relevant parts of $\boldsymbol{k}$-space

$$E_c(\boldsymbol{k}) - E_v(\boldsymbol{k}) = \hbar\omega_g + \frac{\hbar^2\boldsymbol{k}^2}{2\mu} \quad . \tag{3.69}$$

As a further approximation we extend the integration in (3.65) to the entire $\boldsymbol{k}$-space. Then

$$\boldsymbol{M}(\boldsymbol{r}) = \boldsymbol{M}_0(4\pi r_0^2 r)^{-1} e^{-r/r_0} \tag{3.70a}$$

$$r_0 = (2\mu\omega_g/\hbar)^{-1/2} \tag{3.70b}$$

$$r = |\boldsymbol{r}| \quad . \tag{3.70c}$$

Note that the structure in (3.70) is not band-limited and so the use of this result has meaning only if $r_0$ is somewhat larger than the lattice constant. In common semiconductors with a direct energy gap between 1 and 2 eV, $r_0$ is of the order

5 Å. The spherical structure of $\boldsymbol{M}(\boldsymbol{r})$ is a consequence of spherical symmetry in $\boldsymbol{k}$-space of the band energies. Results for ellipsoidal energy surfaces have been derived by Czajkowski and Balslev (1985).

In case of so-called forbidden transitions, $\boldsymbol{p}(0)$ and thereby $\boldsymbol{M}_0$ vanishes. Then, to lowest nonvanishing order in $\boldsymbol{k}$, we have

$$\boldsymbol{p}(\boldsymbol{k}) \propto \boldsymbol{k} \ . \tag{3.71}$$

With isotropic energy bands, $\boldsymbol{M}(\boldsymbol{r})$ then attains the form

$$\boldsymbol{M}(\boldsymbol{r}) \propto \boldsymbol{r}(r + r_0) r^{-3} e^{-r/r_0} \ . \tag{3.72}$$

It should be emphasized that the analytical results (3.70, 72) for $\boldsymbol{M}(\boldsymbol{r})$ have no meaning when $\boldsymbol{r}$ is of the order of one lattice constant or less. For example, the above expressions for $\boldsymbol{M}(\boldsymbol{r})$ are singular at $\boldsymbol{r} = 0$ while a band limited interpolation of any reasonable dipole matrix element $\boldsymbol{m}_{ij}$ is well behaved at $\boldsymbol{r} = 0$. We shall see in the next chapter that reasonable results for the electrodynamics can be obtained even though the characteristic functions $\boldsymbol{M}(\boldsymbol{r})$ and $V_{eh}(\boldsymbol{r})$ in the edge equations are not band-limited.

## 3.7 Irreversible Processes

The edge equations in their form (3.1) are derived from a Hamiltonian and therefore they have the reversible property typical for a dynamically closed system. The derivation of these equations of course involved a number of approximations. An idealization being only approximately fulfilled was, for example, the assumption of a perfect crystal. The same is true for the neglect of thermal fluctuations. Another severe approximation needing improvement is the breaking of the electronic hierarchy by the RPA (3.18, 19). A common characteristic of all these examples is that they refer to interaction with a complex system B, the so called "Bath", and that the state of B is not known in detail. In the first example B represents the imperfections of the crystal. In the second case the thermal lattice fluctuations form the bath, while in the third example B stands for the unknown higher order electronic correlation functions.

The situation described above is standard for the appearance of irreversible processes. We thus face the problem of incorporating irreversible terms into the edge equations to account for the interaction with a complex unknown environment. The simplest way to do this is by a phenomenological guess. Of this type are the relaxation time ansatz in the interband equation or a "Stosszahlenansatz" in the intraband equation. Let us consider four examples.

### 3.7.1 $T_2$-Damping of the Pair Amplitude

Starting from the general structure (2.51) of the interband equation we add a term accounting for the irreversible change of $Y$

$$\dot{Y}_{12} = (\dot{Y}_{12})_{\text{drift}} + (\dot{Y}_{12})_{\text{source}} + (\dot{Y}_{12})_{\text{irr}} \ . \tag{3.73}$$

The simplest guess that can be made about the irreversible term is to assume that it describes a spatially homogeneous linear relaxation to the equilibrium value $Y = 0$. That amounts to an ansatz

$$(\dot{Y}_{12})_{\text{irr}} = -\frac{1}{T_2} Y_{12} \ . \tag{3.74}$$

The symbol $T_2$ for the relaxation time is reminiscent of the fact that (3.74) is a generalization of the corresponding term in the Bloch equations (2.16). The origin of the $T_2$-damping in Bloch's equations is well understood mainly from detailed work on the magnetic resonance case (Slichter, 1980; Redfield, 1955). It is attributable to a dephasing of the transition amplitude occurring under the influence of random local fields. The inhomogeneous part of the dephasing is responsible for the conspicuous photon echo effect (Abella, 1969). Since the ansatz (3.74) for the semiconductor model has so far not been derived from first principles, a justification is mainly based on a success in applications to specific problems. As can be seen by looking into the solutions presented in Chaps. 4, 5 and 7 at least as a first approximation (3.74) seems to work rather well. An exception occurs in Sect. 4.10 on saddle point excitons where a certain need for a nonlocal damping is indicated (Balslev, 1984).

### 3.7.2 Electron-Electron Collisions

The intraband equations must also be extended by irreversible terms. In this way e.g. (2.49) goes over into

$$\dot{C}_{12} = (\dot{C}_{12})_{\text{drift}} + (\dot{C}_{12})_{\text{source}} + (\dot{C}_{12})_{\text{irr}} \ . \tag{3.75}$$

An important contribution to the irreversible term in (3.75) is due to electron interactions that were suppressed by applying the self-consistent field approximation in Sect. 3.2. In many-body theory it is shown (Bohm and Pines, 1953) that the residual interaction that is left after the self-consistent field approximation has been applied, is of short range type. This makes it plausible that the dynamics associated with the residual interaction is adequately treated in terms of a Boltzmann-type collison operator (Kadanoff and Baym, 1962). When formulating the collision operator it is convenient to replace the density matrix $C$ by the equivalent distribution functin $f_e$, through application of the Wigner transformation (6.12). The reason is that a collision process, when described in $(\boldsymbol{r}, \boldsymbol{k})$-phase space, only operates on the momentum $\boldsymbol{k}$. The irreversible change of $f_e$ due to electron-electron collisions then is assumed to be of the well-known form

$$[\dot{f}_e(\boldsymbol{r}, \boldsymbol{k})]_{\text{irr},ee} = R_{ee}^{\text{in}} - R_{ee}^{\text{out}}$$

$$R_{ee}^{\text{in}} = (2\pi)^{-9} \int w_{ee}(\boldsymbol{k}, \boldsymbol{k}_1, \boldsymbol{k}', \boldsymbol{k}_1') f_e(\boldsymbol{r}, \boldsymbol{k}') f_e(\boldsymbol{r}, \boldsymbol{k}_1')$$
$$\times [n_s - f_e(\boldsymbol{r}, \boldsymbol{k})][n_s - f_e(\boldsymbol{r}, \boldsymbol{k}_1)] d^3\boldsymbol{k}_1 d^3\boldsymbol{k}' d^3\boldsymbol{k}_1'$$

$$R_{ee}^{\text{out}} = (2\pi)^{-9} \int w_{ee}(\boldsymbol{k}', \boldsymbol{k}_1', \boldsymbol{k}, \boldsymbol{k}_1) f_e(\boldsymbol{r}, \boldsymbol{k}) f_e(\boldsymbol{r}, \boldsymbol{k}_1)$$
$$\times [n_s - f_e(\boldsymbol{r}, \boldsymbol{k}')][n_s - f_e(\boldsymbol{r}, \boldsymbol{k}_1')] d^3\boldsymbol{k}_1 d^3\boldsymbol{k}_1' d^3\boldsymbol{k}' \tag{3.76}$$

where $n_s$ is the phase space density of states.

The existence of a term (3.76) in the intraband equations provides for intraband thermalization (compare Sect. 6.2). From experience it is known that this thermalization is a very fast process occurring on a time scale of 1 ps or less. From this it is concluded that the influence of electron-electron processes on intraband dynamics must not be neglected. In a treatment of fast processes it has to be considered explicitly, while for slow processes with a characteristic time scale much longer than a picosecond, the fast electron processes can be eliminated by a procedure known as "adiabatic elimination" (van Kampen, 1985). This elimination amounts to introducing a local equilibrium state with parametric dependence on the slow variables (compare also the treatment of stationary transport in Sect. 6.3).

### 3.7.3 Electron-Lattice Scattering

Another important contribution to the intraband irreversible change is electron scattering by impurities and lattice vibrations. The phenomenlogical collision term accounting for these processes is [in contrast to (3.76)] of second order in the electronic distribution function

$$\dot{f}_e(\boldsymbol{r}, \boldsymbol{k})_{\text{irr},eL} = R_{eL}^{\text{in}} - R_{eL}^{\text{out}}$$
$$R_{eL}^{\text{in}} = (2\pi)^{-3} \int w_{eL}(\boldsymbol{k}, \boldsymbol{k}')[n_s - f_e(\boldsymbol{r}, \boldsymbol{k})] f_e(\boldsymbol{r}, \boldsymbol{k}') d^3\boldsymbol{k}'$$
$$R_{eL}^{\text{out}} = (2\pi)^{-3} \int w_{eL}(\boldsymbol{k}', \boldsymbol{k})[n_s - f_e(\boldsymbol{r}, \boldsymbol{k}')] f_e(\boldsymbol{r}, \boldsymbol{k}) d^3\boldsymbol{k}' \ . \tag{3.77}$$

The influence of (3.77) is dominant in transport processes. The momentum relaxation time due to electron lattice collisions as used in Sect. 6.3 is found to be the integral over solid angle

$$\frac{1}{\tau} = \int w[1 - \cos(\boldsymbol{k} \cdot \boldsymbol{k}')] d\Omega_{kk'} \ . \tag{3.78}$$

Typical values of $\tau$ are again 1 ps or less. Much slower is the exchange of energy between the electronic intraband degrees of freedom and the lattice because the electron phonon scattering is almost elastic. Energy relaxation times are therefore many orders of magnitude larger than the momentum relaxation time $\tau$.

### 3.7.4 Incoherent Generation-Recombination Processes

Besides the coherent radiative interband transitions described by the edge equations, there exist also incoherent generation recombination processes. These processes, radiative or nonradiative (e.g. Auger type) are typically much slower than the processes considered in the foregoing paragraphs. There exists therefore no direct competition between coherent and incoherent generation-recombination processes. A strong influence of incoherent processes is found in transport properties (Schöll, 1982, 1983), but that is outside the scope of this review.

Before closing this section we mention an attempt to treat irreversibility in the edge equations from a more microscopic point of view (Frank, 1985). The method is based on the assumption that the parameters in the edge equations contain small fluctuating parts. An example is the phonon induced paramter fluctuation considered in the theory of Raman scattering (Sect. 7.2). But while in Sect. 7.2 attention is focused on the scattered polariton, a theory of irreversible damping asks for the development of the primary wave. Applying an averaging procedure described by Frisch (1968), Frank was able to show that at least in certain limiting cases the $T_2$ damping for $Y$ and the Boltzmann ansatz for $C$ can be related to this more elaborate model. In general the irreversible terms generated by this method look rather complicated, and it is still uncertain whether they contain any new and interesting physics.

# 4. Linear Interband Response of Unbounded Media

## 4.1 Introductory Remarks

In the present chapter we shall study the linear electrodynamics in a semiconductor near the gap frequency. We assume the crystal to be infinite and consider the excitation so low that saturation terms remain negligible. We then enter the rich field of linear excitonic response explored first by Frenkel (1931), Elliott (1957), Gross (1956) and MacFarlane et al. (1957). The fundamental concept in this field is the exciton, being an elementary excitation in which an electron and a hole form a mobile, hydrogenlike pseudo-particle. In the description of the strong coupling between light and direct gap excitons, Hopfield (1958) introduced the convenient concept of a polariton which is a quantized polarization wave with mixed photon and exciton character.

Before going into detail with various theoretical approaches to excitons and excitonic response, let us go through a short survey of the relevant empirical background.

The oldest experimental method for exploring excitons and the higher lying interband continuum is that of an optical transmission spectroscopy on plane parallel samples. The excitonic resonances appear as absorption lines while the excitonic ionization continuum forms a broad interband absorption above the gap frequency. This is demonstrated in Fig. 4.1 showing the spectrum of a thin sample of GaAs. Here 3 lines of the hydrogenic exciton series are resolved (Hornung and Ulbrich, 1982). Up to 7 lines of the yellow series in $Cu_2O$ have been observed in transmission experiments (Nikitine, 1975).

Samples with good geometrical and compositional quality show Fabry-Perot interference in the strongly dispersive frequency region near the exciton lines. The interference patterns give valuable information on the polariton dispersion relation (Ivchenko, 1982; Kiselev et al., 1974). The same is true for the deflection in thin small-angle prisms (Broser et al., 1980).

Preparation of thin samples (one micron or less) with sufficiently perfect geometry, structure and composition is often difficult. Therefore it is common to study the linear response by reflectivity measurements on plane surfaces of thick samples. A good example is shown in Fig. 4.2. Here the spectrum of CuCl in the region $\hbar\omega = 3.2$ to $3.4\,\mathrm{eV}$ is well explained by two excitonic resonances derived from two closely lying band gaps (Mohler, 1970). In addition to the standard

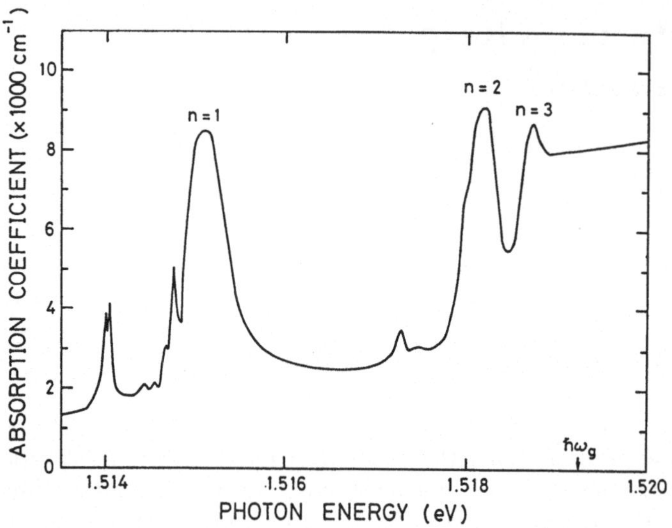


**Fig. 4.1.** Absorption coefficient of GaAs at 1.5 K in the exciton region. The data are obtained from transmission through a 5.7 $\mu$m thick sample. The structure besides the $n = 1, 2, 3$ excitonic lines are due to excitons bound to impurities. From Hornung and Ulbrich (1982)

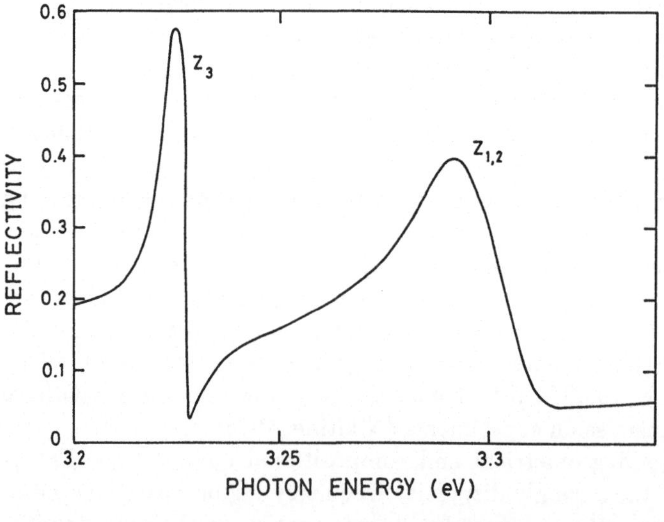


**Fig. 4.2.** Reflectivity at normal incidence of CuCl at 80 K. From Mohler (1970)

measurements of the normal incidence reflectance, a number of refinements have been developed. First to be mentioned is the additional information obtained with oblique incidence (Broser et al., 1978; Pevtsov and Selkin, 1983). Here the ellipsometric technique allows the determination of the phase difference between the two principal reflection coefficients. Another informative geometry is that employed when measuring attenuated total reflectivity (ATR) (Lagois, 1981).

The above experiments for exploring the linear excitonic response are often extended to include a controlled external perturbation such as a magnetic field (Bimberg, 1977), an electric field (Aspnes, 1974) and uniaxial stress (Balslev, 1972).

Since the development of lasers a large variety of new excitonic phenomena have been studied (Hönerlage et al., 1985). Group velocities of polaritons have been measured by time of flight techniques (Segawa et al., 1978; Masumoto, 1979). Resonant Raman scattering by phonons and Brillouin scattering are important new methods for exploring exciton-phonon interaction and polariton dispersion (Yu, 1979; Koteles, 1982). Resonant (electronic) two-photon Raman scattering, two-photon absorption and four-wave mixing (Grun et al., 1982; Hönerlage et al., 1978; Fröhlich et al., 1971) have given accurate information on polariton dispersion relations. An example of this is shown in Fig. 4.3. The last-mentioned nonlinear phenomena are members of a larger group of experiments involving high excitation densities. Another member of this group concerns irreversible high excitation processes giving stimulated emission (Klingshirn and Haug, 1981; Hvam et al., 1983). We shall return to excitonic phenomena at high excitation densities in Chap. 7.

In this introductory section it is appropriate to quote the primary exciton parameters deduced from the spectroscopic experiments mentioned above. In Table 4.1 are given the spectroscopic positions of the lowest excitonic resonances ($\hbar\omega_g - \hbar\omega_x$), the ionization energy $\hbar\omega_x$ of ground state excitons and the excitonic Bohr radius $a_B$. The last-mentioned quantity is calculated from $\hbar\omega_x$ and the static dielectric constant $\varepsilon_s$ determining the screening (see Sect. 3.4).

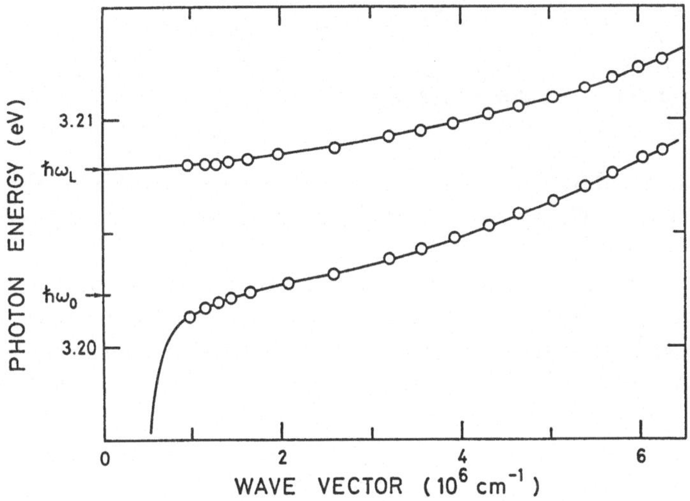


**Fig. 4.3.** Dispersion curves of the $n = 1$ exciton polariton in CuCl measured by biexciton enhanced two-photon Raman scattering

**Table 4.1**

| Material | $\hbar\omega_g - \hbar\omega_x$ [eV] | $\hbar\omega_x$ [meV] | $a_B$ [Å] | $\hbar\omega_{LT}$ [meV] | $\epsilon_s^a$ - |
|---|---|---|---|---|---|
| Ge | 0.898[b] | 1.6[b] | 260 | - | 16 |
| GaAs | 1.515[c] | 4.2[d] | 130 | 0.13[c] | 12.5 |
| GaSb | 0.811[e] | 1.6[e] | 280 | - | 15.7 |
| InP | 1.418[f] | 5.1[g] | 95 | 0.2[f] | 12.6 |
| CdTe | 1.594[h] | 10[i] | 72 | 0.4[h] | 11 |
| ZnSe | 2.800[j] | 18[j] | 50 | 1.3[aa] | 9 |
| ZnO($A$) | 3.377[k] | 60[k] | 15 | 2[k] | 8(⊥) |
| ZnO($B$) | 3.383[k] | 60[k] | 15 | 11[l] | 9(‖) |
| ZnO($C$) | 3.422[k] | 60[k] | 15 | 15[l] | |
| CdS($A$) | 2.553[n] | 28[n] | 30 | 2[m] | 8.4(⊥) |
| CdS($B$) | 2.569[o] | 28[n] | 30 | 1.3[p] | 9.0(‖) |
| CdS($C$) | 2.632[n] | - | - | - | |
| CdSe($A$) | 1.825[q] | 15[r] | 50 | 0.9[q] | |
| CdSe($B$) | 1.851[s] | 16[r] | 47 | - | 9.2(⊥) |
| CdSe($C$) | 2.283[s] | - | - | - | 10.3(‖) |
| CuCl($Z_3$) | 3.203[t] | 190[u] | 7 | 5.4[t] | 7.9 |
| $Cu_2O$* | 2.03[v] | 97[v] | - | - | - |
| Si** | 1.156[x] | 14[x] | 42 | - | 12 |
| Ge** | 0.740[y] | 4.1[z] | 110 | - | 16 |

[a]Madelung (1982)
[b]Johnson (1968)
[c]Schulteis and Balslev (1983)
[d]Sell (1972)
[e]Varfolomeev et al. (1976)
[f]Evangelisti et al. (1976)
[g]Ekardt et al. (1976)
[h]Ulbrich and Weisbuch (1978)
[i]Johnson (1967)
[j]Venghaus (1979)
[k]Hümmer (1973)
[l]Lagois and Hümmer (1975)
[m]Hopfield and Thomas (1963)
[n]Thomas and Hopfield (1959)
[o]Hopfield and Thomas (1963)
[p]Koteles and Winterling (1980)
[q]Kiselev et al. (1973)
[r]Voigt et al. (1979)
[s]Langer et al. (1970)
[t]Sermage et al. (1979)
[u]Goldmann (1977)
[v]Nikitine (1959)
[x]Shaklee and Nahory (1970)
[y]Martin and Schaberg (1979)
[z]Altarelli and Lipari (1976)
[aa]Sermage and Voos (1977)
*Forbidden transitions
**Indirect band edge

Also shown is the longitudinal-transverse splitting $\hbar\omega_{LT}$ associated with the lowest excitonic resonance. The symbols $A$, $B$, $C$, $Z_3$ refer to one-electron sublevels discussed in Sect. 4.9.

Turning to excitonic phenomena outside the framework of coherent linear response, we first mention photoluminescence and cathodoluminescence. Most informative have been spectra involving phonon-assisted exciton recombination (Gross et al., 1971), donor-acceptor pair spectra (Henry et al., 1971), excitons bound to imperfections (Henry et al., 1971; Dean and Herbert, 1979), and bound exciton complexes (Thewalt, 1982). Surprisingly, the zero-phonon free exciton luminescence in most direct-gap materials cannot be interpreted easily. The intensity depends strongly on surface treatment (Sell et al., 1973; Schultheis and Balslev, 1983) and the reabsorption distorts the spectral shape of the emission in a complicated way (Sermage and Voss, 1977). Recent progress in this field has been reported by Koteles et al. (1985).

## 4.2 The Linear Interband Susceptibility

In the present chapter we concentrate on the bulk interband waves in the low excitation regime. These polariton waves are governed by two coupled equations – a relevant constitutive equation and a Maxwellian field equation. It is convenient first to study the constitutive equation alone. In an unbounded medium this defines the linear susceptibility tensor $\underset{\sim}{\chi}(\omega, \boldsymbol{q})$ as a function of frequency $\omega$ and wave vector $\boldsymbol{q}$. In our case $\underset{\sim}{\chi}(\omega, \boldsymbol{q})$ describes the response of a single band edge. Other interband contributions are to be added in order to find the total interband dielectric function of the semiconductor.

The purpose of the present section is to bring the relevant constitutive equation into a form suitable for deriving $\underset{\sim}{\chi}(\omega, q)$. The result for the Frenkel case is already discussed in Sect. 2.3, while the Wannier case treated in this chapter is more complicated. In the low excitation case we can use (3.67) with $C = D = 0$:

$$\dot{Y}_{12} + i\Omega_{eh} Y_{12} = \frac{i}{\hbar} \boldsymbol{M}(\boldsymbol{r}) \boldsymbol{E}(\boldsymbol{R}) + (\dot{Y}_{12})_{\text{irrev.}} \tag{4.1}$$

where $\boldsymbol{r}$ is the relative coordinate

$$\boldsymbol{r} = \boldsymbol{r}_2 - \boldsymbol{r}_1 = \boldsymbol{r}_e - \boldsymbol{r}_h \tag{4.2}$$

and the reference point $\boldsymbol{r}_{12}$ in (3.67) for later convenience is chosen to be the center of mass coordinate

$$\boldsymbol{R} = (m_e \boldsymbol{r}_e + m_h \boldsymbol{r}_h)/(m_e + m_h) \ . \tag{4.3}$$

We have introduced the indices $e$ and $h$ for the electron and the hole, respectively, instead of the indices 1 and 2. The pointlike transition dipole in (3.1a) is replaced by a smeared-out dipole density $\boldsymbol{M}(\boldsymbol{r})$ as discussed in detail in Sect. 3.6.

In the usual relaxation approximation (Sect. 3.7)

$$(\dot{Y}_{12})_{\text{irrev.}} = -\gamma Y_{12} \tag{4.4}$$

where $\gamma$ is the reciprocal of the dephasing time $T_2$. Changing variables of $Y$ one obtains

$$\left(\Omega_{eh} - i\frac{\partial}{\partial t} - i\gamma\right) Y(\boldsymbol{R}, \boldsymbol{r}, t) = \frac{1}{\hbar} \boldsymbol{M}(\boldsymbol{r}) \boldsymbol{E}(\boldsymbol{R}, t) \ . \tag{4.5}$$

The interband polarization contribution $\boldsymbol{P}(\boldsymbol{R}, t)$ is given by (3.67b):

$$\boldsymbol{P}(\boldsymbol{R}, t) = 2 \int \boldsymbol{M}(\boldsymbol{r}) \operatorname{Re}\{Y(\boldsymbol{R}, \boldsymbol{r}, t)\} d^3\boldsymbol{r} \ . \tag{4.6}$$

It is the aim here to derive the complex susceptibility tensor $\chi(\tilde{\omega}, q)$ in the linear relation

$$P_0 = \varepsilon_0 \chi(\tilde{\omega}, \boldsymbol{q}) \boldsymbol{E}_0 \tag{4.7}$$

where $\boldsymbol{P}_0$ and $\boldsymbol{E}_0$ are complex amplitudes of polarization and electric field, respectively, in case of the space-time dependence $\exp(i\boldsymbol{qR} - \tilde{\omega}t)$. $\tilde{\omega}$ stands for the form

$$\tilde{\omega} = \omega + i\eta \tag{4.8}$$

where $\eta$ is a switch-on rate to be taken in the limit $\eta = 0+$.

The combination of complex notations from the Heisenberg equations and in the frequency representation of the response must be handled with great care. For example, Re$\{Y\}$ and Im$\{Y\}$ stand for components of $Y$ determining the polarization and the current, respectively, while Re$\{\chi\}$ and Im$\{\chi\}$ stand for in-phase and out-of-phase components, respectively, with respect to the electric field at a given frequency. Without complex notation (4.5) becomes

$$-\left(\frac{\partial}{\partial t} + \gamma\right) Y_2(\boldsymbol{R}, \boldsymbol{r}, t) + \Omega_{eh} Y_1(\boldsymbol{R}, \boldsymbol{r}, t) = \frac{1}{\hbar} \boldsymbol{M}(\boldsymbol{r}) \boldsymbol{E}(\boldsymbol{R}, t) \tag{4.9}$$

$$\left(\frac{\partial}{\partial t} + \gamma\right) Y_1(R, r, t) + \Omega_{eh} Y_2(\boldsymbol{R}, \boldsymbol{r}, t) = 0 \ . \tag{4.10}$$

In this notation the (real) polarization $\boldsymbol{P}(\boldsymbol{R}, t)$ is given by

$$\boldsymbol{P}(\boldsymbol{R}, t) = 2 \int \boldsymbol{M}(\boldsymbol{r}) Y_1(\boldsymbol{R}, \boldsymbol{r}, t) d^3\boldsymbol{r} \ . \tag{4.11}$$

At this point we can introduce the conventional electrodynamical complex notation without interference with the quantum mechanical one. Then $\boldsymbol{E}$, $\boldsymbol{P}$, $Y_1$ and $Y_2$ are considered complex, and for a harmonic time variation characterized by the frequency $\tilde{\omega}$ we may write

$$\boldsymbol{E}(\boldsymbol{R}, t) = \tfrac{1}{2} \tilde{\boldsymbol{E}}(\boldsymbol{R}) \mathrm{e}^{-i\tilde{\omega}t} + \tfrac{1}{2} \tilde{\boldsymbol{E}}^*(\boldsymbol{R}) \mathrm{e}^{i\tilde{\omega}^* t} \tag{4.12}$$

where $\tilde{\boldsymbol{E}}(\boldsymbol{R})$ is a complex function of space. Eliminating $Y_2(\boldsymbol{R}, \boldsymbol{r}, t)$ from (4.9, 10) and inserting (4.12) we can express the complex representation $Y_1(\boldsymbol{R}, \boldsymbol{r}, t)$ as follows

$$Y_1(\boldsymbol{R}, \boldsymbol{r}, t) = Y_-(\boldsymbol{R}, \boldsymbol{r}) \mathrm{e}^{-i\tilde{\omega}t} + Y_+(\boldsymbol{R}, \boldsymbol{r}) \mathrm{e}^{i\tilde{\omega}^* t} \tag{4.13}$$

$$Y_-(\boldsymbol{R}, \boldsymbol{r}) = Y'_-(\boldsymbol{R}, \boldsymbol{r}) + iY''_-(\boldsymbol{R}, \boldsymbol{r}) \tag{4.14}$$

$$Y_+(\boldsymbol{R}, \boldsymbol{r}) = Y'_+(\boldsymbol{R}, \boldsymbol{r}) + iY''_+(\boldsymbol{R}, \boldsymbol{r}) \tag{4.15}$$

$$(\Omega_{eh} - \tilde{\omega} - i\gamma) Y_-(\boldsymbol{R}, \boldsymbol{r}) = \tfrac{1}{2} \hbar^{-1} \boldsymbol{M}(\boldsymbol{r}) \tilde{\boldsymbol{E}}(\boldsymbol{R}) \tag{4.16}$$

$$(\Omega_{eh} + \tilde{\omega}^* - i\gamma)Y_+(\boldsymbol{R},\boldsymbol{r}) = \tfrac{1}{2}\hbar^{-1}\boldsymbol{M}(\boldsymbol{r})\tilde{\boldsymbol{E}}^*(\boldsymbol{R}) \ . \tag{4.17}$$

Throughout this book the subscripts + and – refer to complex components of wave functions and response functions associated with the decomposition (4.12). With this convention we are allowed to use the complex notation for electric field $[\tilde{\boldsymbol{E}}(\boldsymbol{R})]$ and electron-hole amplitudes $[Y_\pm(\boldsymbol{R},\boldsymbol{r})]$ simultaneously.

Equations (4.16, 17) can be solved if $\tilde{\boldsymbol{E}}(\boldsymbol{R})$ is given and appropriate boundary conditions are specified. For $\tilde{\boldsymbol{E}}(\boldsymbol{R})$ we assume a plane wave

$$\tilde{\boldsymbol{E}}(\boldsymbol{R}) = \boldsymbol{E}_0 e^{i\boldsymbol{q}\cdot\boldsymbol{R}} \ . \tag{4.18}$$

Then, by symmetry arguments $Y_\pm$ attains the form

$$Y_\pm(\boldsymbol{R},\boldsymbol{r}) = u_\pm e^{i\boldsymbol{q}\cdot\boldsymbol{R}} \ . \tag{4.19}$$

The solution $u_\pm(\boldsymbol{r})$ can then be constructed from the Green's functions as demonstrated in detail in Appendix E. Let the Green's functions $G_\pm(\boldsymbol{r},\boldsymbol{r}')$ satisfy the equations:

$$(\Omega_{eh}(q) - \tilde{\omega} - i\gamma)G_-(\boldsymbol{r},\boldsymbol{r}') = \delta(\boldsymbol{r}-\boldsymbol{r}') \tag{4.20a}$$

$$(\Omega_{eh}(q) + \tilde{\omega}^* - i\gamma)G_+(\boldsymbol{r},\boldsymbol{r}') = \delta(\boldsymbol{r}-\boldsymbol{r}') \ , \tag{4.20b}$$

where $\Omega_{eh}(q)$ is the drift operator evaluated for plane waves in the center-of-mass space. As usual we impose boundary conditions on $G_\pm(\boldsymbol{r},\boldsymbol{r}')$ corresponding to the absence of sources at infinity. Then

$$G_\pm(\boldsymbol{r},\boldsymbol{r}')\to 0 \ ; \quad |\boldsymbol{r}|\to\infty \ . \tag{4.21}$$

The solutions in relative space are then

$$u_\pm(\boldsymbol{r}) = \tfrac{1}{2}\int G_\pm(\boldsymbol{r},\boldsymbol{r}')\boldsymbol{M}(\boldsymbol{r}')\boldsymbol{E}_0 d^3\boldsymbol{r}' \ . \tag{4.22}$$

By inspection one can see that $u_-(\boldsymbol{r})$ has a resonant behaviour near $\omega_g$, while $u_+(\boldsymbol{r})$ has little variation with frequency. We shall call $u_-(\boldsymbol{r})$ the "resonant part" and $u_+(\boldsymbol{r})$ the "antiresonant part".

Using (4.11) and the defining equation for the susceptibility $\underset{\approx}{\chi}$:

$$\boldsymbol{P} = \varepsilon_0 \underset{\approx}{\chi} \boldsymbol{E} \tag{4.23}$$

we finally arrive at the resonant and antiresonant parts, $\underset{\approx}{\chi}_-(\omega,q)$ and $\underset{\approx}{\chi}_+(\omega,q)$, of $\underset{\approx}{\chi}$:

$$\underset{\approx}{\chi}_\pm(\omega,\boldsymbol{q}) = \frac{1}{\hbar\varepsilon_0}\int\int \boldsymbol{M}(\boldsymbol{r})\otimes G_\pm(\boldsymbol{r},\boldsymbol{r}')\boldsymbol{M}(\boldsymbol{r}')d^3\boldsymbol{r}'d\boldsymbol{r} \ . \tag{4.24a}$$

The total susceptibility of the band edge will be

$$\chi(\omega, \boldsymbol{q}) = \chi_+^*(\omega, \boldsymbol{q}) + \chi_-(\tilde{\omega}, \boldsymbol{q}) \ . \tag{4.24b}$$

It follows immediately from (4.20–24) that

$$\chi(-\tilde{\omega}, -\boldsymbol{q}) = \chi^*(\tilde{\omega}, \boldsymbol{q}) \tag{4.25a}$$

$$\chi_+(\tilde{\omega}, \boldsymbol{q}) = \chi_-^*(-\tilde{\omega}, \boldsymbol{q}) \ . \tag{4.25b}$$

Note that the symmetry relation (4.25a) for $\chi(\omega, q)$ does not hold for the components $\chi_-$ and $\chi_+$ individually. Note also that $\eta = \mathrm{Im}\{\tilde{\omega}\}$ and $\gamma$ always appear additively in $\underset{\sim}{\chi}_\pm(\omega, \boldsymbol{q})$. Thus, we may put $\eta = 0$ in case of a finite dephasing time $T_2 = \gamma^{-1}$.

The response expressed in (4.24) corresponds to electric dipole transitions. Among the higher-order processes we shall discuss electric quadrupole transitions. In this case the source term in the interband edge equation becomes proportional to the quadrupolar density $\underset{\sim}{Q}(\boldsymbol{r})$ times the spatial derivative of the electric field (Appendix B)

$$(\Omega_{eh} - \omega - i\gamma)Y_-(\boldsymbol{R}, \boldsymbol{r}) = \frac{1}{\hbar}\underset{\sim}{Q}(\boldsymbol{r})\nabla \otimes \boldsymbol{E}(\boldsymbol{R}) \ . \tag{4.26}$$

The resulting response is given by the nonlocal susceptibility

$$\underset{\sim}{\chi}(\omega, \boldsymbol{q}) = \chi_Q^{IV} \boldsymbol{q} \otimes \boldsymbol{q} \tag{4.27}$$

where the fourth rank tensor $\chi_Q^{IV}$ can be calculated by the same methods as used for obtaining (4.24):

$$\chi_-^{IV} \propto \int\int \underset{\sim}{Q}(\boldsymbol{r}) \otimes G_-(\boldsymbol{r}, \boldsymbol{r}')\underset{\sim}{Q}(\boldsymbol{r}')d^3\boldsymbol{r}\, d^3\boldsymbol{r}' \ . \tag{4.28}$$

Note that the nonlocality in (4.27) is fundamentally different from the so-called spatial dispersion in the dipole approximation. The latter is caused by the fact that the Green's function in (4.24a) is the inversion of the operator $(\Omega_{eh}(\boldsymbol{q}) - \omega)$ (Stahl and Uihlein, 1979).

So far we have suppressed the multicomponent character of the electron-hole amplitude appearing in case of band degeneracies. If the edge is degenerate then $Y$ and $\boldsymbol{M}$ should be furnished with an index $\lambda$ counting the pair sublevel. Assuming that $\Omega_{eh}$ does not mix sublevels we obtain instead of (4.24a)

$$\underset{\sim}{\chi}_\pm(\omega, \boldsymbol{q}) = \frac{1}{\hbar\varepsilon_0}\sum_\lambda \int\int \boldsymbol{M}_\lambda(\boldsymbol{r}) \otimes G_\pm(\boldsymbol{r}, \boldsymbol{r}')\boldsymbol{M}_\lambda(\boldsymbol{r}')d^3\boldsymbol{r}'\, d^3\boldsymbol{r} \ . \tag{4.29}$$

## 4.3 Results Neglecting the Electron-Hole Interaction

The calculation of the interband susceptibility is particularly simple if the electron-hole interaction is neglected. The results obtained give a rough impression of the absorption edge and the broad absorption continuum stretching into the ultra-violet region. Furthermore, the analytical results for $\chi(\omega)$ near the gap frequency are so simple that the conceptual difference between a conventional "golden rule" treatment and the coherent-wave approach becomes particularly clear. We shall in this section first reproduce the conventional derivation of the optical response near a semiconductor edge. Thereafter we apply – for comparison – the coherent-wave approach along the lines discussed in Sect. 4.2.

The standard semiclassical approach is obtained by expanding the operator (B.2) in Appendix B to first order in the vector potential $\boldsymbol{A}$. This gives the perturbation operator

$$H = \frac{1}{2m_0}\boldsymbol{p}(\boldsymbol{A}_0 e^{-i\omega t} + \boldsymbol{A}_0^* e^{i\omega t}) \tag{4.30}$$

where $\boldsymbol{p}$ is the momentum operator. Note that the spatial dependence of $\boldsymbol{A}$ is neglected. This gives the $\boldsymbol{q} = 0$ limit of $\chi(\omega, \boldsymbol{q})$. The direct transitions from a full valence band to an empty conduction band occur at a rate (per unit volume) given by

$$W = \frac{\pi e^2}{2\hbar^2 m_0^2}\sum_\lambda \int |\boldsymbol{p}_\lambda(\boldsymbol{k})\cdot\boldsymbol{A}_0|^2 \delta[E_c(\boldsymbol{k}) - E_v(\boldsymbol{k}) - \hbar\omega]d^3\boldsymbol{k} \tag{4.31}$$

where $\boldsymbol{p}_\lambda$ is the interband momentum matrix element for the pair sublevel $\lambda$. Simple energy balance can be used for relating the imaginary part $\chi''(\omega)$ of $\chi(\omega)$ to the power dissipation $\hbar\omega W$ obtained from (4.31). Hence

$$\chi''(\omega) = \frac{\pi^2 e^2}{\varepsilon_0 m_0^2 \omega^2}\sum_\lambda \int |\boldsymbol{p}_\lambda(\boldsymbol{k})\cdot\boldsymbol{e}|^2 \delta[E_c(\boldsymbol{k}) - E_v(\boldsymbol{k}) - \hbar\omega)d^3\boldsymbol{k} \tag{4.32}$$

where $\boldsymbol{e}$ is a unit vector parallel to $\boldsymbol{A}_0$. When $\boldsymbol{p}_\lambda(\boldsymbol{k})$ is nearly constant in relevant parts of $\boldsymbol{k}$-space we may write (Stern, 1963)

$$\chi''(\omega) = \frac{e^2}{4\pi\varepsilon_0 m_0^2 \omega^2}\sum_\lambda |\boldsymbol{p}_\lambda\cdot\boldsymbol{e}|^2 \varrho(\hbar\omega) \tag{4.33}$$

where $\varrho(\hbar\omega)$ is the joint density of states.

Near a direct edge with an isotropic reduced mass $\mu$ one finds

$$\chi''(\omega) = \frac{e^2}{4\pi\varepsilon_0 m_0^2 \omega^2}\sum_\lambda |\boldsymbol{p}_\lambda\cdot\boldsymbol{e}|^2 \left(\frac{2\mu}{\hbar}\right)^{3/2} \mathrm{Re}\{(\hbar\omega - \hbar\omega_g)^{1/2}\}\ . \tag{4.34}$$

When dealing with spin degeneracy only, the sum over $\lambda$ includes two spin-allowed contributions.

The unavoidable non-parabolicity of bands in real semiconductors gives corresponding restrictions on the validity of (4.34). By means of Kane's small-gap approximation (Kane, 1956) one can estimate the influence of non-parabolicity and a variation in $\boldsymbol{k}$-space of $\boldsymbol{p}_\lambda(\boldsymbol{k})$. Here we shall stick to the parabolic approximation valid for $|\omega - \omega_g| \lesssim \omega_g$. Then we may write

$$\chi''(\omega) = \frac{F}{\omega^2}\mathrm{Re}\{(\omega - \omega_g)^{1/2}\} \tag{4.35}$$

where $F$ is a constant.

Assuming (unrealistically) that (4.35) is valid for all frequencies, one can calculate analytically the real part $\chi'(\omega)$ from the Kramers-Kronig relations:

$$\chi''(\omega_0) = \frac{2\omega_0}{\pi}\int\limits_0^\infty \frac{\chi'(\omega)d\omega}{\omega^2 - \omega_0^2} \tag{4.36a}$$

$$\chi'(\omega_0) = \frac{2}{\pi}\int\limits_0^\infty \frac{\omega\chi''(\omega)d\omega}{\omega^2 - \omega_0^2} \ . \tag{4.36b}$$

When (4.35) is inserted into (4.36b) the use of contour integration leads to (Cardona, 1969)

$$\chi'(\omega) = \frac{F}{\omega^2}[2\omega_g^{1/2} - (\omega_g + \omega)^{1/2} - \mathrm{Re}\{(\omega_g - \omega)^{1/2}\}] \ . \tag{4.37}$$

The value of $\chi'(\omega)$ at the gap depends strongly on the high frequency behaviour of $\chi''(\omega)$. However, the shape and strength of the term which is singular at $\omega_g$ is determined by $\chi''(\omega)$ near the absorption edge. Therefore this singular term can be trusted in spite of the severe approximation (4.35). Then $\chi(\omega)$ can be expressed as

$$\chi(\omega) = C(\omega) - \sum_\lambda |\boldsymbol{p}_\lambda \boldsymbol{e}|^2 \frac{e^2}{4\pi\varepsilon_0 m_0^2\omega^2}\left(\frac{2\mu}{\hbar}\right)^{3/2}(\omega_g - \tilde{\omega})^{1/2} \tag{4.38}$$

where $C(\omega)$ is a slowly varying, real function.

It is interesting to compare the mathematical and conceptual differences between the above standard derivation of $\chi$ and the coherent wave approach. In the long wave limit the Green's function $\boldsymbol{G}_-(\boldsymbol{r}, \boldsymbol{r}')$ of the resonance part is given by

$$\left(-\frac{\hbar}{2\mu}\nabla^2 + \omega_g - \tilde{\omega}\right)G_-(\boldsymbol{r}, \boldsymbol{r}') = \delta(\boldsymbol{r} - \boldsymbol{r}') \ . \tag{4.39}$$

When $\boldsymbol{M}_\lambda(r)$ has spherical symmetry it is convenient to use a Green's function $g_0(r,\varrho)$ for a shell source at $|\boldsymbol{r}| = \varrho$, (see Appendix E)

$$\left[-\frac{\hbar}{2\mu}\left(\frac{\partial^2}{\partial r^2}+\frac{2}{r}\frac{\partial}{\partial r}\right)+\omega_g-\tilde{\omega}\right]g_0(r,\varrho)=\frac{\delta(r-\varrho)}{4\pi\varrho^2}\ . \tag{4.40}$$

The allowed solution of (4.40) is

$$g_0(r,\varrho)=\begin{cases}\dfrac{2\mu}{4\pi\hbar\kappa\varrho r}\sinh(\kappa r)\mathrm{e}^{-\kappa\varrho}\ ; & r<\varrho\ ,\\[2mm] \dfrac{2\mu}{4\pi\hbar\kappa\varrho r}\sinh(\kappa\varrho)\mathrm{e}^{-\kappa r}\ ; & r>\varrho\ ,\end{cases} \tag{4.41}$$

where

$$\kappa=\left(\frac{2\mu}{\hbar}(\omega_g-\omega-i\eta)\right)^{1/2}\ . \tag{4.42}$$

With $\boldsymbol{M}_\lambda(r)$ now being a function of $r\equiv|\boldsymbol{r}|$ we obtain from (4.24a) the relation

$$\underset{\approx}{\chi}_-(\omega)=\frac{1}{\hbar\varepsilon_0}\sum_\lambda\int \boldsymbol{M}_\lambda(r)\otimes g_0(r,\varrho)\boldsymbol{M}_\lambda(\varrho)(4\pi)^2r^2\varrho^2d\varrho\, dr\ . \tag{4.43}$$

For $\boldsymbol{M}_\lambda(r)$ we may insert an evanescent spherical wave with strength $\boldsymbol{M}_{\lambda_0}$ and decay length $r_0$ [cf. (3.70a)]. Then

$$\underset{\approx}{\chi}_-(\omega)=\sum_\lambda \boldsymbol{M}_{\lambda_0}\otimes\boldsymbol{M}_{\lambda_0}\frac{\mu}{4\pi\varepsilon_0 r_0\hbar^2}(1+\kappa r_0)^{-2} \tag{4.44a}$$

or in case of isotropic response

$$\chi_-(\omega)=M_0^2\frac{\mu}{4\pi\varepsilon_0 r_0\hbar^2}(1+\kappa r_0)^{-2} \tag{4.44b}$$

as found previously (Stahl, 1979). Near the gap (to first order in $\kappa r_0$) one finds

$$\chi_-(\omega)=M_0^2\frac{2\mu}{4\pi\varepsilon_0\hbar^2}\left\{\frac{1}{2r_0}-\left[\frac{2\mu}{\hbar}(\omega_g-\tilde{\omega})\right]^{1/2}\right\}\ . \tag{4.45}$$

Adding the antiresonance part and inserting values for $r_0$ and $\boldsymbol{M}_{\lambda_0}$ from (3.70) one obtains

$$\begin{aligned}\underset{\approx}{\chi}(\omega)=&\sum_\lambda \boldsymbol{p}_\lambda\otimes\boldsymbol{p}_\lambda\frac{e^2}{4\pi\varepsilon_0 m_0^2\omega_g^2}\left(\frac{2\mu}{\hbar}\right)^{3/2}\\ &\times\left[\tfrac{1}{2}\omega_g-(\omega_g-\tilde{\omega})^{1/2}+\tfrac{1}{2}\omega_g(1+\sqrt{2})^{-2}\right]\ .\end{aligned} \tag{4.46}$$

To first order in $|\omega_g - \tilde{\omega}|^{1/2}$ this is the same as found conventionally [cf. (4.38)]. From (4.46) it is also seen that the antiresonant part (the last term in the square bracket) is real and rather small.

In order to demonstrate the influence of changing the shape of $\boldsymbol{M}(\boldsymbol{r})$ we can compare the above result with that obtained from a shell structure

$$\boldsymbol{M}(\boldsymbol{r}) = \boldsymbol{M}_0 \frac{\delta(r - \varrho_0)}{4\pi \varrho_0^2} \ . \tag{4.47}$$

Then the functions $u_{\pm}(\boldsymbol{r})$ become proportional to $g_0(r, \varrho_0)$, and the susceptibility is found to be

$$\begin{aligned} \chi_-(\omega) &= \frac{1}{\hbar\varepsilon_0} M_0^2 g_0(\varrho_0, \varrho_0) \\ &= M_0^2 \frac{\mu}{4\pi\varepsilon_0 \hbar^2 \varrho_0^2 \kappa} (1 - \mathrm{e}^{-2\kappa\varrho_0}) \ . \end{aligned} \tag{4.48}$$

Expansion of (4.48) near the gap (first order in $\kappa\varrho_0$) yields

$$\chi_-(\omega) = M_0^2 \frac{2\mu}{4\pi\varepsilon_0 \hbar^2} \left\{ \frac{1}{\varrho_0} - \left[ \frac{2\mu}{\hbar} (\omega_g - \tilde{\omega}) \right]^{1/2} \right\} . \tag{4.49}$$

This result is the same as found for the evanescent structure of $M(r)$ if $2r_0 = \varrho_0$.

The above result is an indication of a general applicability of the shell source approximation. Let us therefore explore it in a general framework. For a wide class of electron-hole potentials, the Green's function $g_0(r, \varrho_0)$ is characterized by a finite value at $r = 0$ and a slope jump at $\varrho_0$ proportional to $\varrho_0^{-2}$. As a consequence we have for small $\varrho_0$ :

$$\left| \frac{dg_0}{dr} \right|_{\varrho_0 + 0} \gg \left| \frac{dg_0}{dr} \right|_{\varrho_0 - 0} \ . \tag{4.50}$$

Then the jump condition for $dg_0/dr$ at $\varrho_0$ can be replaced by

$$\left| \frac{\partial g_0}{\partial r} \right|_{r = \varrho_0 + 0} = -\frac{2\mu}{4\pi \varrho_0^2 \hbar} \ . \tag{4.51}$$

Consequently we may write in general

$$\chi_-(\omega) = -\sum_\lambda \boldsymbol{M}_{\lambda_0} \otimes \boldsymbol{M}_{\lambda_0} \frac{\mu}{2\pi\varepsilon_0 \varrho_0^2 \hbar} \frac{u_-}{du_-/dr} \bigg|_{r = \varrho_0} \tag{4.52}$$

where $u_-(r)$ is the allowed solution (regular at infinity) to the equation

$$[\Omega_{eh}(\boldsymbol{q}) - \tilde{\omega}] u_-(r) = 0 \ . \tag{4.53}$$

The two equations (4.52, 53) are useful in many applications. First, they are particularly well suited for a numerical integration in relative space as demonstrated in works by Balslev (1985) on a Debye-Hückel screened electron-hole interaction and by Zimmermann (1985) on a complex (dissipative) electron-hole potential. Second, they also simplify the calculations based on analytical functions. For example, consider electron-hole production in two dimensions. The pertinent solution to the interband equation is in this case

$$u_-(r) \propto K_0(\kappa r)$$

where $K_0$ is the zero order McDonald function (Abramowitz and Stegun, 1965). Expanding $K_0$ and $dK_0/dr$ for small arguments one obtains

$$\chi(\omega) \approx -\frac{M_0^2 \mu}{\pi \varepsilon_0 \hbar} \ln(\kappa \varrho_0)$$

where $\varrho_0$ is the radius of a circular source. This result gives a step function for $\chi''(\omega)$ and a logarithmic singularity at $\omega_g$ for $\chi'(\omega)$ (as also found by conventional methods).

The results obtained in this section can be summarized as follows:

a) In the long wave limit ($q = 0$) the interband waves in relative space are characterized by evanescent spherical waves for frequencies below the gap and propagating (outgoing) waves above the gap.

b) The edge susceptibility $\chi$ consists of a background $\chi(\omega_g)$ and a singular part. The background part is real and is proportional to $M_0^2/\langle r \rangle$ where $\langle r \rangle$ is the typical spatial width of $M(r)$. Consequently, an uncertainty about $\langle r \rangle$ is unimportant as long as $\tilde{\varepsilon}$ is adjustable.

c) Near the gap frequency (to first order in $|\omega - \omega_g|^{1/2}$) the singular part $\chi(\omega) - \chi(\omega_g)$ is proportional to $M_0^2$ independent of the detailed radial dependence of $M(r)/M_0$.

d) The behaviour of $\chi(\omega) - \chi(\omega_g)$ far from the gap frequency does depend on the shape of $M(r)/M_0$ as can be seen from (4.44) and (4.48). None of these results were obtained from dipole densities $M(r)$ derived accurately from (3.65), and they have not been subject to a proper band-limited interpolation. Consequently, the results (4.44) and (4.48) with $\omega$ far from the gap frequency have little physical interest.

e) Unlike the coherent wave approach the conventional result (4.33) involving the joint density of states is valid also far above the band gap. However, as discussed in the next section, an accurate conventional treatment of the electron-hole interaction is based on the effective mass approximation which breaks down far above the gap considered. The extended range of validity for the joint-density-of-states approach can therefore only be trusted if one has made sure that the electron-hole interaction may be neglected.

## 4.4 Elliott's Theory

It was recognized some time ago by Frenkel (1931), Peierls (1931) and Wannier (1937) that the electron-hole interaction neglected in the previous section must be included in a detailed description of interband transitions. The electron-hole interaction modifies the optical response strongly particularly near the fundamental absorption edge. Inspired by experimental results on $Cu_2O$ (Gross, 1956) and the elemental semiconductors Si and Ge (Macfarlane et al., 1957), Elliott developed a very accurate theory (Elliot, 1957) on the optical response of Wannier excitons. He combined the effective mass approximation with a perturbation theory of optical transitions and founded in this way the fruitful field of interband optical response of the semiconductor edge. In this section we shall reproduce the standard Elliott theory in order to give a historical and conceptual introduction to excitonic phenomena.

Let us start by defining the Bloch type two-particle wave function $\Phi_{B\lambda}(\boldsymbol{k}_e, \boldsymbol{k}_h)$. It describes in the many particle system the state characterized by one electron with wave vector $\boldsymbol{k}_e$ in the conduction band and one electron with wave vector $-\boldsymbol{k}_h$ missing in the valence band. $\lambda$ specifies the pair of bands or pair sublevel. From this basis we can construct the wave function in the Wannier representation

$$\Phi_{W\lambda}(\boldsymbol{R}_e, \boldsymbol{R}_h) = \frac{1}{\overline{N}} \sum_{\boldsymbol{k}_e, \boldsymbol{k}_h} \exp\left(i\boldsymbol{k}_e \cdot \boldsymbol{R}_e + i\boldsymbol{k}_h \cdot \boldsymbol{R}_h\right) \Phi_{B\lambda}(\boldsymbol{k}_e, \boldsymbol{k}_h) \tag{4.54}$$

where $\overline{N}$ is the number of lattice cells. In the presence of electron-hole interaction the new eigenstates – the excitons – have wave functions

$$\psi_\lambda(\boldsymbol{q}, \nu) = \sum_\mu \sum_{\boldsymbol{R}_e, \boldsymbol{R}_h} F_{\boldsymbol{q}\nu}^{\lambda\mu}(\boldsymbol{R}_e, \boldsymbol{R}_h) \Phi_{W\mu}(\boldsymbol{R}_e, \boldsymbol{R}_h) \ . \tag{4.55}$$

Here, $\boldsymbol{q}$ and $\nu$ specify the exciton: $\boldsymbol{q}$ is the total wave vector while $\nu$ specifies the internal motion. In what follows we neglect the mixing of pair sublevels by the electron-hole interaction and drop the superscript on the envelope function $F$. The effective mass theory yields (Knox, 1963)

$$F_{\boldsymbol{q}\nu}(\boldsymbol{R}_e, \boldsymbol{R}_h) \rightarrow (\overline{N}\Omega)^{-1/2} f_\nu(\boldsymbol{r})\, e^{i\boldsymbol{q}\cdot\boldsymbol{R}} \ , \tag{4.56}$$

$$\left(-\frac{\hbar}{2\mu}\nabla^2 + V_{eh} - E_\nu\right) f_\nu(\boldsymbol{r}) = 0 \ , \tag{4.57}$$

$$E_{\boldsymbol{q}\nu} = \hbar\omega_g + \frac{\hbar^2 q^2}{2m} + E_\nu \ . \tag{4.58}$$

$V_{eh}$ is the electron-hole potential, $m = m_e + m_h$ is the total mass, $E_{\boldsymbol{q}\nu}$ is the exciton energy, and $E_\nu$ is the energy of the relative electron-hole motion. The arrow in (4.56) stands for the transition from functions of discrete variables to band limited functions of continuous variables (see Appendix A).

As the dielectrically screened Coulomb interaction in most semiconductors varies sufficiently slowly on the scale of a lattice constant, we may insert in (4.57)

$$V_{eh} = -\frac{e^2}{4\pi\varepsilon_0\varepsilon_s r} \tag{4.59}$$

where $\varepsilon_s$ is the static dielectric constant. Then, by analogy with the hydrogen atom, one obtains the following two types of excitons – "closed-orbit" and "open-orbit" excitons. The closed-orbit excitons are characterized by

$$f_\nu(\boldsymbol{r}) = \psi_{n\ell m}(\boldsymbol{r}/a_B) \tag{4.60a}$$

$$E_\nu = -\frac{E_x}{n^2} \;\; ; \;\; n = 1, 2, \ldots \tag{4.60b}$$

where $\psi_{n\ell m}$ is a scaled, normalized hydrogenic wave function while $E_x \equiv \hbar\omega_x$ and $a_B$ are the exciton Rydberg and Bohr radius, respectively:

$$a_B = \hbar^2(4\pi\varepsilon_0\varepsilon_s)/\mu e^2 \ , \tag{4.61a}$$

$$E_x = \hbar\omega_x = \frac{e^2}{8\pi\varepsilon_0\varepsilon_s a_B} \ . \tag{4.61b}$$

Open-orbit excitons have

$$f_\nu(\boldsymbol{r}) = (\overline{N}\Omega)^{-1/2}\psi_{k\ell m} \ , \tag{4.62a}$$

$$k = (2\mu E_\nu/\hbar^2)^{1/2} \ , \tag{4.62b}$$

where $\psi_{k\ell m}$ is a hydrogenic wave function with positive electron energy. As a computational aid Elliott uses solutions in form of standing spherical waves. In spite of this structure we use the name open-orbit excitons. In (4.62a) $k$ is the (real) wave number of the radial dependence in the limit $r \to \infty$. This yields an asymptotic dependence of the form $r^{-1}\sin(kr - \phi)$ where $\phi$ is varying slowly with $r$. Thus $\nu$ stands for $n, \ell, m$ in case of closed orbit excitons and $k, \ell, m$ in case of open orbit excitons.

In the golden rule approach the absorptive part $\chi''$ of the interband susceptibility becomes

$$\chi''(\omega, \boldsymbol{q}) = \frac{\pi e^2}{\varepsilon_0 m_0^2 \omega^2} \sum_\nu |\boldsymbol{p}_{q\nu} \cdot \boldsymbol{e}|^2 \delta(E_{q\nu} - \hbar\omega) \tag{4.63}$$

where $\boldsymbol{p}_{q\nu}$ is the momentum matrix element between the ground state and the exciton specified by $\boldsymbol{q}, \nu$. In the limit in which the $\boldsymbol{p}$-matrix element between the Bloch states involved is considered constant, it is easy to express $\chi''(\omega, \boldsymbol{q})$ in terms of $f_\nu(\boldsymbol{r})$ and $\boldsymbol{p}$ (Elliott, 1957):

$$\chi''(\omega, \boldsymbol{q}) = \frac{\pi e^2}{\varepsilon_0 m_0^2 \omega^2} \sum |f_\nu(0)|^2 |\boldsymbol{p}\cdot\boldsymbol{e}|^2 \delta(E_{\boldsymbol{q}\nu} - \hbar\omega) \ . \tag{4.64}$$

From well-known properties of hydrogenic wave functions (Elliott, 1957) it is found that

$$|f_\nu(0)|^2 = a_B^{-3} n^{-3} \delta_{\ell,0}\delta_{m,0} \tag{4.65a}$$

for closed-orbit excitons and

$$|f_\nu(0)|^2 = \frac{\pi/(ka_B)}{1 - \exp(-2\pi/ka_B)} \tag{4.65b}$$

for open-orbit excitons. Only $s$-like excitons contribute and so the summation over $\nu$ involves only the principal quantum number $n$ and the wave number $k$ of the continuum states. The density of continuum states can be calculated by standard techniques and is proportional to $k$. The final result for $\chi''(\omega, 0)$ then is

$$\chi''(\omega, 0) = \frac{e^2 |\boldsymbol{p}\cdot\boldsymbol{e}|^2}{\varepsilon_0 m_0^2 a_B^3 \omega^2} D(\hbar(\omega - \omega_g)) \tag{4.66}$$

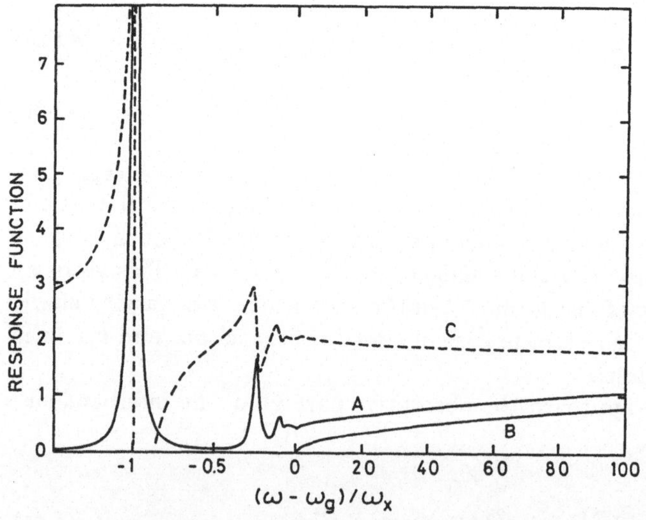


**Fig. 4.4.** Calculated real and imaginary part of the susceptibility near the fundamental absorption edge. Curve $A$ is $\chi''(\omega)$ expressed in (4.66, 67) or (4.75). Curve $B$ is $\chi''(\omega)$ in the absence of electron hole interaction. Curve $C$ is $\chi'(\omega)$ expressed in (4.75). An appropriate broadening is included and the exciton mass is considered infinite. Note the different frequency scale above and below $\omega_g$

where the interband spectral density $D$ is given by

$$D(E_\nu) = \sum_{n=1}^{\infty} n^{-3}\delta(E_\nu - E_x/n^2) + \frac{\frac{1}{2}\theta(E_\nu)/E_x}{1 - \exp\left(-2\pi E_x^{1/2}/E_\nu^{1/2}\right)} \tag{4.67}$$

where $\theta(E_\nu)$ stands for the unit step function.

The spectral density $D(E_\nu)$ is shown in Fig. 4.4 with a little damping included for clarity. The lines below the gap are caused by $s$-like closed orbit excitons. Above the gap the absorption is due to open orbit excitons. This absorption continuum is considerably enhanced compared with the absorption found without electron-hole interaction. It is seen that this Coulomb enhancement is noticeable even $100\,E_x$ above the gap.

So far no analytical expression for $\chi'(\omega)$ has been obtained by a Kramers-Kronig transformation of Elliott's result (4.66, 67). In a restricted range near a specific exciton line, accurate results can be obtained by describing the contributions from other resonances and the continuum by a background susceptibility. At the series limit $\omega = \omega_g$, the contribution to $\chi'(\omega)$ from the closed orbit excitons alone diverges. As we shall see in the next section this divergence is cancelled by the contribution from open orbit excitons [last term in (4.67)].

## 4.5 Coherent Wave Approach to Excitonic Response

We saw in Sect. 4.3 that the susceptibility near the gap frequency can be derived from coherent wave theory without knowing the detailed real-space structure of the dipole density $\boldsymbol{M}(\boldsymbol{r})$. This gives us the freedom to use that structure which gives the simplest calculations. For allowed transitions we use the shell structure with a shell radius given by

$$\varrho_0 = 2r_0 = 2a_B(\omega_x/\omega_g) \ . \tag{4.68}$$

This range of $\boldsymbol{M}(\boldsymbol{r})$ is obtained from (3.65) and the comparison between (4.45) and (4.49).

We apply the shell source approximation leading to (4.52, 53) and include in $\Omega_{eh}$ the dielectrically screened Coulomb potential. Using the isotropy in relative space we obtain

$$\Omega_{eh}(\boldsymbol{q}) = -\frac{\hbar}{2\mu}\left(\frac{\partial^2}{\partial r^2} + \frac{2}{r}\frac{\partial}{\partial r}\right) + \frac{\hbar}{2m}q^2 + \omega_g - \frac{e^2/\hbar}{4\pi\varepsilon_0\varepsilon r} \tag{4.69a}$$

$$(\Omega_{eh} - \omega - i\gamma)u(r) = 0 \tag{4.69b}$$

$$\chi_-(\omega, \boldsymbol{q}) = \left.\frac{-\mu M_0^2}{2\pi\varepsilon_0\varrho_0^2\hbar}\frac{u}{du/dr}\right|_{r=\varrho_0} , \tag{4.69c}$$

with the condition that $u(r \to \infty) = 0$. The solutions to (4.69a, b) can then be expressed as

$$u(r) = e^{-\kappa r} U(1 - (\kappa a_B)^{-1}, 2, 2\kappa r) \ , \tag{4.70}$$

where

$$\kappa = \left[ \frac{2\mu}{\hbar}(\omega_g - \omega - i\gamma) - \frac{\mu}{m} q^2 \right] \ , \tag{4.71}$$

and $U(a, b, z)$ is a confluent hypergeometric function known as one of the Kummer functions. The notation of Abramowitz and Stegun (1965) is adopted. The quantity $\kappa$ will be used often in the following sections: In the $q = 0$ limit, $\kappa$ is essentially real below the gap. It is equal to $a_B^{-1}$ near the lowest excitonic resonance and approaches zero when $\omega \to \omega_g$ (and $\gamma = 0$). Above the gap frequency $\kappa$ is essentially imaginary with $\kappa = -ik$ where $k$ is the asymptotic wave number in relative space. As we are interested in $\chi$ at most a few exciton Rydbergs above and below $\omega_g$ we see that $|\kappa|$ is at most of the order $a_B^{-1}$. As $\varrho_0$ is generally much smaller than $a_B$ we may use throughout that

$$|\kappa| \varrho_0 \ll 1 \ . \tag{4.72}$$

This limit was the basis for choosing a rather artificial structure of $\boldsymbol{M}(\boldsymbol{r})$. The small value of $\varrho_0$ can also be used to express $u(\varrho_0)$ and $du(\varrho_0)/dr$ by simpler functions than the Kummer functions. Expanding $u(r)$ for small arguments we obtain (Abramowitz and Stegun, 1965)

$$u(r) \approx \tfrac{1}{2} a_B r^{-1} - \ln(\kappa r) - \psi(1 - (\kappa a_B)^{-1}) - \tfrac{1}{2} \kappa a_B \ , \tag{4.73a}$$

$$du(r)/dr \approx -\tfrac{1}{2} a_B r^{-2} \ , \tag{4.73b}$$

where $\psi(x) = \Gamma^{-1} d\Gamma/dx$ is the digamma function. As singularities in $u(r)$ are induced by $\kappa$ going to resonance as well as by $r$ going to zero, it is necessary to retain in (4.73a) all orders of $r^{-1}$ between zeroth and first order. Inserting (4.73) into (4.69c) we get

$$\chi_-(\omega, \boldsymbol{q}) = \frac{M_0^2 \mu}{\pi \varepsilon_0 \hbar^2 a_B} [\tfrac{1}{2} a_B \varrho_0^{-1} - \ln(\kappa \varrho_0) - \psi(1 - (\kappa a_B)^{-1}) - \tfrac{1}{2} \kappa a_B] \ . \tag{4.74}$$

This result is expressed slightly differently by Stahl and Balslev (1982). We may explore $\chi(\omega, \boldsymbol{q})$ further by suitable series expansions for the digamma function. Below the gap ($\kappa$ essentially real) the relevant expansion gives the result

$$\chi_-(\omega, \boldsymbol{q}) = \frac{M_0^2 \mu}{\pi \varepsilon_0 \hbar^2 a_B} \left[ \frac{1}{2} a_B \varrho_0^{-1} - \ln(\kappa \varrho_0) + \overline{\gamma} - \sum_{n=1}^{\infty} \left( n \left( n - \frac{1}{\kappa a_B} \right) \right)^{-1} - \frac{1}{2} \kappa a_B \right] , \tag{4.75}$$

where $\overline{\gamma}$ is Euler's constant. Without damping, $\chi_-$ has simple poles for $\kappa a_B = n^{-1}$ with residues proportional to $n^{-3}$, and so the imaginary part of (4.75) becomes identical to Elliott's result for resonances below the gap. Above the gap the relevant expansion of $\psi(x)$ yields (Abramowitz and Stegun, 1965)

$$\chi_-(\omega, \boldsymbol{q}) = \frac{M_0^2 \mu}{\pi \varepsilon_0 \hbar^2 a_B} \left( \frac{1}{2} a_B \varrho_0^{-1} - \ln(\varrho_0 / a_B) + \frac{i\pi}{1 - \exp(-2\pi / k a_B)} \right) , \tag{4.76}$$

where $k = i\kappa$. Equation (4.76) is obtained by expanding the real part of $\chi_-(\omega, q)$ to first order in $k$. The imaginary parts of (4.74) and (4.76) are equal to any order of $k$ (if $k$ is real). Note that the real part $\chi'_-(\omega_g + 0, 0)$ at the gap frequency is well defined although the contribution from closed-orbit excitons taken alone would diverge. In the simultaneous treatment of the resonances and the continuum this divergence is removed as seen also in Fig. 4.4. The value

$$\chi'(\omega_g + 0, 0) = \frac{M_0^2 \mu}{2\pi \varepsilon_0 \hbar^2 a_B} [a_B \varrho_0^{-1} - 2\ln(\varrho_0 / a_B)] \tag{4.77}$$

is almost the same as without electron-hole interaction [cf. (4.49)].

Comparing the imaginary part $\chi''_-(\omega, 0)$ from the coherent wave approach with Elliott's results, there is a complete agreement in spite of the very different theoretical procedure. This aspect will be discussed further in the next section.

It is appropriate also to consider "forbidden" transitions mentioned at the end of Sect. 3.6. In this case $\boldsymbol{M}(\boldsymbol{r})$ has inversion symmetry, and in the same spirit as for allowed transitions we may use a $p$-like shell source

$$\boldsymbol{M}(\boldsymbol{r}) \propto \boldsymbol{r} \delta(r - \varrho_0) \ . \tag{4.78}$$

The regular solutions to the wave equation in relative space then become $p$-like and can be expressed as (cf. Appendix E)

$$u(\boldsymbol{r}) = z \, \mathrm{e}^{-\kappa r} U(2 - (\kappa a_B)^{-1}, 4, 2\kappa r) \ , \tag{4.79}$$

where $z$ is one of the coordinates in relative space. The resulting susceptibility given by

$$\chi_- \propto \left. \frac{u/z}{\partial(u/z)/\partial r} \right|_{r=\varrho_0} \tag{4.80}$$

has simple poles for $(\kappa a_B)^{-1} = 2, 3, 4\ldots$ . The imaginary part of (4.80) becomes identical to the result of a conventional treatment (Elliott, 1957; McLean, 1960) and agrees with experiments on $Cu_2O$ (Nikitine, Grun and Sieskind, 1961).

In the electric dipole approximation the $n = 1$ line ($\kappa a_B = 1$) is missing. However, quadrupole transitions forming $n = 1$ excitons are allowed if the quadrupole density $\underset{\approx}{Q}(r)$ mentioned in Sect. 4.2 has $s$-like symmetry. The nonlocal response characteristic for quadrupole transitions (see Sect. 4.2) is observed for the weak $n = 1$ absorption line in $Cu_2O$ (Grun, 1962).

## 4.6 What Is New?

In the last section we saw that the coherent wave approach reproduces in great detail the results for $\chi''(\omega, 0)$ from Elliott's theory. In view of this satisfactory agreement we want in this section to discuss the structural differences between the two approaches.

### 4.6.1 $\chi'$ and $\chi''$

A remarkable property of the edge equation (4.1) is that it allows a simple simultaneous calculation of both the real and imaginary parts of the susceptibility. The coherent wave treatment replaces the standard procedure (based on stationary states, golden rule, and Kramers-Kronig transform) by a real space integration of a wave equation for a macroscopic bilocal field.

### 4.6.2 Continuous and Discrete Spectra

We have also seen that the coherent wave approach provides a simple, combined treatment of the excitonic resonances and the continuum. There is no need for the mathematically complicated distinction between localized and delocalized wave functions with all the corresponding bookkeeping concerning normalization, density of states etc.

### 4.6.3 The Residual Epsilon

The simultaneous treatment of the response of open-orbit and closed-orbit excitons means that the residual dielectric constant $\tilde{\varepsilon}$ is not the same as the background dielectric constant $\varepsilon_b$ used when treating a single resonance (Hopfield and Thomas, 1963). The total dielectric function near the band gap is

$\tilde{\varepsilon}$ plus the contribution $\chi$ from the lowest edge. However, the "lowest edge" can not be defined precisely at higher interband energies, and so $\tilde{\varepsilon}$ is model dependent. The most important quantity related to $\tilde{\varepsilon}$ is the width of $\boldsymbol{M}(\boldsymbol{r})$ as discussed in Sect. 3.6. In other words the uncertainty in our knowledge on the width of $M(r)$ can be absorbed in $\tilde{\varepsilon}$. This is also the case for the antiresonance part of $\chi$ being as structureless as the response of higher interband transitions.

### 4.6.4 The Structure of $Y$

A conceptual difficulty of Elliott's theory arises from the fact that the bare (uncoupled) exciton problem has as stationary continuum states only *standing* waves in relative space. In contrast to this, the coherent-wave treatment gives directly the appealing structure of an outgoing spherical wave in relative space when electron-hole pairs are created by light with a frequency above $\omega_g$. Also below the gap there are fundamental differences between Elliott's excitonic wave functions and the coherent waves in relative space. Elliott's solution is characterized by a certain size of excitons depending on the principal quantum number. The coherent-wave amplitude in relative space $u(r)$ has a continuously varying spatial width defined for example as the reciprocal logarithmic derivative of $u(r)$ for $r \to \infty$. This width is $\kappa^{-1}$ which depends only on frequency (for $q = 0$) and coincides with the exciton radius only at resonance. Near the origin of the relative space $u(r)$ differs significantly from conventional exciton wave functions unless the electron-hole system is driven at resonance. The difference is due to the fact that in the coherent wave theory one considers wave functions driven by a source. For a shell source at $r = \varrho_0$ the electric field amplitude is proportional to $(-du/dr)^{-1}$ at $\varrho_0$, while the polarization is porportional to $u(\varrho_0)$. Instructive examples of solutions $u(r)$ are shown in Fig. 4.5. Note that

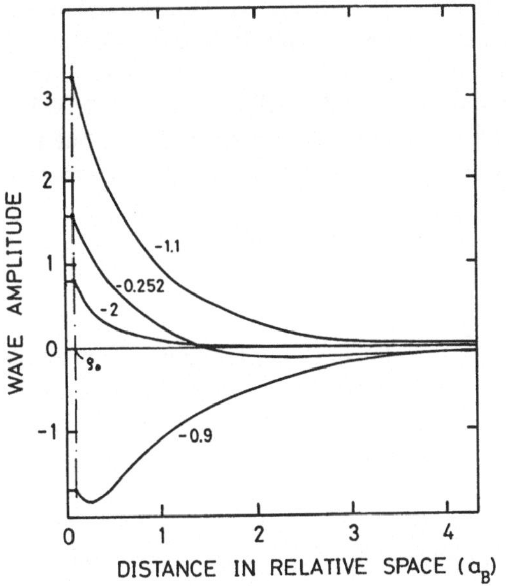


**Fig. 4.5.** Relative space amplitude of Wannier exciton polaritons for a shell structure of the dipole density $M(r)$. The frequency of each curve is given as the value of $(\omega - \omega_g)/\omega_x$. The shell radius is $0.1\, a_B$. The curves are normalized to have the same slope on the outer surface of the shell

all curves for $r>\varrho_0$ are solutions of the homogeneous wave equation. They are normalized to have the same electric field (same slope at $\varrho_0+0$). Then the amplitude $u(\varrho_0)$ represents the susceptibility. The shell source approximation gives the characteristic jump in the slope of $u(r)$ at $\varrho_0$. For more realistic structures of the dipole density $\boldsymbol{M}(\boldsymbol{r})$ this jump is smoothed out. However, this change does not invalidate the shell source approximation as useful for obtaining insight into the relative space properties of the wave amplitude $Y$. Near resonance the excitonic wave function $F_\nu(\boldsymbol{r})$ and the relative space part $u(\boldsymbol{r})$ of $Y(\boldsymbol{R},\boldsymbol{r})$ are solutions of the same wave equation. In spite of this similarity the two functions have different physical interpretations. $|F_\nu(\boldsymbol{r})|^2$ is a probability density and the phase of $F_\nu(\boldsymbol{r})$ has no physical meaning. In contrast, $u(\boldsymbol{r})$ is not subject to normalization and the phase of $u(\boldsymbol{r})$ has a physical meaning. Let us briefly trace back how the polarization and the polarization current are given by the complex function $u(\boldsymbol{r})$. Recalling that we have been dealing only with the resonance part we can construct $Y_-(\boldsymbol{R},\boldsymbol{r},t)$ :

$$Y_-(\boldsymbol{R},\boldsymbol{r},t)=Y_0u(\boldsymbol{r})\exp\left(i\boldsymbol{q}\cdot\boldsymbol{R}-i\omega t\right) \ . \tag{4.81}$$

Then the (real) resonance part of the polarization is

$$\boldsymbol{P}_-(\boldsymbol{R},t)=2\boldsymbol{M}_0\,\mathrm{Re}\{Y_-(\boldsymbol{R},\varrho_0,t)\} \tag{4.82}$$

while the corresponding polarization current is represented by the imaginary part of $Y_-(\boldsymbol{R},\varrho_0,t)$.

### 4.6.5 Other $V_{eh}$ Can Be Handled

A convenient property of the coherent wave treatment is that the response can be derived directly without the detour via stationary states. This is particularly useful when the electron-hole interaction is not the simple Coulomb potential. As mentioned in Sect. 4.3 the response can be derived by a simple radial integration of the homogeneous edge equation with the subsequent use of (4.52). This method can handle not only different radial dependence of $V_{eh}$, but also dissipative interactions described by a complex potential $V_{eh}$ (Zimmermann, 1985).

### 4.6.6 The Frequency Range

So far we have only used the edge equations for deriving the response near the gap frequency. An important question is: Can the coherent-wave approach be used in a wider frequency range than Elliott's theory? Several uncertainties appear far above and below the gap. In addition to non-parabolicity and additional band edges there is an inherent uncertainty by using functions $M(r)$, $V_{eh}(r)$ which are not band-limited. As shown for example by Adler (1962) and by Sham (1975), the inclusion of higher-lying band pairs is essential when treating the steep part of the Coulomb interaction. The inaccuracy of the use of non-band-limited functions in a two-band model is difficult to estimate in detail. It is intuitively clear that such errors become important for rapid spatial variation of the solution $Y$. As this variation occurs on a length scale of the order $|\kappa|^{-1}$ it is clear that the coherent-wave approach is best suited for describing the near-band gap response. It should be noted that Elliott's theory is also based on band-limited wave functions and should not be applied when the envelope wave function varies considerably on the scale of a lattice constant. All we know at present is that both approaches break down when $|\kappa|^{-1}$ is somewhere between $a_B$ and the lattice constant.

## 4.7 Exciton Polaritons

In the preceding sections we have described the bulk excitonic phenomena in terms of the susceptibility $\chi(\omega, \boldsymbol{q})$ with $\omega$ and $\boldsymbol{q}$ as free parameters. Considering also Maxwell's equations we arrive at crystalline excitations in the form of polariton waves with interrelated $\omega$ and $\boldsymbol{q}$. The relevant coupled equations are obtained by combining the interband edge equation with a field equation in which the magnetic field is eliminated from Maxwell's equations:

$$\left[-\frac{\hbar^2}{2m}\nabla_R^2 - \frac{\hbar^2}{2\mu}\nabla_r^2 + V_{eh} + \hbar(\omega_g - \omega - i\gamma)\right] Y_-(\boldsymbol{R}, \boldsymbol{r}) = \tfrac{1}{2}\boldsymbol{M}(\boldsymbol{r}) \cdot \boldsymbol{E}(\boldsymbol{R}) \qquad (4.83)$$

$$-\varepsilon_0 c^2 \nabla_R \times \nabla_R \times \boldsymbol{E}(\boldsymbol{R}) = -2\omega^2 \int Y_-(\boldsymbol{R}, \boldsymbol{r}) \boldsymbol{M}(\boldsymbol{r}) d^3\boldsymbol{r} - \omega^2 \varepsilon_0 \tilde{\varepsilon} \boldsymbol{E}(\boldsymbol{R}) \ , \qquad (4.84)$$

where $\tilde{\varepsilon}$ is the residual dielectric constant (including also the antiresonant part of the edge response). The right hand side of (4.84) is the time derivative of the total displacement current, split up into two parts, the resonant edge part and the residual part.

For bulk solutions of (4.83, 84) the bilocal structure of $Y$ is irrelevant. The electrodynamics in the bulk is uniquely determined by the dielectric function

$$\varepsilon(\omega, \boldsymbol{q}) = \tilde{\varepsilon} + \chi_-(\omega, \boldsymbol{q}) \ . \qquad (4.86)$$

With this response function the plane wave solutions to (4.84) must fulfill the condition:

$$\frac{c^2}{\omega^2} \boldsymbol{q} \times \boldsymbol{q} \times \boldsymbol{E}(\boldsymbol{R}) = -\varepsilon(\omega, \boldsymbol{q}) \boldsymbol{E}(\boldsymbol{R}) \ . \qquad (4.87)$$

As usual we distinguish between two types of polariton waves, the transverse modes with

$$\nabla_R \cdot \boldsymbol{E}(\boldsymbol{R}) = 0 \ , \tag{4.87a}$$

$$\frac{c^2}{\omega^2} q^2 = \varepsilon(\omega, \boldsymbol{q}) \ , \tag{4.87b}$$

and the longitudinal modes with

$$\nabla \times \boldsymbol{E}(\boldsymbol{R}) = 0 \ , \tag{4.88a}$$

$$\varepsilon(\omega, \boldsymbol{q}) = 0 \ . \tag{4.88b}$$

When exploring the $\omega - \boldsymbol{q}$ dependence of the above polariton modes, two cases are of interest: To find for a real $\omega$ the generally complex wave vector $\boldsymbol{q}$, or to find for a given real $\boldsymbol{q}$ the complex frequency $\omega$. The former problem arises when waves are excited via the surface by a source with given frequency. The latter case is typical for Raman experiments where the scattering geometry selects a real wave vector for which the imaginary part of the frequency determines the life time broadening.

We now turn to the response function $\chi_-(\omega, \boldsymbol{q})$ derived for Wannier excitons in Sect. 4.5. Inserting (4.75) into (4.87,88) one finds polariton dispersion curves shown in Fig. 4.6 for real frequencies.

Vacuum photons couple primarily to transverse polaritons, and so we concentrate in the following on transverse solutions to (4.84) with a dispersion relation given by (4.87).

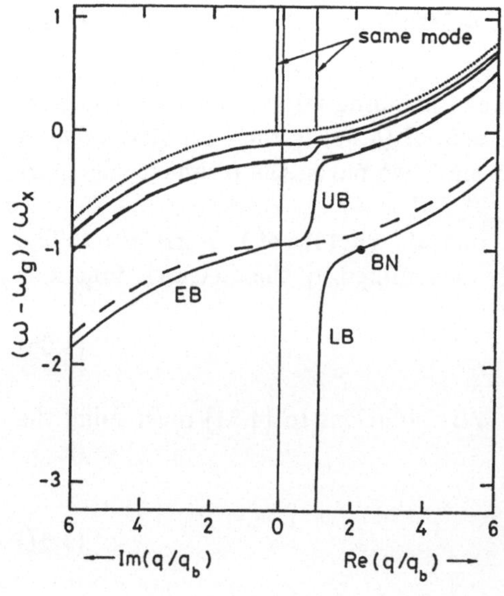


**Fig. 4.6.** Polariton dispersion curves for excitonic resonances in the Wannier model. *Full curves* are transverse modes, *dashed curves* are longitudinal modes, and the *dotted curve* is the series limit

In the coherent-wave theory, the polariton has a structure in both ordinary (center of mass) space and the hyperspace of the relative coordinate. The dependence $u_-(r)$ of the electron-hole amplitude in relative space is asymptotically given by

$$u(r) \propto r^{-1} e^{-\kappa r} \tag{4.89}$$

for $r \to \infty$, independent of the electron hole interaction. Below the gap frequency, both $\kappa$ and the polarization wave vector $q$ are essentially real. Then the relative space can be said to be totally reflecting (in the limit $\gamma = 0$). This property of relative space is changed when the polariton frequency exceeds $\omega_g$. Then $\kappa$ is imaginary and $q$ becomes complex. The spherical wave in relative space becomes outgoing, and the energy necessary for this wave propagation is supplied by the spatially damped center of mass wave. The structure of an outgoing wave in relative space is consistent with the corresponding particle description in which the electron and the hole fly apart with relative kinetic energy $\hbar(\omega - \omega_g)$.

For visualizing the structure of $Y$ above and below the gap frequency it is instructive to examine a one-dimensional exciton model. As will be discussed in Sect. 5.4, a suitably chosen potential $V_{eh}(x)$ and dipole density $M(x)$ for a one dimensional model lead to

$$Y_- \propto e^{iqX} e^{-\kappa|x|} \tag{4.90}$$

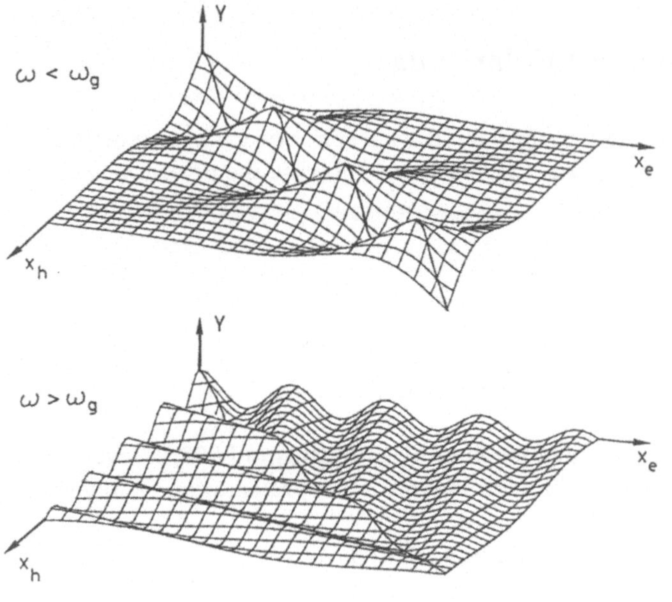


**Fig. 4.7.** The real part of the electron-hole amplitude $Y(x_e, x_h)$ for a one-dimensional exciton model discussed in Sect. 5.4. The wave amplitude is expressed in (4.90). The amplitude along the diagonal $x_e = x_h$ is proportional to the polarization

where $x = x_e - x_h$ is the relative coordinate and $X$ is the center of mass coordinate. The behaviour of the real part $Y'$ in the $x_e, x_h$ plane is shown in Fig. 4.7 for $\omega < \omega_g$ ($\kappa$ and $q$ real) and for $\omega > \omega_g$ ($\kappa$ imaginary, $q$ complex).

When the dephasing time $T_2$ is long, we may write

$$q = q(\gamma = 0) + i\gamma \left.\frac{\partial q}{\partial \omega}\right|_{\gamma=0} \tag{4.91}$$

because $\partial q/\partial\omega \approx \partial q/\partial(i\gamma)$. For modes which propagate in the absence of damping ($q(\gamma = 0)$ real) we obtain

$$\mathrm{Im}\{q\} = \alpha/2 = \gamma/v_g \tag{4.92}$$

where $\alpha$ is the absorption coefficient and $v_g$ is the group velocity. Thus, in this approximation there is a very simple relation between the spatial decay of the polariton wave and the group velocity.

As seen from Fig. 4.6, Wannier exciton polaritons with a finite mass $m$ can have infinitely many different wave vectors $\boldsymbol{q}$ for a given frequency. This kind of nonlocal response is called "spatial dispersion" (Agranovich and Ginzburg, 1966). The fact that two or more polariton modes can be excited through a surface by a single external plane wave gives severe problems in describing the photon-polariton coupling through a surface. Chapter 5 is devoted to this problem.

## 4.8 Single Resonance Polaritons

A dielectric function is often considered as a sum of oscillator contributions some of which may have spatial dispersion. In this case the polariton spectrum is given by

$$-(c^2/\omega^2)\boldsymbol{q}\times\boldsymbol{q}\times\boldsymbol{E} = \varepsilon(\omega, \boldsymbol{q})\boldsymbol{E} \tag{4.93}$$

$$\varepsilon(\omega, \boldsymbol{q}) = 1 + \sum_n \frac{f_n\omega_n^2}{\omega_n^2 - \omega^2 - i\Gamma_n\omega + \beta_n q^2} \ . \tag{4.94}$$

Having assumed an unbounded medium when setting up (4.93, 94) we can just as well express the response in a real-space-time representation as follows

$$\ddot{\boldsymbol{P}}_n - \beta_n\nabla^2\boldsymbol{P}_n + \Gamma_n\dot{\boldsymbol{P}}_n + \omega_n^2\boldsymbol{P}_n = f_n\omega_n^2\varepsilon_0\boldsymbol{E} \tag{4.95}$$

$$-c^2\nabla\times\nabla\times\boldsymbol{E} - \ddot{\boldsymbol{E}} = \sum_n \ddot{\boldsymbol{P}}_n/\varepsilon_0 \ . \tag{4.96}$$

If $\omega_n$, $\Gamma_n$, $f_n$, $\beta_n$ describe the full response of Wannier excitons including the continuum, then the oscillator equations (4.95, 96) and the coherent wave polariton equations (4.83, 84) represent the same system and give the same bulk response. The oscillator equations have a reduced spatial dimensionality, but must handle many components $\boldsymbol{P}_n$ of the crystalline excitation. The elimination in (4.95, 96) of the hyperspace of the relative coordinate resembles the hydrodynamic approximation in intraband dynamics (see Sect. 6.5).

The use of the oscillator representation is particularly useful when the attention is focused on one resonance far from others. Then for a single resonance with polarization contribution $\boldsymbol{P}_0$ :

$$\ddot{\boldsymbol{P}}_0 - \beta_0 \nabla^2 \boldsymbol{P}_0 + i\Gamma_0 \dot{\boldsymbol{P}}_0 + \omega_0^2 \boldsymbol{P}_0 = f_0 \omega_0^2 \varepsilon_0 \boldsymbol{E} \ , \tag{4.97}$$

$$-c^2 \nabla \times \nabla \times \boldsymbol{E} - \varepsilon_b \ddot{\boldsymbol{E}} = \ddot{\boldsymbol{P}}_0 / \varepsilon_0 \ , \tag{4.98}$$

where $\varepsilon_b$ is a background dielectric constant including the response from the remaining oscillators. Equations (4.97, 98) are the Hopfield-Thomas polariton equations which are useful for visualizing the response by a ball-and-spring model (Hopfield and Thomas, 1963). However, it must be kept in mind that this "hydrodynamic" approximation is closely linked to the assumption that the polariton waves meet no obstacles such as surfaces, impurities, phonons or other polaritons which act on the electron and the hole individually. This limitation of the applicability of (4.95–98) is not present in the coherent wave approach because the perturbations of the polariton propagation are usually well described in the configuration space of the electron and the hole as discussed in detail in Chaps. 5–7.

In the spirit of an oscillator model we may evaluate the coherent wave result (4.75) as follows:

$$\varepsilon(\omega, \boldsymbol{q}) = \tilde{\varepsilon} + \chi_-(\omega, \boldsymbol{q}) \approx \tilde{\varepsilon} + \Delta\chi - \frac{M_0^2 \mu}{\pi \varepsilon_0 \hbar^2 a_B [1 - (\kappa a_B)^{-1}]} \ , \tag{4.99}$$

where all terms in (4.75) except for the $n = 1$ term of the sum are lumped into a background contribution $\Delta\chi$. In a frequency range near the resonance at $\omega_0 = \omega_g - \omega_x$, $\Delta\chi$ can be considered as constant, and the expression (4.99) can be transformed into

$$\begin{aligned} \varepsilon(\omega, \boldsymbol{q}) &= \tilde{\varepsilon} + \Delta\chi + \frac{2\omega_x M_0^2 \mu}{\pi \varepsilon_0 \hbar^2 a_B (\omega_0 - \omega - i\gamma + \hbar q^2/2m)} \\ &= \varepsilon_b + \frac{\varepsilon_b \omega_{LT}}{\omega_0 - \omega - i\gamma + \hbar q^2/2m} \ . \end{aligned} \tag{4.100}$$

In the last step we have introduced the longitudinal-transverse splitting $\omega_{LT}$ defined by the condition that $\varepsilon(\omega, 0) = 0$ for the frequency $\omega_L = \omega_0 + \omega_{LT}$ (without damping). In terms of quantities in the coherent-wave theory $\varepsilon_b$ and $\omega_{LT}$ are given by

$$\varepsilon_b = \tilde{\varepsilon} + \Delta\chi \tag{4.101}$$

$$\omega_{LT} = \frac{M_0^2}{\pi\varepsilon_0\varepsilon_b\hbar a_B^3} \ . \tag{4.102}$$

Typical values of $\omega_{LT}$ and $M_0^2$ are given in Table 4.1 (p. 58).

We may then rewrite the Hopfield-Thomas equations for transverse polaritons with frequency $\omega$ as follows [since $\omega_0^2 - \omega^2 \approx 2\omega_0(\omega_0 - \omega)$]

$$(\omega_0 - \omega - \frac{\hbar}{2m}\nabla^2 - i\gamma)\boldsymbol{P}_0 = \omega_{LT}\varepsilon_b\varepsilon_0\boldsymbol{E} \tag{4.103}$$

$$(\nabla^2 c^2/\omega^2 + \varepsilon_b)\varepsilon_0\boldsymbol{E} = -\boldsymbol{P}_0 \ , \tag{4.104}$$

where

$$\omega_{LT} = \omega_0 f_0/2\varepsilon_b \tag{4.105a}$$

$$\gamma = 2\Gamma_0 \tag{4.105b}$$

$$\hbar/m = \beta_0/\omega_0 \ . \tag{4.105c}$$

It is interesting to express the energy flux density of a polariton wave in terms of the fields. When applying the above notation to the results of Bishop and Maradudin (1976) and Selkin (1977) one obtains the energy flux density

$$\boldsymbol{S} = c^2\varepsilon_0\boldsymbol{E}\times\boldsymbol{B} - \frac{\hbar}{2m\omega_{LT}\varepsilon_0\varepsilon_b}\nabla\boldsymbol{P}_0 \otimes \boldsymbol{P}_0 \ . \tag{4.106}$$

In (4.106), $\boldsymbol{E}$, $\boldsymbol{P}_0$ and the magnetic induction $\boldsymbol{B}$ are real vector fields. The first term on the r.h.s. of (4.106) is the Poynting vector, i.e. the energy flux density obtained in the absence of spatial dispersion. The second term is the crystalline contribution to the energy flux density.

In case of a single resonance it is simple to evaluate the real-space dielectric function $\varepsilon^r(\omega, \boldsymbol{r})$ defined by

$$\varepsilon^r(\omega, \boldsymbol{r}) = \int \varepsilon(\omega, \boldsymbol{q}) e^{i\boldsymbol{q}\cdot\boldsymbol{r}} d^3\boldsymbol{q} \ . \tag{4.107}$$

$\varepsilon^r(\omega, \boldsymbol{r})$ occurs in the response equation

$$\boldsymbol{P}^\omega(\boldsymbol{r}) = \int (\varepsilon^r(\omega, \boldsymbol{R} - \boldsymbol{r}') - \delta(\boldsymbol{r} - \boldsymbol{r}'))\varepsilon_0\boldsymbol{E}^\omega(\boldsymbol{r}')d^3\boldsymbol{r}' \tag{4.108}$$

between the total electric polarization $\boldsymbol{P}^\omega$ and the electric field $\boldsymbol{E}^\omega$. Inserting (4.100) into (4.107) one gets [cf. Maradudin and Mills (1973)]

$$\varepsilon^r(\omega, \boldsymbol{r}) = \varepsilon_b\delta(\boldsymbol{r}) - \frac{\varepsilon_b\omega_{LT}m}{2\pi\hbar r}e^{-\kappa_e r} \tag{4.109}$$

where

$$\kappa_e = [2m(\omega_0 - \omega - i\gamma)/\hbar]^{1/2} \ . \tag{4.110}$$

The spatial width of $\varepsilon^r(\omega, \boldsymbol{r})$ is of the order $(\mathrm{Re}\{\kappa_e\})^{-1}$. This quantity becomes infinite above resonance when $\gamma = 0$ (wave-like nonlocality). This type of long range nonlocal response is characteristic for spatial dispersion.

## 4.9 Band Edge Degeneracies and Crystal Anisotropy

The basis for the development of the band-edge model was an isotropic array of two-level systems with a lower $p$-like state and an upper $s$-like state. A more realistic situation must include spin-orbit coupling, exchange interaction, crystal anisotropy, and band structure pecularities (Sect. 3.5).

If we include spin, but still neglect the above complexities, the electron-hole pair basis is a 12-fold degenerate $s_{1/2} - p_{3/2}$ manifold. This pair level is influenced by the following perturbations

a) simple excitonic effects (see Sect. 4.4, 5),
b) crystal field,
c) spin-orbit coupling,
d) short-range exchange,
e) long range exchange, and
f) $\boldsymbol{k}$-space anisotropy of the band structure.

These perturbations have been studied extensively in the last 30 years. Pioneering contributions and valuable reviews are given by Dresselhaus et al. (1955), Hopfield (1960), Balderschi and Lipari (1973), Kane (1956, 1975), Cho (1976, 1979), Altarelli and Lipari (1976), Rössler (1979), and Ivchenko (1982). We shall briefly discuss the resulting excitonic structure in two cases: (i) Hexagonal crystal symmetry such as that found in wurtzite materials (CdS, ZnO, etc.) and (ii) Cubic symmetry without inversion symmetry as in zincblende materials (GaAs, CuCl, etc.).

### 4.9.1 Wurtzite Structure

The splitting due to the above mentioned interactions b–e) on an exciton level is shown in Fig. 4.8. For clarity the interactions are applied one after the other with the largest perturbations applied first. It is seen that the crystal field and the spin-orbit coupling split up the 12-fold $s_{1/2} - p_{3/2}$ level into three 4-fold exciton levels denoted as $A$, $B$ and $C$ excitons. Further splitting into singlet and triplet states is caused by the short range exchange interaction. The long range (non-analytical) part of the exchange interaction is included by self-consistency requirements on the electromagnetic fields (Sect. 3.2). As a result, the singlet states split up into longitudinal and transverse levels.

The exciton structure becomes rather simple in case of weak coupling in the effective mass Hamiltonian between singlet and triplet states and between

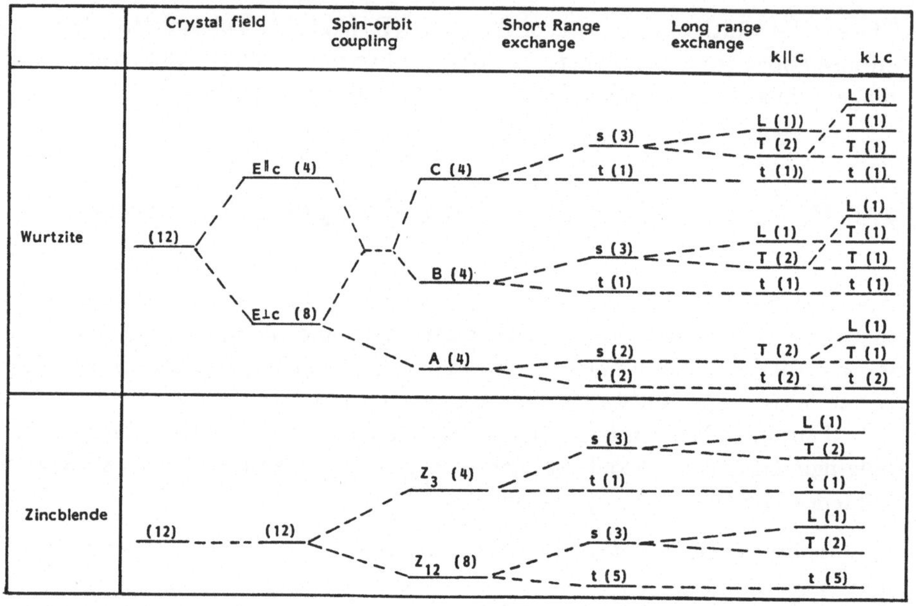


**Fig. 4.8.** Splitting of a $p_{3/2} - s_{1/2}$ pair state in wurtzite and zincblende type crystals. $s$=singlet, $t$=triplet, $L$=longitudinal, $T$=transverse. $k$ and $E$ stand for directions of electric field and wave vector and $c$ stands for the $c$-axis direction. Numbers in parentheses are degeneracies

the levels $A, B, C$ mutually. Then the electrodynamics can be described by three decoupled amplitudes $\boldsymbol{Y}_A$, $\boldsymbol{Y}_B$ and $\boldsymbol{Y}_C$ considered as vectors. The three components in $\boldsymbol{Y}_B$ or $\boldsymbol{Y}_C$ describe the amplitude of the three singlet levels, while $\boldsymbol{Y}_A$ is lying in the plane perpendicular to the $c$ axis since the $A$ level has only two singlet states. There is no need to define amplitudes for the 4 triplet states. In this notation the dipole densities can be described as second rank tensors so that

$$\hbar\left(\Omega_{eh} - i\frac{\partial}{\partial t}\right)\boldsymbol{Y}_\lambda(\boldsymbol{R},\boldsymbol{r},t) = \underset{\sim}{\boldsymbol{M}}_\lambda(\boldsymbol{r})\boldsymbol{E}(\boldsymbol{R},t); \quad \lambda = A,\ B,\ C \tag{4.111}$$

$$\boldsymbol{P}_\lambda(\boldsymbol{R},t) = 2\,\mathrm{Re}\left\{\int \underset{\sim}{\boldsymbol{M}}_\lambda(\boldsymbol{r})\boldsymbol{Y}_\lambda(\boldsymbol{R},\boldsymbol{r},t)d^3\boldsymbol{r}\right\} \quad ; \quad \lambda = A,\ B,\ C\ . \tag{4.112}$$

In a coordinate system with the $c$-axis as coordinate axis, $\underset{\sim}{\boldsymbol{M}}_\lambda(\boldsymbol{r})$ is diagonal with principal values $M_\lambda^{||}(\boldsymbol{r})$, $M_\lambda^{\perp}(\boldsymbol{r})$. As the $A$-exciton is forbidden for $E||c$, $M_A^{||}(\boldsymbol{r}) = 0$. (The above notation is valid for so-called positive spin-orbit coupling. In ZnO the strong admixture of the $\alpha$-states in the valence band gives as an exception of negative spin-orbit coupling. Then the symmetry of the (lowest) $A$-level and the $B$-level is reversed and so, in principle, $M_B^{||}(\boldsymbol{r}) = 0$, $M_A^{||}(\boldsymbol{r}) \neq 0$ in ZnO. However, $M_A^{||}(\boldsymbol{r})$ is rather small in this material).

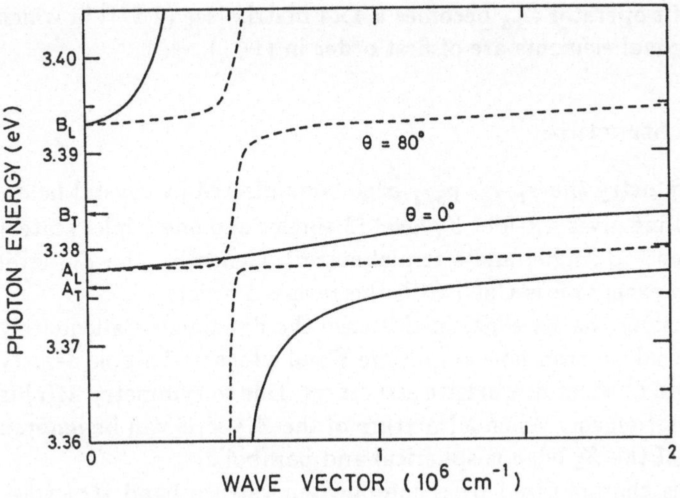


**Fig. 4.9.** Dispersion curves calculated for $A$ and $B$ exciton polaritons in ZnO. Subscripts $L$ and $T$ used on the vertical axis stand for longitudinal and transverse energies, respectively. $\theta$ is the angle between the polariton wave vector and the $c$ axis. After Hümmer and Gebhardt (1978)

The splitting into three levels with anisotropic dipolar coupling gives rise to optical anisotropy. The dispersive effect is a resonance enhanced birefringence (double refraction). The dispersion curves of extraordinary (mixed mode) and ordinary $A(n = 1)$ and $B(n = 1)$ polaritons in ZnO are shown in Fig. 4.9 (Hümmer and Gebhardt, 1978). The mixed mode polaritons are characterized by a longitudinal-transverse splitting which depends on the propagation direction. It is seen that the longitudinal frequencies $(A_L, B_L)$ are independent of propagation direction.

In a coherent-wave approach the optical anisotropy derived from the above tensorial structure of $M(\boldsymbol{r})$ is expressed by the relation

$$P_\lambda^{\pm}(\boldsymbol{R}) = \frac{1}{\hbar} \int \underset{\sim}{M}_\lambda(\boldsymbol{r}) G_{\pm}(\boldsymbol{r}, \boldsymbol{r}') \underset{\sim}{M}_\lambda(\boldsymbol{r}') \boldsymbol{E}(\boldsymbol{R}) d^3\boldsymbol{r}\, d^3\boldsymbol{r}' \tag{4.113}$$

where $G_{\pm}(\boldsymbol{r}, \boldsymbol{r}')$ are Green functions obeying (4.20). The structure of the response given in (4.113) is based on spherical, parabolic bands and isotropic electron-hole interaction. The most important deviations from this simple picture are caused by anisotropy of the translational mass tensor (Broser et al., 1978) and by $\boldsymbol{k}$-linear terms in the valence band structure. The $\boldsymbol{k}$-linear terms give rise to complicated admixture of the triplet states. As a result, the reflectivity spectrum of the $B(n = 1)$ exciton in CdS has an anomalons shoulder (see Fig. 5.1) (Mahan and Hopfield, 1964; Broser et al., 1979) and exhibits optical activity ($\boldsymbol{q}$-linear antisymmetric components of $\underset{\sim}{\chi}$) (Ivchenko, 1982). In the coherent wave treatment of such effects, one should include all four sublevels of the $B$-edge. The drift operator $\Omega_{eh}$ becomes a 4×4 matrix [see (3.57)] in which some of the off-diagonal elements are of first order in $(i\nabla_h)$.

### 4.9.2 Zincblende Structure

In case of cubic symmetry the $s_{1/2} - p_{3/2}$ edge is unaffected by crystal fields. The spin-orbit coupling gives a 4-fold $Z_3$ level (3 singlet and one triplet states) and an 8-fold $Z_{12}$ level (3 singlet and 5 triplet states). Generally, the $Z_{12}$ level is the lower one. An exception is CuCl with the reversed order.

Let us first consider the $Z_3$ edge. In this case the dynamics is adequately described by a vectorial electron-hole amplitude $\boldsymbol{Y}$ and a tensorial dipole density $\underset{\sim}{M}(\boldsymbol{r})$ as for the $B$ and $C$ edges in wurtzite structures. Due to symmetry $\underset{\sim}{\dot{M}}(\boldsymbol{r})$ is a scalar times the unit tensor. When admixture of the $Z_{12}$ level can be ignored the band structure of the $Z_3$ edge is spherical and parabolic.

The $Z_{12}$ edge is characterized by a complicated valence band structure. The 4×4 effective mass Hamiltonian matrix for the $\Gamma_8$ valence band produces a structure with light and heavy holes and with warped energy surfaces (Dresselhaus et al., 1955):

$$E_v(\boldsymbol{k}) = -Ak^2 \pm [B^2k^4 + C^2(k_x^2k_y^2 + k_y^2k_z^2 + k_z^2k_x^2)]^{1/2} \ . \tag{4.114}$$

In this expression, $\boldsymbol{k}$-linear terms are neglected. It is generally impossible to describe the interband dynamics of a $Z_{12}$ edge as products of functions of relative and center-of-mass coordinates (Dresselhaus, 1956; Kane, 1975), and so no analytical excitonic wave functions have been found. Some insight on the center-of-mass motion and on polariton effects can be achieved by approaches developed by Kane (1975), Altarelli and Lipari (1977), Cho (1976), Rössler (1979), and Hönerlage et al. (1975). If we assume the relative space wave function to be essentially independent of the total wave vector $\boldsymbol{q}$, then one can start with a $\boldsymbol{q} = 0$, $n = 1$ exciton manifold with 8 states: $|2+\rangle, |20\rangle$ with $\Gamma_3$ symmetry, $|1+\rangle, |1-\rangle$ and $|2-\rangle$ with $\Gamma_4$ symmetry, $|x\rangle$, $|y\rangle$, $|z\rangle$ with $\Gamma_5$ symmetry. The sublevel structure of the $\boldsymbol{q} = 0$ basis is shown in Fig. 4.8. The effective mass Hamiltonian $H_q$ to be considered is (Rössler, 1979)

$$H_q = \begin{Bmatrix} |2+\rangle & |20\rangle & |1+\rangle & |1-\rangle & |2-\rangle & |x\rangle & |y\rangle & |z\rangle \\ S+2V & 0 & 0 & 0 & 0 & 0 & 0 & C \\ 0 & S-2V & 0 & 0 & C & 0 & 0 & 0 \\ 0 & 0 & S-V & 0 & 0 & -U & -iC & 0 \\ 0 & 0 & 0 & S-V & 0 & iC & U & 0 \\ 0 & C & 0 & 0 & S+2V & 0 & 0 & 0 \\ 0 & 0 & -U & -iC & 0 & S+V & 0 & 0 \\ 0 & 0 & iC & U & 0 & 0 & S+V & 0 \\ C & 0 & 0 & 0 & 0 & 0 & 0 & S-2V \end{Bmatrix} \tag{4.115}$$

This form is valid for $\boldsymbol{q} \| \boldsymbol{z}$ and $\boldsymbol{z}$ chosen as the (111) or the (100) crystallografic direction. The element $C$ is proportional to $|\boldsymbol{q}|$ and is induced by the $\boldsymbol{k}$-linear terms in the valence band structure. $S$, $V$ and $U = \sqrt{3}V$ are quadratic in $|\boldsymbol{q}|$; $S$ describes the "average" exciton dispersion while $V$ is the relevant heavy-hole light-hole splitting parameter. It is seen that the terms $U$ and $C$ transfer dipole coupling from the singlet to the triplet states. Consequently the dispersion curves must include the previously forbidden triplet states. The extra polariton branches and the $\boldsymbol{q}$-linear components in the dispersion curves in CuBr are clearly seen in Fig. 4.10. As shown here, two-photon Raman scattering enhanced by biexcitons gives dispersion curves which agree in detail with theory.

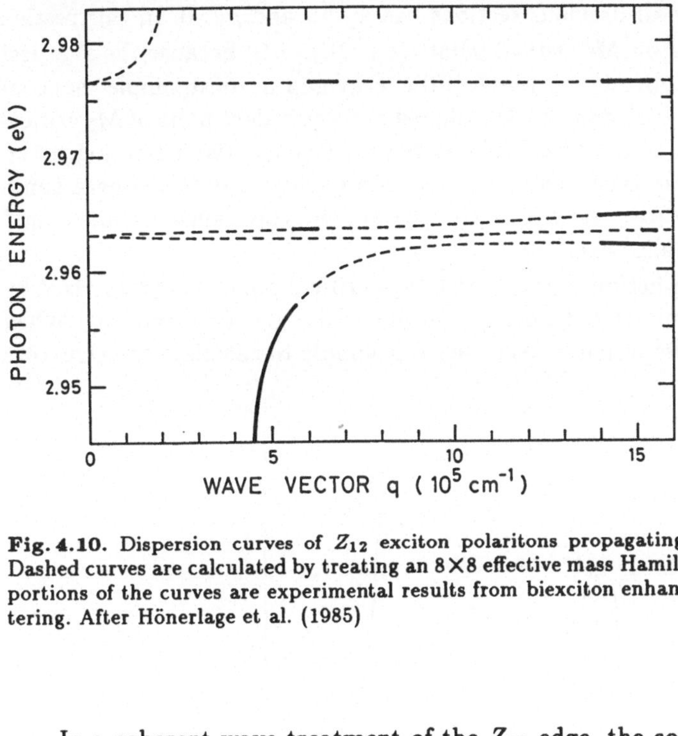


**Fig. 4.10.** Dispersion curves of $Z_{12}$ exciton polaritons propagating along [110] in CuBr. Dashed curves are calculated by treating an 8×8 effective mass Hamiltonian. The fully drawn portions of the curves are experimental results from biexciton enhanced hyper Raman scattering. After Hönerlage et al. (1985)

In a coherent-wave treatment of the $Z_{12}$ edge, the constitutive equations attain the form (3.57) discussed in Sect. 3.5. The drift operator becomes an 8×8 matrix $\Omega_{eh}^{\lambda\mu}$. In the absence of a simple center-of-mass coordinate one can use the transformation: $\boldsymbol{r} = \boldsymbol{r}_e - \boldsymbol{r}_h, \boldsymbol{r}' = (\boldsymbol{r}_e + \boldsymbol{r}_h)/2$, and insert plane wave solutions in midpoint space. If the corresponding wave vector $\boldsymbol{q}$ is chosen to be along high symmetry directions, then $\Omega_{eh}^{\lambda\mu}(\boldsymbol{q})$ has the same structure as $H_q$ in (4.115) with $S$ containing the gap frequency and the electron-hole interaction, and $S$, $V$, $U$, and $C$ containing terms linear and quadratic in $\boldsymbol{q}$ and $(i\nabla_r)$. Green's functions for such an operator have not been developed so far.

## 4.10 Saddle Point Excitons

The optical properties of semiconductors well above the energy gap are influenced by additional critical points in the joint density of states, i.e., spectral points with $\hbar\omega = E_c - E_v$ and $\nabla_k(E_c - E_v) = 0$ (Cardona, 1969) (where $E_c$ and $E_v$ stand for the energy of a conduction band and a valence band, respectively). The critical points are classified according to the notation that a $M_\nu$ critical point has $\nu$ negative principal values of the tensor $\nabla_k \otimes \nabla_k(E_c - E_v)$. In the following, suppose such a "gap" to be parabolic with the band structure

$$E_c(\boldsymbol{k}) - E_v(\boldsymbol{k}) = \hbar\omega_g + \frac{\hbar(k_x^2 + k_y^2)}{2\mu_\perp} + \frac{\hbar k_z^2}{2\mu_\parallel} \tag{4.116}$$

expressed in a suitable coordinate system in reciprocal space. The $M_0$ critical point treated in all previous sections has $\mu_\perp > 0$ and $\mu_\parallel > 0$. In this section we shall concentrate on $M_1$ critical points ($\mu_\perp > 0, \mu_\parallel < 0$) because the sharpest spectral structure above the fundamental energy gap in most simple semiconductors, the $E_1', E_1' + \Delta_1'$ doublet, is assigned to 8 equivalent pairs of $M_1$ critical points on the $L$ line in the first Brillouin zone (Cardona, 1969) (see Fig. 4.11). The doublet structure is caused by the spin-orbit splitting of the valence band. Electron-hole pair states near $M_1$ or $M_2$ critical points are called "saddle point" excitons or "hyperbolic" excitons.

The dielectric function near a saddle type critical point is characterized by an absorption continuum both below and above the gap frequency $\omega_g$. When the electron-hole interaction is neglected it is simple to calculate the structure

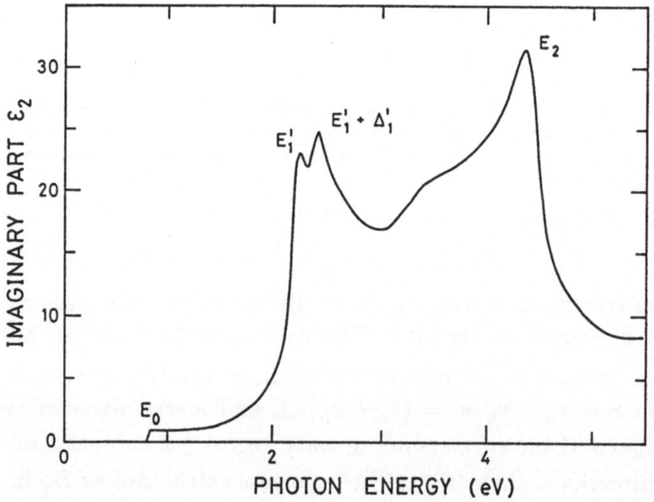


**Fig. 4.11.** Imaginary part $\varepsilon_2$ of the dielectric function of Ge at 100 K. The data above 1.5 eV are obtained from ellipsometry by Vina et al. (1984)

of $\varepsilon(\omega)$. It consists of a smooth background plus a singular contribution $\Delta\chi$ given by

$$\Delta\chi = -C(\omega - \omega_g + i\gamma)^{1/2}; \ \mu_\perp > 0; \ \mu_\| < 0 \ , \tag{4.117a}$$

$$\Delta\chi = -C \ln(\omega_g - \omega - i\gamma); \ \mu_\perp > 0; \ \mu_\| = 0 \ , \tag{4.117b}$$

where $C$ and $\gamma$ are real, positive constants (Cardona, 1969).

The experimental spectra can not be fitted to (4.117) using real values of $C$. This indicates the necessity of including the electron-hole interaction. A crude method for describing excitonic effects is to use (4.117) with $C$ as a complex constant $C = |C|e^{i\phi}$. This is correct in the case of an infinitely short range of $V_{eh}$ (Toyozawa et al., 1967). Experiments agree well with (4.117) when $\omega_g, \gamma, |C|$ and the phase $\phi$ are freely adjustable (Vina, Logothetides and Cardona, 1984), but this agreement seems to be a coincidence in view of the large Coulomb enhancement estimated from many body calculations (Hanke and Sham, 1980). The use of (4.117) also gives an error of the order one exciton Rydberg in the determination of $\omega_g$.

In a restricted range of mass ratio, namely $\mu_\perp/\mu_\| \ll 1$, it is possible to include the Coulomb interaction by using a pseudo-two-dimensional effective mass approach (Veliki and Sak, 1966; Kane, 1969). Here the exciton problem is solved exactly in the limit $\mu_\perp/\mu_\| \to 0$ giving a line spectrum with one predominant line at $\omega_g - 4\omega_\perp$ where $\hbar\omega_\perp$ is the transverse exciton Rydberg energy

$$\hbar\omega_\perp = \frac{\mu_\perp e^2}{2(4\pi\varepsilon_0\varepsilon)^2\hbar^2} \ .$$

Above $\omega_g$ there is a Coulomb-enhanced absorption continuum. By perturbative methods one can explore the influence of a finite mass ratio. Kane (1969) finds in this way a peak in $\varepsilon''$, 2 to 4 transverse exciton Rydbergs below $\omega_g$.

The coherent-wave approach can in principle handle hyperbolic exciton polaritons in an unrestricted range of mass ratios. With a positive constant $g \equiv -\mu_\perp/\mu_\|$ we are faced with the following equations for the response [c.f. (4.20, 24)]

$$\hbar(\Omega_{eh} \pm \omega) G_\pm(\boldsymbol{r}, \boldsymbol{r}') = \delta(\boldsymbol{r} - \boldsymbol{r}') \tag{4.118a}$$

$$\Omega_{eh} = -\frac{\hbar}{2\mu}\left(\frac{\partial^2}{\partial x^2} + \frac{\partial^2}{\partial y^2} - g\frac{\partial^2}{\partial z^2}\right) - \frac{e^2}{4\pi\varepsilon_0\varepsilon r\hbar} + \hbar\omega_g \tag{4.118b}$$

$$\underline{\chi}_\pm = \frac{1}{\varepsilon_0}\int \boldsymbol{M}(\boldsymbol{r}) \otimes G_\pm(\boldsymbol{r}, \boldsymbol{r}') \boldsymbol{M}(\boldsymbol{r}) d^3\boldsymbol{r}\, d^3\boldsymbol{r}' \tag{4.118c}$$

$$\chi = \chi_+ + \chi_- \ . \tag{4.118d}$$

The mixture of spherical and cylindrical symmetry of $\Omega_{eh}$ in (4.118b) represents a serious mathematical problem. It seems impossible to express $G_{\pm}(\boldsymbol{r}, \boldsymbol{r}')$ in terms of known analytical functions. A naive estimate of the structure of $G_{-}(\boldsymbol{r}, \boldsymbol{r}' \approx 0)$ can be obtained by examining the limit $|\boldsymbol{r}| \to \infty$. Introducing spherical coordinates $(r, \theta, \phi)$ and neglecting terms of order $r^{-1}$ and $r^{-2}$ one finds

$$G_{-}(\boldsymbol{r}, \boldsymbol{r}' \approx 0) \propto r^{-1} e^{-F(\theta) r} \ , \tag{4.119a}$$

$$F(\theta) = [(\omega_g - \tilde{\omega})/\omega_{\perp}]^{1/2} (\sin^2\theta - g\cos^2\theta)/a_B \ , \tag{4.119b}$$

where $a_B$ is the transverse exciton Bohr radius. For $\omega > \omega_g$, $G_{-}(\boldsymbol{r}, \boldsymbol{r}' \approx 0)$ has the character of an outgoing wave along the directions $\theta = \pi/2$ (perpendicular to the $z$-axis) and an evanescent behaviour along the $z$-axis ($\theta = 0$). This asymptotic dependence of $G_{-}$ extended to entire relative space is illustrated in Fig. 4.12 where $\mathrm{Re}\{G_{-}(\boldsymbol{r}, \boldsymbol{r}' \approx 0)\}$ is shown as a function of $z$ and $\varrho \equiv (x^2 + y^2)^{1/2}$ for $g = 0.16$. Note that the singularity on the cone $\mathrm{tg}\theta = \sqrt{g}$ is highly artificial because the $r^{-1}$ terms neglected when deriving (4.119) become important on this cone. The asymptotic solution (4.119) for $r \to \infty$ is of little practical use because the response is determined by the behaviour at small relative distances.

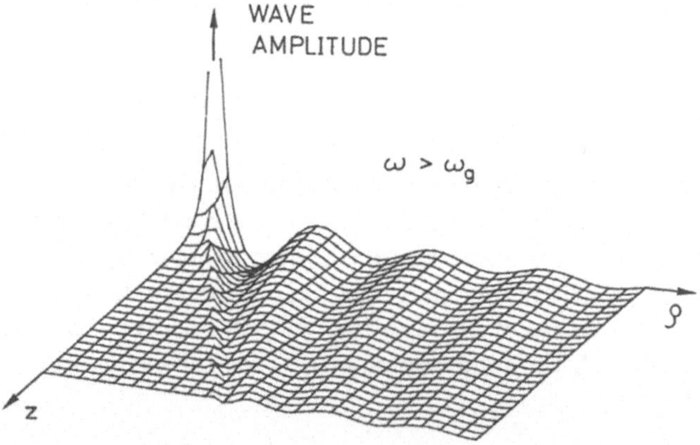


**Fig. 4.12.** Real part of the relative space Green's function for a saddle point exciton in the crude approximation (4.119). The amplitude is drawn for $g = -\mu_{\perp}/\mu_{\parallel}$ and a frequency above the "gap" frequency $\omega_g$. The relative space coordinate $\varrho$ is equal to $(x^2 + y^2)^{1/2}$

Turning to a proper treatment of (4.118) we note that the simplest step is the calculation of $\boldsymbol{M}(\boldsymbol{r})$ by an integration in $\boldsymbol{k}$-space of band structure characteristics [see (3.65)]. In this step it is important to realize that the parabolic approximation (4.116) with negative $\mu_{\parallel}$ is inadequate without a cut-off in $\boldsymbol{k}$-space. A cut-off at the zone edge or a positive $k_z^4$ term in (4.116) must be included to avoid unphysical singularities in the integrand of (3.65). However, it is pointless to search for a realistic structure of $\boldsymbol{M}(\boldsymbol{r})$ until a solution $G_{\pm}(\boldsymbol{r},\boldsymbol{r}')$ to (4.118a, b) is within reach, and that must await further work.

An approximate treatment of (4.118) can be based on the assumption that the response is insensitive to the detailed structure of $\boldsymbol{M}(\boldsymbol{r})$. We have already seen that the radial dependence of $\boldsymbol{M}(\boldsymbol{r})$ is unimportant for the singular contribution to the response at frequencies near $M_0$ critical points (Sects. 4.3, 5). Dealing with an anisotropic problem one can speculate that both the angular dependence of $\boldsymbol{M}(\boldsymbol{r})$ and the relative space part $u(\boldsymbol{r})$ of the electron-hole amplitude are similarly unimportant near the origin. The irrelevance of the detailed behaviour of $u(\boldsymbol{r})$ and $\boldsymbol{M}(\boldsymbol{r})$ for small $|\boldsymbol{r}|$ is related to the irrelevance of the band structure at large distances in reciprocal space from the critical point. With this in mind it seems an acceptable approximation to assume an anisotropic shell source

$$M(r) = M_0 f(\theta)\delta(r - r_0)/(4\pi r_0)^2 \tag{4.120}$$

with considerable freedom to chose the angular function $f(\theta)$, where $\theta = \arccos(z/r)$. $f(\theta)$ is normalized over the full solid angle. A particularly simple calculation scheme is obtained by choosing a relative space wave function $u(\boldsymbol{r})$ such that

$$(\Omega_{eh} - \omega)u(r,\theta) = 0 \quad \text{with} \tag{4.121}$$

$$u(r,\theta) = 0\ ; \quad r \to \infty\ , \tag{4.122a}$$

$$u(r_0,\theta) = u_0\ ; \quad r \le r_0\ . \tag{4.122b}$$

If we assume $f(\theta)$ to have the same angular dependence as $(\partial u/\partial r)^{-1}$ evaluated at $r_0$, then (4.118c) becomes (Balslev, 1984)

$$\chi_- \propto \int \left[\frac{u}{\partial u/\partial r}\right]_{r=r_0} \sin\theta\, d\theta\ . \tag{4.123}$$

The method gives reasonable results when compared with experiments and with Kane's model (Balslev, 1984) but should be explored further for several reasons: (a) The above backward procedure where the propagator $\Omega_{eh}$ determines the angular structure of the source should be justified by firm arguments; (b) The inclusion of damping essential when replacing $\infty$ in (4.122a) by some great distance is nontrivial (Balslev, 1984) and should be studied in detail; (c) The band structure (4.116) and thereby $\omega_{eh}$ in (4.118b) is not realistic for $k_z^2$ larger than say $2\mu_{||}\omega_g/\hbar$, and so a stabilizing term should be incorporated in the effective mass treatment e.g. a $\partial^4/\partial z^4$ term in $\Omega_{eh}$ corresponding to a $k_z^4$ term in the band structure.

# 5. Exciton Polaritons in Half-Space Geometry

## 5.1 Introductory Remarks

A large class of optical experiments can only be properly interpreted if there is a well defined coupling between bulk polaritons and external photons. Near excitonic resonances, the understanding of this coupling is troublesome for two reasons, namely the "dead-layer" problem and the "additional boundary condition" (ABC). The dead (or exciton-free) layer is the region below a surface with noticeable distortion of the exciton polariton structure. The dead layer thickness is of the order $a_B$ which is not always negligible compared with the relevant polariton wave lengths. Consequently, the dead layer acts like an optical coating. The ABC is the condition necessary when the bare mechanical excitation has propagating character as is the case if the total exciton mass is finite. The need for an ABC can be inferred from the fact that a finite exciton mass allows more than one polariton with common frequency (and polarization direction) to propagate with different wave vectors. As can be seen from Fig. 4.4 there is in principle a need for infinitely many ABCs, one for each principal quantum number in the exciton series.

In the case of an isolated spatially dispersive resonance there are for each frequency $\omega$ two wave vectors $q_1$ and $q_2$ of polaritons with common propagation direction and polarization. In a normal incidence reflectivity measurement on a crystal filling the half-space $z>0$, we have

$$E_x = E_0(e^{iz\omega/c} + \tilde{r}e^{-iz\omega/c}) \qquad z<0 \tag{5.1}$$

$$E_x = E_1 e^{iq_1 z} + E_2 e^{iq_2 z} \qquad z>0\ . \tag{5.2}$$

It is clearly seen that the Maxwellian boundary conditions ($E_x$ and $\partial E_x/\partial z$ continuous) are insufficient for determining the unknowns $\tilde{r}$, $E_1/E_0$, and $E_2/E_0$.

Attempts to bypass these problems by fitting theories with $a_B \to 0$ and $m \to \infty$ to experiments are rarely successful and the oscillator parameters ($\Gamma$, $\omega_0$, $\omega_{LT}$) obtained are unreliable. Such a simple approach is particularly poor for explaining the famous "spike" at $\omega_L$. This is an observed subsidiary peak at the longitudinal frequency in the reflectivity of CdS and many other semiconductors with lower band gap. As an example of such a spike, Fig. 5.1 shows the reflectivity of CdS.

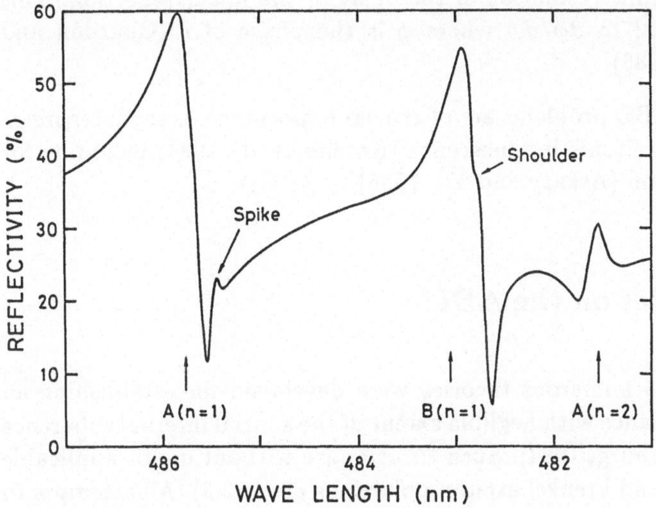


**Fig. 5.1.** Reflectivity spectrum of CdS for $E \perp c$. The spike at the $A(n=1)$ structure and the reversal of the $A(n=2)$ structure are assigned to dead layer phenomena. The shoulder in the $B(n=1)$ structure is due to $k$-linear terms in the band structure (Sect. 4.9). From Evangelisti et al. (1974)

In our presentation we shall divide the theories on the half-space problem into 3 groups: (i) early theories in which the dead-layer problem is ignored or accounted for by phenomenological fitting parameters (Sect. 5.2), (ii) theories based on boundary conditions for electronic wave functions (Sect. 5.3), and (iii) the coherent wave approach (Sect. 5.4,5).

The results to be explained by these theories involve not only the normal incidence reflectivity $|\tilde{r}|^2$ in (5.1), but also any kind of experiments where exciton polaritons in a crystal couple to external photons via the surface. Below are listed five categories of experiments used for gaining more information on dead-layer and ABC problems:

a) Ellipsometric measurements with normal incidence on wurtzite materials for obtaining information on the phase of the complex ratio $\tilde{r}(E \| c)/\tilde{r}(E \perp c)$ (Pevtsov et al., 1980).
b) Oblique incidence experiments (Sect. 5.6) giving two principal reflectivities $|\tilde{r}_s|^2$, $|\tilde{r}_p|^2$ and, with ellipsometry, the phase of $\tilde{r}_s/\tilde{r}_p$ (Pevtsov and Selkin, 1983; Broser et al., 1978).
c) Propagation of surface polaritons at frequencies between $\omega_0$ and $\omega_L$ studied by attenuated total reflection (Sect. 5.6) (Lagois and Fischer, 1979, 1982).
d) Efficiency measurements on resonance Brillouin and Raman scattering (Sects. 7.2,3) (Brenig et al., 1978; Ulbrich and Weisbuch, 1978; Yu, 1979).

e) Picosecond time resolution of the delay of the linear reflection. This delay is equal to $\partial\phi/\partial\omega$ where $\phi$ is the phase of $\tilde{r}$ (Gourdon and Lavallard, 1985).

The dead-layer and ABC problems are of crucial importance in the interpretation of zero-phonon excitonic luminescence (Koteles et al., 1984) including for example time resolution (Askary and Yu, 1985).

## 5.2 Early Theories on the ABC

In the period 1957–75 numerous theories were developed for establishing an ABC for a single resonance with neglible extent of the spatial internal coherence (Agranovich and Ginzburg, 1984). Such theories are without doubt applicable to phonon polaritons and Frenkel exciton polaritons (Sect. 2.3). All attempts to handle the Wannier case in a similar manner remained unsatisfactory because of the interference between dead layer effects and the ABC problem.

Neglect of the dead-layer phenomena in most of the early theories leads to an ABC formulated as

$$P_{ox} + \tilde{\gamma}\frac{\partial P_{ox}}{\partial z} = 0 \ ; \quad z = 0+ \ , \tag{5.3}$$

where $P_{ox}$ is the resonant polarization (directed along the $x$-axis) and $\tilde{\gamma}$ is a parameter which in the general case is complex. Deutsche and Mead (1965) have shown that any condition involving $P_{ox}$ alone can be brought into the form (5.3) on the basis of causality. More complicated ABCs involving $E_x$, $P_{ox}$ and their spatial derivatives were suggested by Akhmediev and Yatsishen (1978).

It is interesting to relate (5.3) to the energy flux density at $z = 0$. The $z$-component is given by (4.106)

$$S_z = \frac{1}{2}\left[c^2\varepsilon_0(E_x^* B_y + E_x B_y^*) - \frac{i\hbar\omega}{2\varepsilon_0\varepsilon_b\omega_{LT}m} \times\left(P_{ox}^*\frac{\partial P_{ox}}{\partial z} - P_{ox}\frac{\partial P_{ox}^*}{\partial z}\right)\right] \tag{5.4}$$

where $E_x$, $B_y$ and $P_{ox}$ are complex amplitudes. The first term is continuous at $z = 0$ thanks to the Maxwellian boundary conditions. The second term is zero in vacuum. Consequently $S_z$ is discontinuous unless $\tilde{\gamma}$ in (5.3) is real (or if $|\tilde{\gamma}|\rightarrow\infty$) (Bishop and Maradudin, 1976).

The oldest ABC is that of Pekar (1957) who suggested $\tilde{\gamma} = 0$ in (5.3). The ABC of Fuchs and Kliewer (1969) applied to metallic response had $|\tilde{\gamma}|\rightarrow\infty$. A crude "ball-and-spring" model suggested by Skettrup and Balslev (1971) involves a detuned surface layer with bulk-like nonlocal coupling between this layer and the rest of the crystal. The model favors Pekar's ABC unless the detuned surface layer is very thin. In the beginning of the seventies several groups of physicists developed theories leading to the results

$$P^{\omega}(r)+\varepsilon_0 E^{\omega}(r)=\int\limits_{z'>0}(\varepsilon^r(\omega,r-r')+R_e\varepsilon^r(\omega,r^+))\varepsilon_0 E(r')d^3r' \ , \quad (5.5a)$$

$$r^+=(x-x',\ y-y',\ z+z') \ . \quad (5.5b)$$

$\varepsilon^r$, $P^{\omega}$ and $E^{\omega}$ have the same meaning as in (4.108) and $R_e$ is a parameter specific of the model. The so-called dielectric model leads to the result $R_e=0$ (Birman and Sein, 1972; Maradudin and Mills, 1973; Agarwal et al., 1974). On the basis of a microscopic argument concerning conservation of excitonic amplitude, Zeyher et al. (1972) suggest $|R_e|=1$. The response given in (5.5) is equivalent to the boundary condition (5.3) (Zeyher et al., 1972; Ivchenko, 1982; Halevi and Fuchs, 1984) if

$$\tilde{\gamma}=-\frac{1+R_e^*}{1-R_e^*}[2m(\omega-\omega_0)/\hbar]^{-1/2} \ . \quad (5.6)$$

Thus, in the dielectric model $\tilde{\gamma}$ is imaginary (discontinuous energy flux) above resonance. In the model of Zeyher and coworkers (1972) $\tilde{\gamma}$ is imaginary below resonance unless $R_e=1$ or $-1$ (corresponding to the Pekar and the Fuchs-Kliewer ABC, respectively).

None of the above theories can be applied consistently to the Wannier case. To see this, we can compare two lengths, namely $a_B$ and

$$|\kappa_e|^{-1}=|2m(\omega_0-\omega)/\hbar|^{-1/2} \ .$$

$a_B$ is the typical length of the internal polariton structure while $|\kappa_e|^{-1}$ is the decay length or the wave length (divided by $2\pi$) of the real space response function $\varepsilon^r(\omega,r)$ [see (4.109)]. In this comparison a relevant value of $\omega_0-\omega$ is $2\omega_{LT}$, the typical width of excitonic spectral structures. Then

$$a_B|\kappa_e|=\left(\frac{2m\omega_{LT}}{\mu\omega_x}\right)^{1/2} \ . \quad (5.7)$$

This quantity is of the order unity in most semiconductors, and so the distortion depth below the surface and the relevant spatial scale of the nonlocal response are generally comparable for excitons in semiconductors.

An indication of pronounced effects of the finite exciton radius is the success of the Hopfield-Thomas (HT) model (Hopfield and Thomas, 1963). Here the dead layer is assumed homogeneous with $P_0=0$ and the Pekar ABC is applied to the interface between the dead layer and the bulk. Then the amplitude reflection $\tilde{r}$ in (5.1) is given by (Evangelisti et al., 1974A)

$$\tilde{r} = \frac{r_b + r_0 e^{i\theta}}{1 + r_b r_0 e^{i\theta}} \tag{5.8a}$$

$$r_b = (1 - \varepsilon_b^{1/2})/(1 + \varepsilon_b^{1/2}) \tag{5.8b}$$

$$r_0 = (\varepsilon_b^{1/2} - n_0)/(\varepsilon_b^{1/2} + n_0) \tag{5.8c}$$

$$n_0 = \frac{c}{\omega}(q_1 q_2 + \varepsilon_b \omega^2/c^2)/(q_1 + q_2) \tag{5.8d}$$

$$\theta = 2\ell\varepsilon_b^{1/2}\omega/c \ . \tag{5.8e}$$

Here $\ell$ is the dead layer thickness and $\theta$ is the round-trip retardation in the dead layer. In the HT model $\ell$ is an adjustable parameter attaining values of the order $a_B$. An estimate of $\ell$, however, is possible from theories developed much later (see Sect. 5.3). The HT model has several attractive features: It is consistent with respect to energy flux and length scales, it is computationally easy to handle and it gives a reasonable fit to the experiments. For example it reproduces the spike at $\omega_L$ (Hopfield and Thomas, 1963; Sell et al., 1973). It also accounts for the reversal of the reflectivity spectrum of the $n = 2$ excitonic resonance (Evangelisti et al., 1974) (cf. Fig. 5.1). The dead layer thickness $\ell$ is usually a much more important fitting parameter than for example the mass $m$ and $\tilde{\gamma}$ in (5.3), and so the anomalies caused by spatial dispersion are often blurred by dead-layer effects (Grosman et al., 1975).

The microscopic justification of the HT model was originally based on the image charge forces which are repulsive below the surface of a dielectric (Hopfield and Thomas, 1963). However, as will be discussed later, the dominant contribution to the depletion of the excitonic polarization near the surface is the steep potential barrier at $z = 0$ experienced individually by the electrons and holes. These contact forces (as well as the image charge forces) give an effective dead-layer thickness scaling as the exciton radius. One would therefore expect $\ell/(na_B)$ to be independent of the material and the principal quantum number $n$. Usually, values of $\ell/(na_B)$ between 2 and 3 fit the experiments well. However, there are significant exceptions: $\ell/(na_B) \simeq 5$ has been found for mixed mode polaritons in ZnO (Hümmer and Gebhardt, 1978) (see Sect. 5.8) and for $n = 2$ excitons in CdS (Evangelisti et al., 1974A).

A frequently studied refinement of the HT model is the use of an inhomogeneous dead layer, i.e. a stepwise or continuous variation of $\omega_0$, $\omega_{LT}$ and/or $\Gamma$. For example, Ekhardt et al. (1979) use an enhanced value of the damping $\Gamma$ in a transition layer, and Lagois (1981) introduces adjustable eigenfrequencies $\omega_0$ in a 3-layer sandwich below the dead layer. The increased number of fitting parameters in such models gives better fits to experiments. The approach by Lagois (1981) is particularly interesting in view of the adiabatic approximation discussed in Sects. 5.3 and 5.7, because the assumption of a spatially varying resonance frequency can be given a microscopic explanation.

## 5.3 The Adiabatic Approximation

In the period 1976–81 a series of theories was developed on basis of half-space boundary conditions on the two-particle excitonic wave function (Sakoda, 1976; Balslev, 1978; Mattis and Beni, 1978; D'Andrea and Del Sole, 1979, 1984; Zeyher, 1981). In very different ways these works use half-space excitonic solutions as building blocks for constructing the optical response and explore in this manner the nature of the dead layer. What these theories have in common with the coherent-wave approach is the insight that the boundary condition at the surface has to be applied to the wave function rather than the polarization.

In the present section we shall discuss one of the above theories, namely that based on the adiabatic approximation (Sakoda, 1976; Balslev, 1978).

Let us first neglect spatial dispersion ($m^{-1} = 0$) and rewrite (4.57) in the form

$$\left(-\frac{\hbar^2}{2\mu}\nabla^2 + V_{eh}(\boldsymbol{r}, Z) + E_\nu(Z)\right) f_\nu(\boldsymbol{r}, Z) = 0 \tag{5.9}$$

where $Z$ is the $z$ component of the center of mass coordinate and $V_{eh}(\boldsymbol{r}, Z)$ is the two-particle potential with three contributions: The direct Coulomb interaction, the image charge interaction and a cut-off potential experienced by each particle at the surface. By treating $Z$ as a parameter in (5.9) we are dealing with an adiabatic approximation also used for separating electronic and vibronic motions in solids and molecules. In an Elliott-like treatment it is reasonable to assume that $E_\nu(Z)$ and $|f_\nu(0, Z)|^2$ define, respectively, a spatially dependent resonance frequency and an oscillator strength. Then, the $n = 1$ excitonic susceptibility is

$$\chi(\omega, Z) = \frac{\varepsilon_b \omega_{LT}(Z)}{\omega_0(Z) - \omega - i\gamma} \ , \quad \text{where} \tag{5.10}$$

$$\omega_0(Z) = \omega_g - E_1(Z)/\hbar \ , \tag{5.11a}$$

$$\omega_{LT}(Z) = \omega_{LT}(\infty)|f_1(0, Z)|^2/|f_1(0, \infty)|^2 \ . \tag{5.11b}$$

Note that $\chi(\omega, Z)$ in (5.10) describes a local, but inhomogeneous response so that

$$\left(\frac{c^2}{\omega^2}\frac{\partial^2}{\partial Z^2} + \varepsilon_b\right) E(Z) = -\chi(\omega, Z) E(Z) \ . \tag{5.12}$$

Having neglected spatial dispersion there is no need for an ABC when matching solutions to (5.12) for $z>0$ with electro-magnetic waves in the vacuum (for $z<0$).

Let us now return to (5.9) and specify $V_{eh}(\boldsymbol{r}, Z)$. The image charge potential $V_s$ is given by

$$V_s(\boldsymbol{r}, Z) = \frac{\varepsilon_s - 1}{\varepsilon_s + 1} \frac{e^2}{4\pi\varepsilon_0\varepsilon_s} \left\{ \frac{1}{4z_e} + \frac{1}{4z_h} - \frac{1}{[(x_e - x_h)^2 + (y_e - y_h^2) + (z_e + z_h)^2]^{1/2}} \right\} , \tag{5.13}$$

where

$$(x_e, y_e, z_e) = \boldsymbol{r}_e = \boldsymbol{R} + \frac{m_h}{m}\boldsymbol{r} \tag{5.14a}$$

$$(x_h, y_h, z_h) = \boldsymbol{r}_h = \boldsymbol{R} - \frac{m_e}{m}\boldsymbol{r} . \tag{5.14b}$$

The cut-off forces can be included by the boundary conditions

$$f_\nu(\boldsymbol{r}, Z) = 0 \ ; \quad z_e \leq 0 \tag{5.15a}$$

$$f_\nu(\boldsymbol{r}, Z) = 0 \ ; \quad z_h \leq 0 . \tag{5.15b}$$

In the relative space these two conditions represent two infinitely high potential barriers forming a two-dimensional potential well. In the limit $\mu/m \to 0$ one of the walls [say the electron wall leading to (5.15a)] is much closer to the origin than the other.

Equation (5.8) must be solved with the boundary conditions (5.15) and the potential

$$V_{eh}(\boldsymbol{r}, Z) = V_s(\boldsymbol{r}, Z) - \frac{e^2}{4\pi\varepsilon_0\varepsilon r} . \tag{5.16}$$

In case of band bending there are additional terms to be considered as discussed in Sect. 5.7.

Both $V_s$ and the cut-off forces tend to reduce the amplitude of $f_\nu(\boldsymbol{r}, Z)$ near the walls of the potential well. Most efficient are the cut-off forces, and so we may, as a crude approximation neglect $V_s$. Then the ground state at $Z = 0$ is characterized by (Satpathy, 1983)

$$f_1(r, 0) = z\mathrm{e}^{-r/2a_B} \tag{5.17a}$$

$$E_1(0) = -\hbar\omega_x/4 \tag{5.17b}$$

corresponding to a $p$-state in half-space. The influence of the cut-off forces has roughly an exponential dependence on $Z$ with $a_B$ as the typical decay length, and so a simple estimate would be

$$\omega_0(Z) = \omega_0(\infty) + \tfrac{3}{4}\omega_x e^{-Z/a_B} \tag{5.18a}$$

$$\omega_{LT}(Z) = \omega_{LT}(\infty)(1 - e^{-Z/a_B}) \; . \tag{5.18b}$$

Equation (5.18a) shows that a surface layer is detuned upwards corresponding to a repulsive potential for the excitonic center-of-mass motion. When $\omega$ is near $\omega_0(\infty)$ the resonance polarization is reduced in a surface layer. This reduction represents the dead layer. It is usually less important that $\omega_{LT}(Z)$ is reduced in the dead layer. A crude assumption (Balslev, 1978) is that the effective dead layer is the region in which

$$|\omega_0(Z) - \omega_0(\infty)| > \omega_{LT} \; . \tag{5.19}$$

From (5.17a) we obtain the dead layer thickness $\ell$

$$\ell = a_B \ln \frac{3\omega_x}{4\omega_{LT}} \; . \tag{5.20}$$

The estimate (5.20) compares surprisingly well with the dead-layer thickness obtained by fitting a HT model to experiments, as seen in Table 5.1. An exception to this trend is the case of GaAs having an exceptionally large exciton Bohr radius. We shall return to this case later.

**Table 5.1.** Estimate of exciton-free surface layer for various substances

| Material | $\omega_x/\omega_{LT}$ | $a_B$ [Å] | $\ell$ (theor.) [Å] | $\ell$ (exp.) [Å] |
|---|---|---|---|---|
| GaAs | 32 | 130 | 413 | 290[b] |
| InP | 26 | 95 | 282 | 260[c] |
| ZnO ($\theta_{k,c} = 80°$) | 180[a] | 15 | 74 | 85[d] |
| ZnSe | 14 | 50 | 117 | 130[e] |
| CdS ($k\parallel c$) | 14 | 30 | 71 | 110[f] |
| CdS ($k\perp c$) | 14 | 30 | 71 | 70[f,g] |
| CdSe | 17 | 50 | 127 | 150[g] |

[a] Mixed mode $B$ excitons (Balslev, 1981)
[b] Sell et al. (1973)
[c] Evangelisti et al. (1974)
[d] Hümmer and Gebhardt (1978)
[e] Lagois (1981)
[f] Hopfield and Thomas (1963)
[g] Evangelisti et al. (1974A)

Turning to a more refined application of the adiabatic approximation we note that the excitonic "potential" $\hbar\omega_0(Z)$ and the strength $\omega_{LT}(Z)$ can be calculated by appropriate variational methods (Sakoda, 1976; Balslev, 1978). Figure 5.2 shows the result of finding the ground state of (5.8) for two extreme cases: $\mu/m \ll 1$ is widely applicable. It is characteristic of this limit that the influence of the image charge forces is small (see Fig. 5.2). A response which

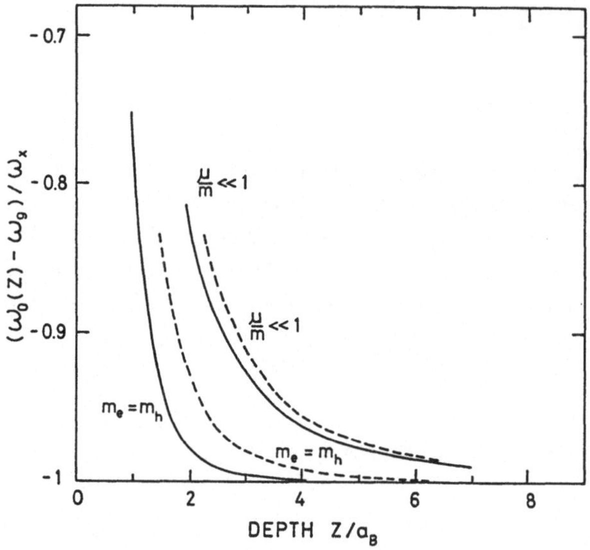


**Fig. 5.2.** Adiabatic potential $\hbar\omega_0(Z)$ for the ground state exciton. *Full curves* are calculated without the image charge potential (5.13) while the *dashed curves* are calculated with an image charge potential using $\varepsilon_s = 8.0$. After Balslev (1978)

is nonlocal and inhomogeneous at the same time requires that the real-space dielectric function must be written as a function $\varepsilon^r(\omega, \boldsymbol{r}, \boldsymbol{r}')$ giving the electric displacement at $\boldsymbol{r}$ from an electric field at $\boldsymbol{r}'$. The theories of D'Andrea and Del Sole (1984) and Zeyher (1981) describe the half-space response by such a dielectric function and find the reflectivity by an integral equation. In contrast to this the adiabatic approximation is suitable for real-space differential equations in the spirit of Thomas and Hopfield (1963):

$$\left(\omega_0(Z) - \omega - i\gamma - \frac{\hbar}{2m}\frac{\partial^2}{\partial Z^2}\right)\boldsymbol{P}_0(Z) = \omega_{LT}(Z)\varepsilon_0\varepsilon_b\boldsymbol{E}(Z) \tag{5.21a}$$

$$\left(\frac{c^2}{\omega^2}\frac{\partial^2}{\partial Z^2} + \varepsilon_b\right)\varepsilon_0\boldsymbol{E}(Z) = -\boldsymbol{P}_0(Z) \ . \tag{5.21b}$$

This form does not have a continuous energy flux in the limit $\gamma = 0$ unless $\omega_{LT}(Z)$ is constant. Sakoda (1976) finds more complicated expressions with terms in (5.21a) proportional to $P$ or $\partial P/\partial Z$ times appropriate spatial derivatives of $\omega_{LT}(Z)$. It is not clear if such terms improve the accuracy. They may be of the same order as the inherent errors of the adiabatic approximation.

When solving (5.21) for $Z>0$ and matching the solutions to the vacuum waves there is still a need of an ABC. This is not provided by the adiabatic approximation. However, the detuning at $Z \approx 0$ is so pronounced that conditions allowing $P(Z = 0)$ to be finite require very high computational precision in a numerical integration of (5.21). It is therefore reasonable (but not so far rigorously proved) that the Pekar ABC should be applied at $Z = 0$ in the adiabatic approximation.

The logarithmic slope of $\omega_0(Z)$ in Fig. 5.2 is of the order $a_B^{-1}$ while it is infinite in the HT model. Comparing these two cases with each other and with experiments it is concluded that (a) the less steep excitonic potential fits the experiments better (Balslev, 1981); (b) the HT model underestimates the transmission coefficient above $\omega_L$ from vacuum to the lower branch polariton (Fig. 5.3); (c) the HT model generally overestimates the damping $\gamma$ and $\omega_{LT}$ (Balslev, 1981). An interesting comparison between the HT model and the adiabatic approximation is displayed in Fig. 5.4. Here is shown the required frequency dependence of $\ell/a_B$ for an HT model to give a reflectivity spectrum identical to that of a representative adiabatic potential $\omega_0(Z)$.

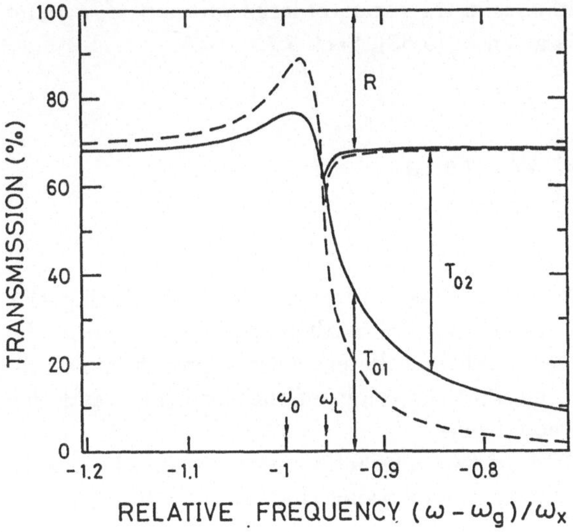


**Fig. 5.3.** The reflectivity $R$ and the energy transmittances $T_{01}$ and $T_{02}$ from vacuum to lower and upper branch polaritons, respectively. *Full curves* show the result from the adiabatic approximation while the *dashed curves* are from the HT-model. The parameters used are typical for GaAs

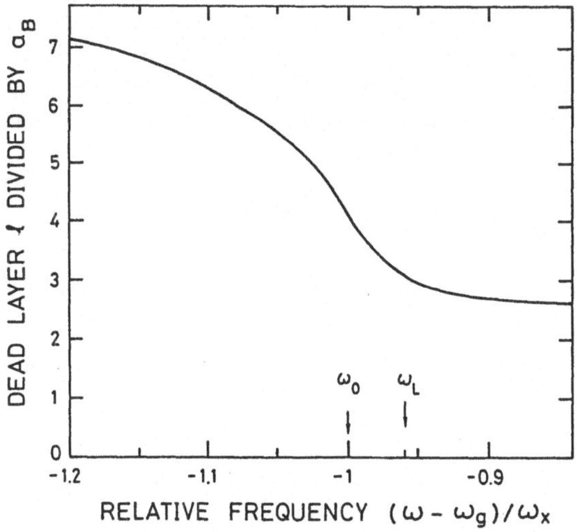


**Fig. 5.4.** Frequency dependence of the dead layer thickness to be assumed in the HT model in order to give the same reflectance $|\bar{r}|^2$ as that of an excitonic potential given by $\omega_0(z) = \omega_0 + \omega_x \exp(1.2 - z/a_B)$. Values for $\varepsilon_b$, $\omega_x$, $\omega_g$, $m$ etc. are typical for GaAs

Two other interesting applications of the adiabatic approximation will be mentioned here. First, the influence of an electric field can be included in the potential in (5.9) as will be discussed in Sect. 5.7. Second, when solving (5.8) near the $n = 2$ resonance, it is found that $V_{eh}(\mathbf{r}, Z)$ splits the $2p2s$ level which is more or less degenerate in the bulk. Both levels are optically active and produce extra structure in the reflectivity spectrum (Balslev, 1983).

It was shown recently (D'Andrea and Del Sole, 1985) that the variational results in Fig. 5.2 are rather inaccurate compared with the analytical results of Satpathy (1985) obtained in the limit $m^{-1} = 0$. However, general conclusions regarding the convergence properties of the variational procedure applied to the adiabatic approach must be postponed until other trial functions have been explored. A new trial function suggested by Schultheis and Balslev (1983) is significantly better than the old ones in the region of large values of $Z/a_B$ and $m/\mu$. This new trial function is given in (5.65), Sect. 5.7.

## 5.4 Coherent Interband Waves in One-Dimensional Half-Space

It is very instructive to study surface effects of polaritons by reducing the dimensionality to one, thereby simplifying the mathematics (Stahl, 1981). In this section it will be demonstrated that the coherent wave approach in a semi-infinite one-dimensional system clarifies the origin of the dead layer and the role of the electronic boundary conditions.

First we consider the motion of two interacting particles on the $z$-axis stretching from $-\infty$ to $\infty$. We choose the following characteristic functions in the edge equation:

$$M(z) = M_0\delta(z) \tag{5.22}$$

$$V_{eh}(z) = -2\hbar\omega_x a_B \delta(z) \ . \tag{5.23}$$

This choice gives a well-defined response with parameters $M_0$, $\omega_x$ and $a_B$ analogous to a three-dimensional system. Then the one-dimensional polariton equations become:

$$\left(\omega_g - \omega - i\gamma - \frac{\hbar}{2m}\frac{\partial^2}{\partial Z^2} - \frac{\hbar}{2\mu}\frac{\partial^2}{\partial z^2} - 2\omega_x a_B \delta(z)\right) Y(Z, z) = \tfrac{1}{2} M_0 \delta(z) E(Z) \tag{5.24}$$

$$\left(\frac{c^2}{\omega^2}\frac{\partial^2}{\partial Z^2} + \tilde{\varepsilon}\right)\varepsilon_0 E(Z) = -2M_0 Y(Z, 0) = -P(Z) \tag{5.25}$$

$$z = z_e - z_h \tag{5.26a}$$

$$Z = (z_e m_e + z_h m_h)/m \ . \tag{5.26b}$$

A bulk solution to (5.24,25) has the structure

$$Y(Z,z) = Y_0 e^{iqZ} e^{-\kappa|z|} \quad . \tag{5.27}$$

Then, from (5.24,25)

$$\kappa^2 = (\omega_g - \omega - i\gamma)\omega_x^{-1} a_B^{-2} + \mu q^2/m \tag{5.28a}$$

$$\chi \equiv \frac{P_-}{\varepsilon_0 E} = \frac{M_0}{\varepsilon_0 a_B \hbar\omega_x}(\kappa a_B - 1)^{-1} \tag{5.28b}$$

$$c^2 q^2/\omega^2 = \chi + \tilde{\varepsilon} \quad . \tag{5.28c}$$

The susceptibility at $q = 0$ has a single resonance at $\omega = \omega_g - \omega_x$ and an absorption continuum for $\omega > \omega_g$. Near the resonance the decay length in relative space is $a_B$. Eliminating $\chi$ and $q$ from (5.28) one obtains a cubic equation in $\kappa$. One of the three solutions has negative real part of $\kappa$ which is incompatible with the requirement $Y(Z, |z| \to \infty) = 0$ for bulk solutions. The remaining two solutions are allowed. For $\gamma = 0$ and $\omega < \omega_g$, $q^2$ is real which corresponds to a totally reflecting relative space. One of the modes, the photon-like one, becomes weakly damped for $\omega > \omega_g$. The structure of $Y(Z,z)$ is shown in Fig. 4.6, and the dispersion curves are shown in Fig. 5.5.

Let us now confine the two particles to the positive $z$-axis by an infinitely steep and high potential barrier at $z_e = 0$ and $z_h = 0$. The cases with and without spatial dispersion shall be discussed separately.

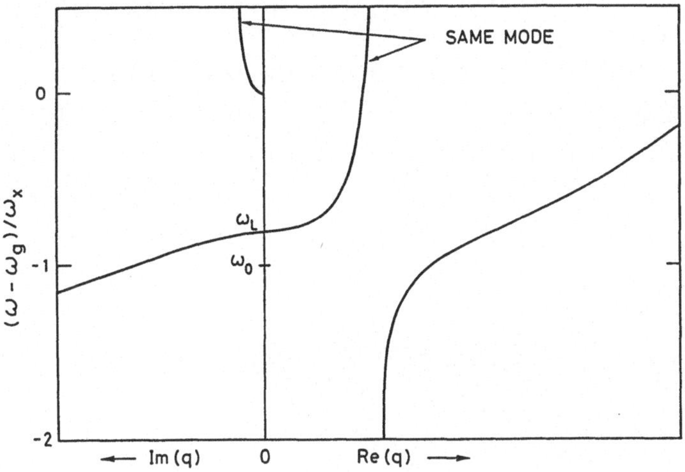


**Fig. 5.5.** Dispersion curves of polaritons of the one-dimensional exciton model

### 5.4.1 No Spatial Dispersion

We insert in (5.24–26) $m_h^{-1} = m^{-1} = 0$ and $m_e = \mu$. The relevant boundary conditions for $Y(Z,z)$ are then

$$Y(Z,z) = 0 \ ; \quad z = -Z \tag{5.29a}$$

$$Y(Z,z) = 0 \ ; \quad z \to \infty \ , \tag{5.29b}$$

and the solution has the form

$$Y(Z,z) = \begin{cases} Y(Z,0)\mathrm{e}^{-\kappa z} \ ; & z>0 \ , \\ Y(Z,0)\dfrac{\sinh(\kappa z + \kappa Z)}{\sinh(\kappa Z)} \ ; & -Z<z<0 \ , \end{cases} \tag{5.30}$$

where $\kappa$ does not depend on $q$. $Y(Z,0)$ is defined by the equations (Stahl, 1981)

$$\frac{d^2E}{dZ^2} = -\frac{\omega^2}{c^2}[\tilde{\varepsilon} + \chi(\omega,Z)]E \tag{5.31}$$

$$Y(Z,0) = \frac{\varepsilon_0 E(Z)\chi(\omega,Z)}{2M_0} \tag{5.32}$$

$$\chi(\omega,Z) = \frac{M_0^2}{a_B\varepsilon_0\hbar\omega_x}\{\tfrac{1}{2}\kappa a_B[\coth(\kappa Z)+1]-1\}^{-1} \ . \tag{5.33}$$

The structure of $Y$ as a function of $z_e$, $z_h$ is shown in Fig. 5.6.

The response (5.33) is local and inhomogeneous. $\chi$ is zero at the surface at any frequency and achieves the bulk value [proportional to $(\kappa a_B - 1)^{-1}$] at a depth of a few times $\kappa^{-1}$.

It is interesting to calculate the frequency profile $\omega_0(Z)$ for which

$$\chi^{-1}(\omega_0(Z),Z) = 0 \ . \tag{5.34}$$

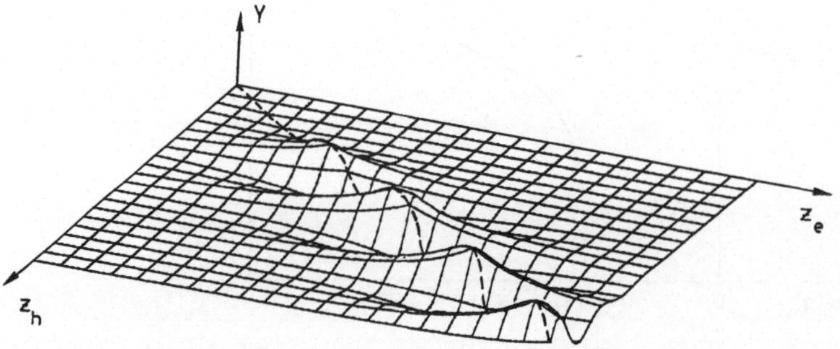


**Fig. 5.6.** Real part of $Y(z_h, z_e)$ near resonance for the one-dimensional exciton model in the half-space geometry. The result is valid for the limit $m_e/m_h = 0$

Then in the spirit of the adiabatic approximation, $\omega_0(Z)$ is the local resonance frequency. From (5.33,34) we find

$$\omega_0(Z) = \omega_g - \lambda^2(Z)\omega_x \tag{5.35}$$

where $\lambda(Z)$ satisfies the equation

$$\lambda(Z)[\coth(\lambda(Z)Z/a_B) + 1] = 2 \ . \tag{5.36}$$

The function $Z(\omega_0)$ inverse to that in (5.35) is

$$Z(\omega_0) = -\frac{a_B}{2}\left(\frac{\omega_g - \omega_0}{\omega_x}\right)^{-1/2} \ln\left[1 - \left(\frac{\omega_g - \omega_0}{\omega_x}\right)^{1/2}\right] \ . \tag{5.37}$$

$\omega_0(Z)$ reaches the gap frequency at $Z = a_B/2$ and for $0<Z<a_B/2$ there is no solution to (5.34). The overall structure of $\omega_0(Z)$ for $Z>a_B/2$ is similar to that of the 3-dimensional model (see Fig. 5.2).

A calculation along the lines of Sect. 5.3 (the adiabatic approximation) is also very simple for $m^{-1} = 0$. We are then faced with the problem

$$\left(-\frac{\hbar}{2\mu}\frac{\partial^2}{\partial z^2} - 2\omega_x a_B \delta(z) + \omega_g - \omega_0(Z)\right) f(Z,z) = 0 \tag{5.38a}$$

$$f(Z,z) = 0 \ ; \quad z = -Z \tag{5.38b}$$

$$f(Z,z) = 0 \ ; \quad z \to \infty \ . \tag{5.38c}$$

After lengthy, but straightforward calculations it can be shown that (5.38) for $Z>a_B/2$ gives the same $\omega_0(Z)$ as obtained by (5.34) and that the oscillator strength is proporitonal to

$$\omega_{LT}(Z) \propto f^2(Z,0) = [a_B\lambda^{-2}(Z) - \tfrac{1}{2}Z\sinh^{-2}(\lambda(Z)Z/a_B)]^{-1} \ . \tag{5.39}$$

For $Z \to (a_B + 0)/2$, $\omega_{LT}(Z)$ goes to zero. For $Z<a_B/2$, (5.38) has no bound states.

In the limit of infinite hole mass treated here, the adiabatic approximation is exact. This fact is not obvious from the above findings because the result (5.33) of the coherent wave theory is not equal to the form

$$\chi(\omega, Z) = \begin{cases} \dfrac{\tilde{\varepsilon}\omega_{LT}(Z)}{\omega_0(Z) - \omega - i\gamma} \ ; & Z>a_B/2 \\ 0 \ ; & Z<a_B/2 \end{cases} \tag{5.40}$$

found in the adiabatic treatment. The reason for the discrepancy between (5.33) and (5.40) is that the absorption continuum is neglected in (5.40). If this continuum is included in an Elliott-type treatment of (5.38) one arrives at the coherent wave result (5.33) (Balslev, 1987).

## 5.4.2 Spatial Dispersion

For a finite total mass $m$ there exist no simple half-space solutions to (5.24,25). Two elaborate calculation schemes have been explored.

First, one can transform (5.24,25) into difference equations which are linear in the amplitudes $Y$ and $E$ of evenly spaced points in the configuration space (Balslev and Stahl, 1982).

Secondly, Gotthard, Stahl and Czajkowski (1984) have developed a semianalytical method based on the Lebedev-Kontorowicz transformation (Lebedev, 1973). The method uses the fact that the boundary value problem separates in polar coordinates $\varrho = (Z^2 + z^2)^{1/2}$, $\phi = \arctan z/Z$. The superposition of separable solutions can then be written as

$$Y_i(\varrho, \phi) = \int_0^\infty K_{i\lambda}(\varrho)\tilde{P}(\lambda)\frac{\sinh(\lambda\phi_i - \lambda\phi)}{\sinh \lambda\phi_i}d\lambda \ , \tag{5.41}$$

where $K_{i\lambda}(\varrho)$ is the McDonald function with imaginary index (Abramowitz and Stegun, 1965), $i = 1,2$ correspond to $\phi<0$ and $\phi>0$, respectively, $\phi_i$ are domain boundaries and $\tilde{P}(\lambda)$ is a weighting function. Apart from a numerical matrix inversion the Lebedev transformation procedure is an analytical method.

Both calculation procedures give well defined reflectivity curves on basis of the parameters $\omega_g$, $\omega_x$, $\omega_{LT}$, $\gamma$, $m$, $a_B$ and $\tilde{\varepsilon}$. In other words, the wave equations, the boundary conditions on $Y$ and the Maxwellian boundary conditions provide a situation where no extra assumption or parameters are necessary. The dead-layer and ABC problems are handled simultaneously. Furthermore, the calculated reflectivity contains all essential features of the experimental results on GaAs and CdS (see Fig. 5.7).

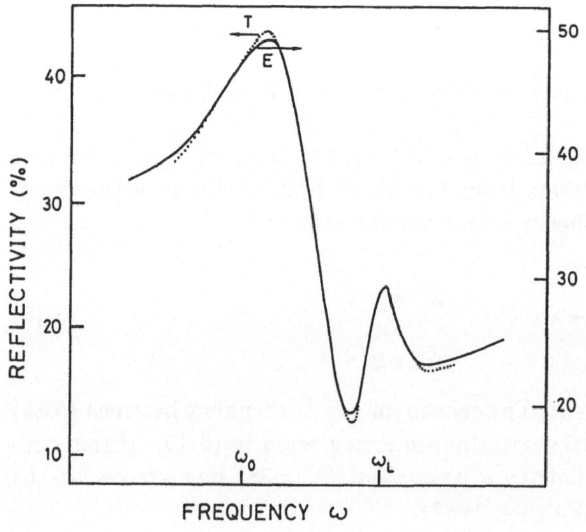


**Fig. 5.7.** The reflectivity of CdS near the $A(n = 1)$ exciton for $E \perp c$, $k \| c$. Curve $E$ is the experimental result of Hopfield and Thomas (1963), curve $T$ is the calculated result from the one-dimensional model (Gotthard et al., 1984). Note the different reflectivity scales

## 5.5 Normal Incidence Reflectivity of a Three-Dimensional Crystal

Let us first set up the coherent wave polariton equations and the boundary conditions for a three-dimensional half-space geometry at normal incidence. When the translational and rotational symmetry is used, then the 6 dimensions in the initial problem can be reduced to three. In this discussion we choose $Z, z, \varrho$ as independent spatial variables. $Z$ and $z$ are the $z$ components (perpendicular to the surface) of the center of mass and relative coordinate, respectively, while $\varrho = \sqrt{x^2+y^2}$, see Fig. 5.8. For $Z>0$ (inside the crystal) and $\varrho>0$ we may write

$$\Omega_{eh} = \omega_g - \frac{\hbar}{2\mu}\left(\frac{\partial^2}{\partial z^2} + \frac{\partial^2}{\partial \varrho^2} + \frac{1}{\varrho}\frac{\partial}{\partial \varrho}\right) - \frac{\hbar}{2m}\frac{\partial^2}{\partial Z^2} - \frac{2\omega_x a_B}{(\varrho^2+z^2)^{1/2}} \tag{5.42a}$$

$$(\Omega_{eh} - \omega - i\gamma)Y(Z,z,\varrho) = \tfrac{1}{2}M((\varrho^2+z^2)^{1/2})E(Z) \tag{5.42b}$$

$$\left(\frac{c^2}{\omega^2}\frac{\partial^2 E}{\partial Z^2} + \tilde{\varepsilon}\right)\varepsilon_0 E(Z) = -2\iint M((\varrho^2+z^2)^{1/2})\, Y(Z,z,\varrho) 2\pi\varrho\, d\varrho\, dz \; . \tag{5.43}$$

Here $Y$ stands for the resonant components. Image charge forces are neglected.

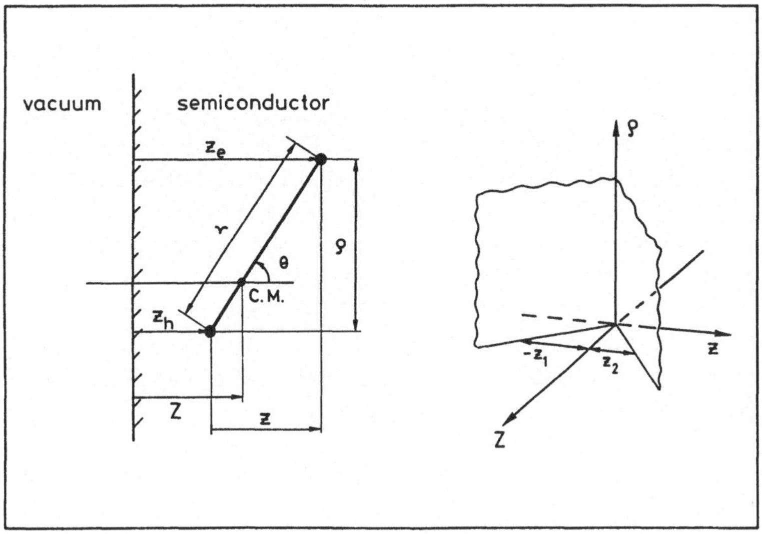


**Fig. 5.8.** Spatial coordinates in three-dimensional half-space dynamics of excitons

The boundary conditions are

$$Y = 0 \; ; \quad z \geq z_1(Z) = \frac{m_h}{m} Z \tag{5.44}$$

$$Y = 0 \; ; \quad z \leq z_2(Z) = -\frac{m_e}{m} Z \tag{5.45}$$

$$\partial Y / \partial \varrho = 0 \; ; \quad \varrho = 0 \tag{5.46}$$

$$|Y| < \infty \; ; \quad \varrho = 0 \tag{5.47}$$

$$Y = 0 \; ; \quad \varrho \rightarrow \infty \tag{5.48}$$

where $z_1(Z)$ and $z_2(Z)$ are relative $z$-coordinates of points on the planes given by $z_e = 0$ and $z_h = 0$, respectively. The relevant domain is shown in Fig. 5.8 with mutually perpendicular axes for $Z$, $z$ and $\varrho$.

In addition to conditions (5.44–48) one must impose the Maxwellian boundary conditions at $Z = 0$ and radiation conditions depending on the external sources. In a reflection experiment with vacuum at $Z<0$ the Maxwellian conditions and the definition of the amplitude reflection coefficient $\tilde{r}$ can be combined so that $\tilde{r}$ is given by $E$ and $dE/dZ$ evaluated at $Z = 0+$ [c.f. (5.1a)]:

$$\tilde{r} = \frac{iE\omega/c - dE/dZ}{iE\omega/c + dE/dZ} \; ; \quad Z = 0_+ \; . \tag{5.49}$$

This relation holds for the solution to (5.42–48) characterized by an absence of sources at $Z \rightarrow \infty$. The latter condition is, for finite damping $\gamma$, equivalent to the boundary condition

$$Y(Z_{\max}, z, \varrho) = 0 \; ; \quad Z_{\max} \rightarrow \infty \; . \tag{5.50}$$

It is clearly seen that the mathematical complexity has increased considerably compared with the one-dimensional case. For example, it is not clear whether the semianalytical method used by Gotthard et al. (1984) can be further developed to cope with the three dimensional case. The new aspect of the situation compared with the one-dimensional case is the appearance of the extra variable $\varrho$. But since the boundary conditions associated with this variable are so simple, one can conclude that the three-dimensional problem (5.42–50) is equally well posed in the sense that there is still no need for extra conditions.

A numerical treatment of (5.42–48) has been reported by Gotthard (1985). He replaces the radiation condition (5.50) by requiring that $Y$ is a sum of two right-going bulk modes for $Z \geq 8a_B$. The two bulk modes are those with smallest $q$ (upper and lower branch, see Fig. 4.6). Gotthard breaks up the numerical treatment into a boundary value problem in the $\varrho, z$ plane and a numerical integration on the $Z$ axis, and so in its mathematical structure it is related to the adiabatic approximation. Gotthard's method involves some approximations concerning the second difference proportional to $m^{-1}$. The results obtained are challenging by their detailed agreement with experiments also at oblique incidence (see Sect. 5.6) (Stahl, 1985).

In the search for more accurate solutions to (5.42–48) than those of Gotthard one could use standard numerical methods developed for elliptical boundary value problems. These methods in three dimensions are known to be extremely demanding of computer time and core memory, and so before one starts such a program it is sensible to explore any kind of faster numerical methods, e.g. iterative or variational procedures. It is also appropriate to reexamine the alternative theoretical approaches to the three-dimensional half-space problem: The adiabatic approximation is valid in a restricted range of mass ratios and – as learned from the one-dimensional model – it neglects the dispersive influence of higher lying absorption. In the latter respect the method of D'Andrea and Del Sole (1984) is also an approximation. Among the alternatives should also be mentioned the solution of Bethe-Salpeter equations in a multi-layer structure (Zeyher, 1981). This method could be developed further for a proper comparison with other approaches.

## 5.6 Oblique Incidence Reflection and Surface Polaritons

In the efforts to understand the surface structure of exciton polaritons it is very instructive to extend the experiments to include geometries other than that of normal incicende. In this section we shall discuss two such extensions namely oblique incidence reflectivity (OIR) and attenuated total reflection (ATR). In both cases the wave propagation has a wave-vector component $q_x$ parallel to the surface. All modes inside ($z>0$) and outside ($z<0$) the medium have this common $x$ component of the wave vector. If $q_x<\omega/c$ we are dealing with OIR since the modes in vacuum propagate at an angle $\beta = \arcsin(q_x c/\omega)$ with the surface normal (see Fig. 5.9A,B). If $q_x>\omega/c$ the waves in vacuum are evanescent in the $z$ direction (see Fig. 5.9C). Surface polaritons have this vacuum wave vector and can be excited across an air gap between the sample and a prism of glass or a similar material. The internal reflection in the prism is then studied in the spectral region of the excitonic resonances.

Let us assume that the electron-hole amplitude has three components which can be combined into a vector $\boldsymbol{Y}$ (see Sect. 4.9). Then

$$\hbar(\Omega_{eh} - \omega)\boldsymbol{Y}_-(Z,\boldsymbol{r}) = \underset{\sim}{\boldsymbol{M}}(\boldsymbol{r})\boldsymbol{E}_t(Z)/2 \tag{5.51}$$

$$\begin{aligned}&[(-\varepsilon_0 c^2 \nabla_R \times \nabla_R \times) + \varepsilon_0 \tilde{\varepsilon}\omega^2]\boldsymbol{E}_t(Z)\mathrm{e}^{iq_x X}\\ &\quad = -2\omega^2 \mathrm{e}^{iq_x X}\int \underset{\sim}{\boldsymbol{M}}(\boldsymbol{r})\boldsymbol{Y}_-(Z,\boldsymbol{r})d^3r \ ,\end{aligned} \tag{5.52}$$

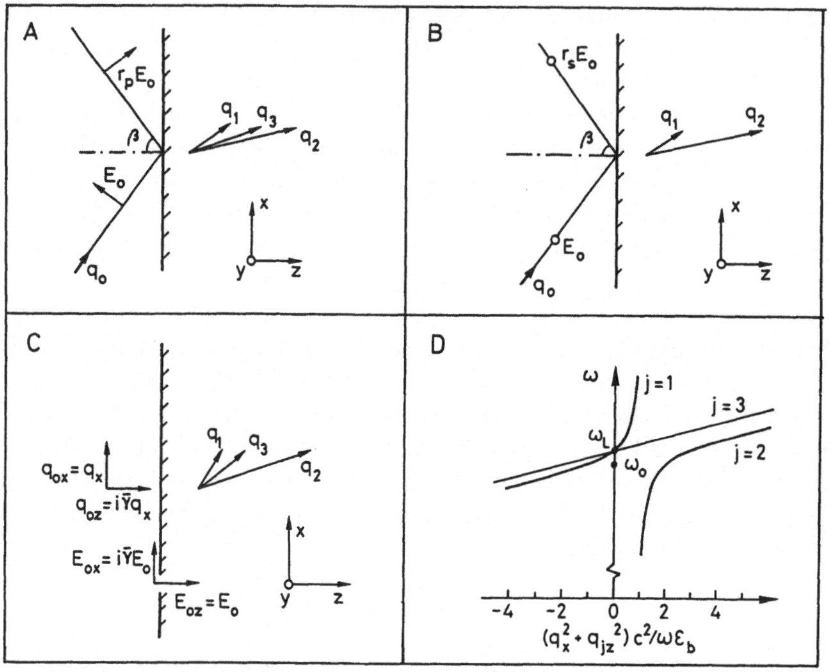


**Fig. 5.9.** Electric fields, wave vectors and polariton dispersion curves relevant for the discussion of oblique incidence reflection and attenuated total reflection. For clarity the vectors $\boldsymbol{q}_1$, $\boldsymbol{q}_2$, $\boldsymbol{q}_3$ are drawn without taking into account that $q_{1z}$ and/or $q_{2z}$ are sometimes imaginary

where $\boldsymbol{Y}_-(Z,\boldsymbol{r})$ and $\boldsymbol{E}_t(Z)$ are fields without the common factor $\exp(iq_xX - i\omega t)$. $\boldsymbol{E}_t$ is the transmitted electric field inside the medium. The boundary conditions for $\boldsymbol{Y}_-(Z,\boldsymbol{r})$ are:

$$\boldsymbol{Y}_-(Z,\boldsymbol{r}) = 0 \ ; \quad z = mZ/m_e \tag{5.53a}$$

$$\boldsymbol{Y}_-(Z,\boldsymbol{r}) = 0 \ ; \quad z = -mZ/m_h \tag{5.53b}$$

$$\boldsymbol{Y}_-(Z,\boldsymbol{r}) = 0 \ ; \quad |\boldsymbol{r}|\to\infty \tag{5.53c}$$

$$\boldsymbol{Y}_-(Z,\boldsymbol{r}) = 0 \ ; \quad Z\to\infty \quad \gamma\neq 0 \ . \tag{5.53d}$$

In the case of a single, spatially dispersive resonance there are generally three bulk modes excited, namely two transverse modes (wave vectors $\boldsymbol{q}_1$, $\boldsymbol{q}_2$) and a longitudinal mode (wave vector $\boldsymbol{q}_3$). The solutions $q_{jz}(j = 1,2,3)$ are real or imaginary depending on $\omega$ and $q_x$ as can be seen from Fig. 5.9D. Note that there is only one solution for $q_z$ in the absence of spatial dispersion.

### 5.6.1 Oblique Incidence

Let $\beta$ be the angle of incidence and let the amplitudes $E_i$ and $E_r$ characterize the incident and reflected wave, respectively:

$$E_i = E_0 \exp\left[iq_x x + iz(\omega/c)\cos\beta - i\omega t\right] \; ; \quad z<0 \tag{5.54}$$

$$E_r = \tilde{r} E_0 \exp\left[iq_x x - iz(\omega/c)\cos\beta - i\omega t\right] \; ; \quad z<0 \tag{5.55}$$

$$E_t = E_t(Z)\exp\left(iq_x x - i\omega t\right) \; ; \quad z>0 \; . \tag{5.56}$$

There are two principal values $\tilde{r}_s$ and $\tilde{r}_p$ of the amplitude reflectivity, corresponding to the incident electric field perpendicular and parallel, respectively, to the plane of incidence (the $x, z$ plane) (see Fig. 5.9A,B).

For the $s$ polarization the Maxwellian boundary conditions are

$$E_{0y}(1+\tilde{r}_s) = E_{ty}(0) \; , \tag{5.57a}$$

$$E_{0y}(1-\tilde{r}_s) = E'_{ty}(0)c/i\omega \; . \tag{5.57b}$$

In this case the longitudinal mode is not excited. For the $p$ polarization

$$-E_0(1-\tilde{r}_p)\sin\beta = \tilde{\varepsilon}E_{tz}(0) \tag{5.58a}$$

$$E_0(1+\tilde{r}_p)\cos\beta = E_{tx}(0) \tag{5.58b}$$

$$E_0(1-\tilde{r}_p) = -\sin\beta E_{tz}(0) + E'_{tz}(0)c/i\omega \; . \tag{5.58c}$$

In (5.57,58) $E'$ means the derivative with respect to $z$.

The set of equations and conditions (5.51–58) is in principle sufficient to determine $\tilde{r}_s$ and $\tilde{r}_p$, but as in the case of normal incidence the computational problems have prevented and exact treatment up to now. Gotthard (1984, 1985) has used two types of approximate solutions, one based on one-dimensional dynamics in relative space (Gotthard, 1984) and one based on difference equations solved approximately as discussed in Sect. 5.5. In both cases the excitation of the longitudinal wave was not considered. This may introduce errors for the $p$ polarization.

A conventional treatment of oblique incidence on a spatially dispersive medium must be based on ABC's for all components of the resonance polarization $P_0$. A common form is (Pekar, 1957; Kliewer and Fuchs, 1968; Maradudin and Mills, 1973; Halevi and Fuchs, 1978):

$$P_{0z} + \tilde{\gamma}_{\|}\frac{\partial P_{0z}}{\partial z} = 0 \qquad z = 0_+ \tag{5.59a}$$

$$P_{0\alpha} + \tilde{\gamma}_{\perp}\frac{\partial P_{0\alpha}}{\partial t} = 0 \qquad z = 0_+ \qquad \alpha = x, y \; . \tag{5.59b}$$

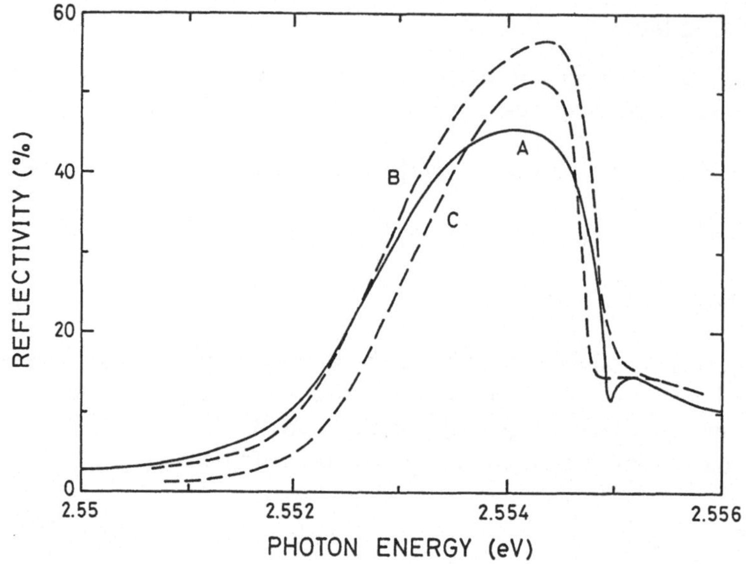


**Fig. 5.10.** Oblique incidence reflectivity near the $A(n=1)$ exciton in CdS for $E\perp c$, surface normal $\perp c$, and an incident angle of 78°. Curve $A$ is measured by Broser et al. (1979), Curve $B$ is calculated from the HT model with parameters obtained from the normal incidence spectrum (Broser et al., 1978). Curve $C$ is calculated by D'Andrea and Del Sole (1984)

$\tilde{\gamma}_{\|}$ and $\tilde{\gamma}_{\perp}$ are fitting parameters or constants defined by the model. In a conventional treatment (5.58a) should be replaced by

$$-E_0(1-\tilde{r}_p)\sin\beta = \varepsilon_b E_{tz}(0) + P_{0z}(0)/\varepsilon_0 \ . \tag{5.60}$$

Reasonable fits to spectra of ZnSe (zincblende structure) are obtained by considering $\tilde{\gamma}_{\|}$, $\tilde{\gamma}_{\perp}$ as freely adjustable complex quantities or by using $\tilde{\gamma}_{\|} = \tilde{\gamma}_{\perp} = 0$ in connection with a dead layer (Pevtsov et al., 1980). Broser et al. (1978, 1979) treat the complicated situation arising with oblique incidence reflection from an anisotropic structure, namely CdS. They use $\tilde{\gamma}_{\|} = \tilde{\gamma}_{\perp} = 0$ and assume a dead layer of usual thickness. These extensive studies demonstrate the shortcomings of the Pekar-Hopfield-Thomas approach because the normal incidence oscillator parameters cannot reproduce the oblique incidence spectra (see Fig. 5.10). The model of D'Andrea and Del Sole (1984) does not give a substantially better fit (Fig. 5.10) but it contains no adjustable dead-layer thickness.

### 5.6.2 Attenuated Total Reflection

In ATR the otherwise total internal reflection in a prism is measured when the studied semiconductor crystal is close to the reflecting surface (Otto, 1974; Tokura et al., 1981). The air gap between the prism and the sample is of the order 1 $\mu$m. This gives a measurable coupling to surface polaritons. The wave vector in the surface plane is $q_x = n_s(\omega/c)\sin\alpha$ where $\alpha$ is the angle of incidence in the prism and $n_s$ is the refractive index of the prism material. As $q_x$ is made larger than $\omega/c$, the vacuum wave-vector is

$$q_0 = q_x(1, 0, -i\overline{\gamma}) \tag{5.61}$$

where $\overline{\gamma}$ is real (see Fig. 5.9C). In the absence of spatial dispersion there is a single real solution $\omega(q_x)$ corresponding to the dispersion relation of a conventional surface polariton. $\omega(q_x)$ lies between the transverse and the longitudinal frequency of the resonance (Agranovich, 1984; Lagois and Fischer, 1979, 1982). In the presence of spatial dispersion there are generally no surface waves with real $\omega$ and $q_x$. The reason is that the surface modes couple to at least one propagating bulk mode radiating energy into the crystal. The mode responsible for this is the upbending lower branch polariton. It is seen that $q_{2Z}^2 > 0$ unless $q_x = n_s(\omega/c)\sin\alpha$ is larger than the background wave vector $\varepsilon_b^{1/2}\omega/c$. However, in prism materials transparent at $\omega \approx \omega_0$, $n_s$ turns out to be smaller than $\varepsilon_b^{1/2}$, so in practice no excitable surface polariton can have all three wave vectors with $q_z$ imaginary.

Usually the spatial dispersion is small enough to give rather weak damping of surface polaritons. This gives a well-defined dispersion relation $\mathrm{Re}\{\omega(q_x)\}$. It is important to note that the calculated result for $\mathrm{Re}\{\omega(q_x)\}$ depends on the ABC-dead-layer model used (Lagois and Fischer, 1982).

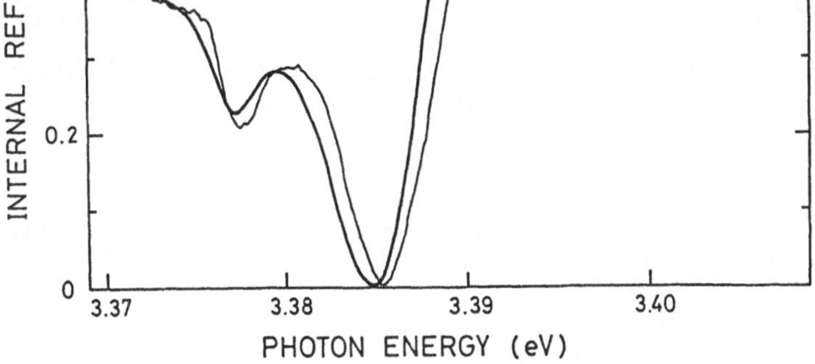

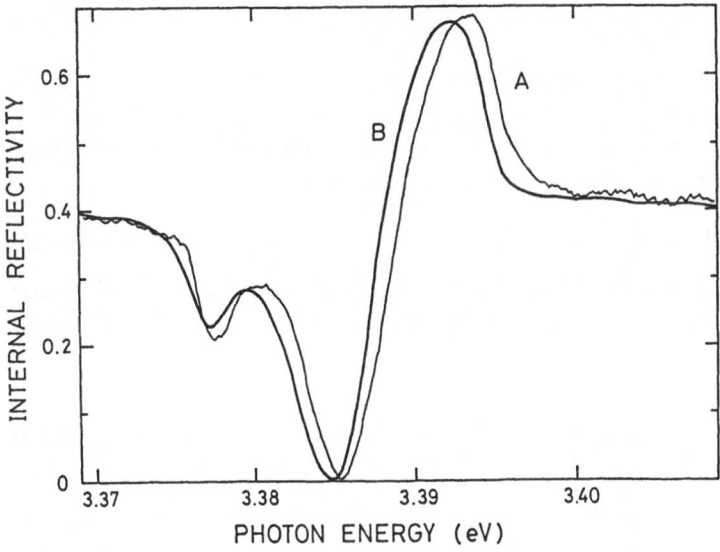

**Fig. 5.11.** Spectra of attenuated total reflection near the $A, B(n = 1)$ exitons in ZnO for $E \perp c$. Curve $A$ is experimental while curve $B$ is calculated from the HT model. The parameters of the HT model are taken from fits to normal incidence reflectivity spectra. From Lagois (1981)

Extensive experimental studies on ATR are reported by Lagois (1981), who studied ZnO and included normal incidence measurements on the same samples. The ATR and normal incidence spectra could not be described in detail by a Pekar-Hopfield-Thomas model [$\tilde{\gamma}_{\|} = \tilde{\gamma}_{\perp} = 0$ in (5.59) and a dead layer]. This is seen in Fig. 5.11 showing an experimental ATR spectrum and a spectrum calculated for a HT-model with parameters $\omega_0, \omega_{LT}, \gamma, \varepsilon_b, \ell$ obtained from normal incidence. Lagois (1981) achieved full agreement between the two types of experiments by extending the HT model to include 3 additional surface layers with different values of $\omega_0$. As the resulting depth dependence indicates the existence of band bending we shall postpone the discussion of Lagois' results to Sect. 5.7.

## 5.7 Band Bending

The reflectivity of direct gap semiconductors near the exitonic resonances is very sensitive to surface treatment and external perturbations. In GaAs, for example, the spectra of $|\tilde{r}|^2$ change drastically with optical illumination, application of very small transverse electric fields and exposure to air (Sell et al., 1973; Fischbach et al., 1976; Lagois et al., 1978; Schultheis and Balslev, 1983). As we shall see in this section such effects are readily explained by the influence of surface electric fields on the excitonic motion. Such fields are due to chemical or structural imperfections with strong influence on the electrochemical potential. Usually the electric field $\boldsymbol{F}$ has an exponential depth dependence, for example

$$\boldsymbol{F} = (0, 0, F_0 \mathrm{e}^{-z/d}) \quad z>0 \ . \tag{5.62}$$

$F_0$ is the maximum field and $d$ is the Debye-Hückel screening length in case of nondegenerate statistics. The self-consistent potential is

$$\Phi = dF_0 \mathrm{e}^{-z/d} \quad z>0 \tag{5.63}$$

corresponding to a total band bending of $-edF_0$. In this section we shall discuss the influence on the interband half-space dynamics of such a field. The drift operator becomes

$$\Omega_{eh} = \omega_g - \frac{\hbar}{2m_e}\nabla_e^2 - \frac{\hbar}{2m_h}\nabla_h^2 + \frac{1}{\hbar}[eF_0 d(\mathrm{e}^{-z_e/d} - \mathrm{e}^{-z_h/d}) + V_{eh}] \tag{5.64}$$

where $V_{eh}$ is given by (5.16). Because of mathematical and computational complexities a rigorous solution of (5.64) must wait for further development. Some understanding of the band-bending effects on the $n = 1$ excitonic resonances can be achieved by applying the adiabatic approximation. As discussed in Sect. 5.3 this method consists of two steps: (a) calculation by variational methods of the excitonic surface potential $\hbar\omega_0(Z,d,F_0)$ and (b) numerical integration of the polariton equations.

A result of the first step is shown on Fig. 5.12. The trial function used here is

$$f_{trial}(\mathbf{r},Z) = (1-e^{-z_e/a_B})(1-e^{-z_h/a_B}) \times (\psi_{1s}(\mathbf{r}) + \alpha(Z)\psi_{2s}(\mathbf{r}) + \beta(Z)\psi_{2p}(\mathbf{r})) \tag{5.65}$$

where $\psi_{1s}$, $\psi_{2s}$, and $\psi_{2p}$ are scaled hydrogenic wave functions. (5.65) has the correct behaviour near the boundaries in relative space. Minimizing the energy expectation value by varying $\alpha(Z)$, $\beta(Z)$ one can arrive at a good estimate of $\omega_0(Z,d,F_0)$ of the ground state exciton. Note that in Fig. 5.12 the energy shift due to the electric field is negative, corresponding to the quadratic Stark effect. This negative shift dominates for $z>6a_B$ while the cut-off forces give almost the same upward shift of $\omega_0(Z)$ as that without a field (c.f. Fig. 5.2). The depth of influence of the cut-off forces is greatest when the light mass particle is pushed towards the surface (e.g. $F_0>0$; $m_e<m_h$). It is interesting to compare these calculations with the depth-dependent resonance frequency found experimentally by fitting a 4 layer sandwich model to ATR and simple reflectivity of ZnO (Lagois, 1981). It is seen that there is a close agreement between Lagois' empirically determined dependence $\omega_0(z)$ and the calculated one for $|F_0| \approx 0.2\,F_I$, where $F_I$ is the ionization field

$$F_I = \hbar\omega_x/(ea_B) \; . \tag{5.66}$$

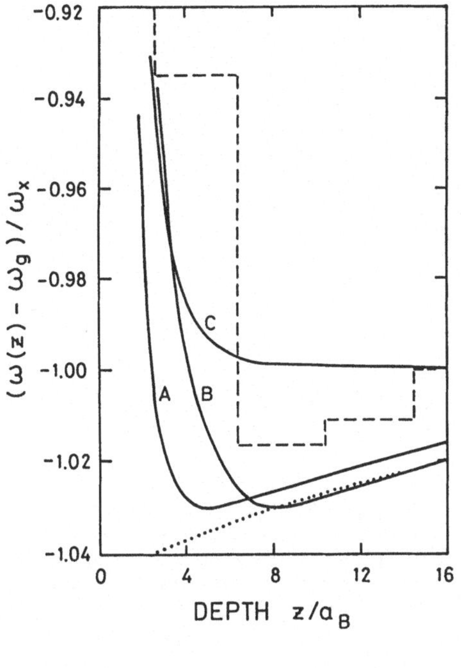


**Fig. 5.12.** Depth dependence of the resonance frequency. Curve *A,B,C* are calculated from the adiabatic approximation, *C* is without band bending, *A* and *B* are calculated with surface electric fields of $-0.2\,F_I$ and $+0.2\,F_I$, respectively (positive into the crystal): The *dotted curve* corresponds to the Stark shift calculated from the local field. From Schultheis and Balslev (1983). The *dashed curve* is the result from fitting a four layer model to ATR/normal incidence experiments on the *B* exciton in ZnO. From Lagois (1981)

(Lagois explained the upshift of $\omega_0(Z)$ close to the surface by higher-order field effects calculated by Blossey (1971), important for $|F_0| \gtrsim F_I$. However, as seen in Fig. 5.12, there is no need for this assumption. In fact, the validity of Blossey's theory close to a surface is rather questionable.)

The next calculational step, viz. the numerical integration of the polariton equations, is necessary when the adiabatic treatment is to be compared with unprocessed reflectivity data. Figure 5.13 shows the calculated spectra for several values of $F_0$ between $-0.28\,F_I$ and $0.28\,F_I$. The excitonic parameters ($\omega_0$, $\omega_x$, $\omega_{LT}$, $m$, $\varepsilon_b$) used here are those of GaAs. Taking into account the broadening effect of lateral variations of $F_0$ and $d$, there is good agreement between the observed spectra of GaAs (Curve A, Fig. 5.13) and calculated results with $|F_0|$ somewhere between 0.2 and $0.3\,F_I$, both spectra having double minima. The oscillatory behaviour is due to complicated interference effects occuring when the Stark shift $(-9\hbar\omega_x F_0^2/8F_I^2)$ is of the order $-\omega_{LT}$.

The dependence of the excitonic reflectivity spectra on various surface conditions has been the subject of many experimental works. To be mentioned here are studies of heat treated samples (Evangelisti et al., 1974A), of optical bias (Lagois et al., 1978) (Curves B and C, Fig. 5.13), of small transverse fields (Lagois et al., 1978), and of surfaces freshly cleaved in liquid helium (Schultheis and Balslev, 1983). The effects are all to be assigned to changes of the band-

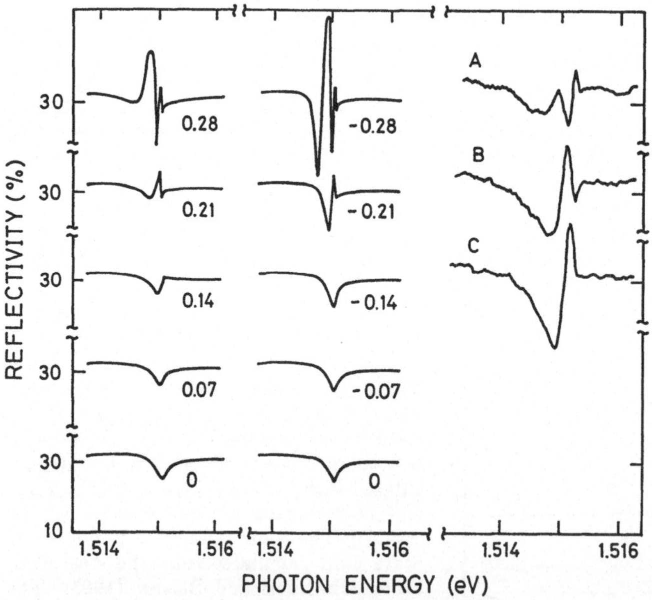


**Fig. 5.13.** Reflectivity spectra of GaAs. The 10 curves to the left are calculated using the adiabatic approximation for different surface electric fields. The field penetration depth $d$ (see text) is $40\,a_B$ while values of $F(0)/F_I$ are given in the figure. From Schultheis and Balslev (1985). Curves $A,B,C$ are experimental results from eptaxial GaAs with different intensities of a secondary illumination (optical bias): Curve $A$ corresponds to zero optical bias curve $C$ to maximum optical bias. From Lagois et al. (1978)

bending, but in most cases the mechanisms behind the relation between this band-bending and the actual surface treatment are not understood in detail.

On the other hand, if the quantitative influence of surface electric fields on the spectra could be fully explored, then the excitonic reflectivity can be developed as a tool for characterizing the surface in terms of local fields.

## 5.8 Pseudo-Two-Dimensional Edge Dynamics

Somewhat related to the half-space geometry is the situation where electrons and holes are confined to a thin layer in a semiconductor. This occurs in metal-oxide-semiconductor (MOS) inversion layers, in heterojunctions, and in quantum wells. Here the electronic dynamics can often be considered as pseudo-two-dimensional in the sense that there is a pronounced quantization in the motion perpendicular to the layer. In this case the electron-hole amplitude takes the form

$$Y(\mathbf{r}_h, \mathbf{r}_e) = f_m(z_h, Z_e) y_m(x_h, y_h, x_e, y_e) \tag{5.67}$$

where $m$ counts the pair sublevels of the confinement quantization. Then the two-dimensional (2D) edge equations for $y_m$ contain gap frequencies $\omega_{gm}$ assigned to the pair sublevels.

2D excitons have been studied in multiple or single quantum wells since the pioneering work by Dingle et al. (1974). The most common structure explored is a layer of GaAs ($\approx 100$ Å) between thick layers of $Al_xGa_{1-x}As$ ($x \approx 0.2$–$0.5$), the latter material having the highest band gap. As reviewed thoroughly by Chemla and Miller (1985), the 2D excitons have interesting optical properties. The absorption spectrum is given by

$$\begin{aligned}\alpha(\omega) \propto \sum_{m,n} \left(n - \frac{1}{2}\right)^{-3} \delta\left(\omega - \omega_{gm} + \omega_x \left(n - \frac{1}{2}\right)^{-2}\right) \\ + \frac{2\theta(\omega - \omega_{gm})}{1 + \exp\left[2\pi\omega_x^{1/2}/(\omega - \omega_{gm})^{1/2}\right]}\end{aligned} \tag{5.68}$$

where $\theta$ is the unit step function and $\hbar\omega_x$ is the 3D exciton Rydberg energy. This expression is valid for a simple band edge. In real quantum wells the valence band structure gives rise to two series, a heavy exciton and a light exciton series. Generally only $n = 1$ lines are resolved.

There are three remarkable properties of 2D excitons to be mentioned here; (a) Their binding energy is four times as large as that of 3D excitons [see (5.68)]; this allows excitonic effects to be studied and used at room temperature. (b) They show large electroabsorption via the Stark effect (Miller et al., 1984), and (c) they give rise to resonance non-linear response at intensities much lower than for bulk excitations (Chemla et al., 1985).

So far the coherent-wave approach has not been applied to the above pseudo-two-dimensional systems. The potential areas of applications are resonance nonlinear response and polariton effects for wave propagation along the 2D structure.

Intraband 2D dynamics is mainly studied in MOS inversion layers and heterojunctions. The most spectacular phenomena are Shubnikov-de Haas oscillations and the quantum Hall effect (von Klitzing et al., 1980; Ando et al., 1982). Of interest are also 2D plasmons which, unlike 3D plasmons, have zero frequency in the long wave limit (Stern, 1967; Chaplik, 1972). They can be studied experimentally by a grating coupler technique, i.e. an optical coupling through a metallic, periodic pattern (Allen et al., 1977). In the detailed study of 2D plasmons the application of a magnetic field is essential [see for example Batke et al., (1985)].

# 6. Intraband Dynamics

## 6.1 Density Matrices and Distribution Functions

Intraband processes in the two band semiconductor are either motion of electrons of the conduction band or of holes of the valence band. In this section we shall concentrate on conduction-band electrons. An appropriate level for the description of intraband dynamics is the kinetic level. The quantum version of kinetic theory is based on the equation of motion for the single particle density matrix

$$C(\mathbf{r}_1,\mathbf{r}_2,t) = \langle c^\dagger(\mathbf{r}_1,t)c(\mathbf{r}_2,t)\rangle \ . \tag{6.1}$$

Evidently $C$ is hermitian

$$C(\mathbf{r}_1,\mathbf{r}_2) = C^*(\mathbf{r}_2,\mathbf{r}_1) \ . \tag{6.2}$$

Sometimes it is useful to consider $C$ as a function of the two variables

$$\mathbf{r} = (\mathbf{r}_1+\mathbf{r}_2)/2 \qquad \mathbf{r}' = \mathbf{r}_1-\mathbf{r}_2 \ . \tag{6.3}$$

Then the notation

$$\tilde{C}(\mathbf{r},\mathbf{r}') = C(\mathbf{r}_1,\mathbf{r}_2) \tag{6.4}$$

will be used.

The decomposition of $\tilde{C}$ into real and imaginary parts coincides with the decomposition into symmetric and antisymmetric parts with respect to $\mathbf{r}'$

$$\tilde{C} = \tilde{C}' + i\tilde{C}'' \tag{6.5}$$

$$\tilde{C}'(\mathbf{r},-\mathbf{r}') = \tilde{C}'(\mathbf{r},\mathbf{r}') \qquad \tilde{C}''(\mathbf{r},-\mathbf{r}') = -\tilde{C}''(\mathbf{r},\mathbf{r}') \ . \tag{6.5a}$$

Observable quantities related to $C$ are the electron density

$$n(\mathbf{r}) = C(\mathbf{r},\mathbf{r}) = \tilde{C}'(\mathbf{r},0) \tag{6.6}$$

and the transport current density

$$j_n(r) = \frac{i\hbar}{2m_e}(\nabla_1 - \nabla_2)C\bigg|_{r_1=r_2=r} = -\frac{\hbar}{m_e}\nabla_{r'}\tilde{C}''\bigg|_{r'=0} . \tag{6.7}$$

Electronic contributions to the electric charge density $\varrho$ and to the electric conduction current $\boldsymbol{J}$ are derived from $n$ and $\boldsymbol{j}_n$

$$\varrho = -en \qquad \boldsymbol{J} = -e\boldsymbol{j}_n . \tag{6.8}$$

The equation of motion for the density matrix in the absence of source terms and irreversible processes is according to (3.1b)

$$\dot{C} + i\left[\frac{\hbar}{2m_e}(\nabla_1^2 - \nabla_2^2) + \frac{e}{\hbar}(\Phi(\boldsymbol{r}_1) - \Phi(\boldsymbol{r}_2))\right]C = 0 \tag{6.9}$$

or if expressed in coordinates $\boldsymbol{r}, \boldsymbol{r}'$:

$$\dot{\tilde{C}} + i\frac{\hbar}{m_e}\nabla_r\nabla_{r'}\tilde{C} + \frac{ie}{\hbar}\left[\Phi\left(\boldsymbol{r} + \frac{1}{2}\boldsymbol{r}'\right) - \Phi\left(\boldsymbol{r} - \frac{1}{2}\boldsymbol{r}'\right)\right]\tilde{C} = 0 . \tag{6.10}$$

In the special case $\boldsymbol{r}' = 0$, (6.10) yields the conservation law

$$\dot{n} + \nabla\cdot\boldsymbol{j}_n = 0 . \tag{6.11}$$

Instead of the density matrix $\tilde{C}$ one can also use the phase-space distribution function $f_e$. (The index "e" refers to electrons: the corresponding quantity for holes will be denoted as $f_h$). The distribution function is derived from $\tilde{C}$ by the Wigner transformation

$$f_e(\boldsymbol{r}, \boldsymbol{k}) = \int \tilde{C}(\boldsymbol{r}, \boldsymbol{r}')e^{i\boldsymbol{k}\cdot\boldsymbol{r}'}d^3\boldsymbol{r}' . \tag{6.12a}$$

The inverse transformation is

$$\tilde{C}(\boldsymbol{r}, \boldsymbol{r}') = (2\pi)^{-3}\int f_e(\boldsymbol{r}, \boldsymbol{k})e^{-i\boldsymbol{k}\cdot\boldsymbol{r}'}d^3\boldsymbol{k} . \tag{6.12b}$$

The description of the electronic system by $C$ or by $f_e$ is equivalent. Typical observables, like $n$ or $\boldsymbol{j}_n$ can therefore just as well be derived from $f_e$ :

$$n(\boldsymbol{r}) = (2\pi)^{-3}\int f_e(\boldsymbol{r}, \boldsymbol{k})d^3\boldsymbol{k} , \tag{6.13a}$$

$$\boldsymbol{j}_n(\boldsymbol{r}) = (2\pi)^{-3}\int \frac{\hbar\boldsymbol{k}}{m_e}f_e(\boldsymbol{r}, \boldsymbol{k})d^3\boldsymbol{k} . \tag{6.13b}$$

Let us translate the equation of motion for $\tilde{C}$ (6.10) into the corresponding equation for $f_e$. With the abbreviation

$$V = -e\Phi$$

(6.10) becomes

$$\dot{\tilde{C}} + \frac{i\hbar}{m_e}\nabla_r\nabla_{r'}\tilde{C} - \frac{i}{\hbar}\left[V\left(r+\frac{1}{2}r'\right) - V\left(r-\frac{1}{2}r'\right)\right]\tilde{C} = 0 . \tag{6.14}$$

The potential difference we express by a series expansion

$$V\left(r+\frac{1}{2}r'\right) - V\left(r-\frac{1}{2}r'\right) = \sum_n \frac{1}{n!}[1-(-1)^n]\left(\frac{r'}{2}\nabla\right)^n V(r) . \tag{6.15}$$

Let us for a moment assume that in (6.15) all terms but the lowest nonvanishing term, i.e., $n = 1$, can be neglected. The insertion of (6.15) into (6.14) then yields

$$\dot{\tilde{C}} + \frac{i\hbar}{m_e}\nabla_r\nabla_{r'}\tilde{C} - \frac{i}{\hbar}r'\nabla V(r)\tilde{C} = 0 . \tag{6.16}$$

Applying to (6.16) the Wigner transformation (6.12) one finds the equation of motion of $f_e$

$$\dot{f}_e + \frac{\hbar k}{m_e}\nabla_r f_e - \frac{\nabla V}{\hbar}\nabla_k f_e = 0 . \tag{6.17}$$

This equation is well known in transport theory as the collisionless Boltzmann-equation. Also the name Wlassow-equation is in use for (6.17) (Landau, Lifschitz, 1981). The type of motion described by (6.17) is called a drift motion. As is shown in many textbooks (Landau, Lifschitz, 1981; Becker, Sauter, 1968), (6.17) can be derived from the idea that $f_e$ is a phase-space density distribution of classical particles moving under the influence of a force

$$F = -\nabla V . \tag{6.18}$$

$f_e(r, k)$ then describes the particle density in the phase space region with position $r$ and momentum $p = \hbar k$. The approximation leading from (6.14) to (6.16) therefore had the effect of making the density matrix equivalent to a classical phase-space distribution. Let us now look into the corrections to this approximation.

The next nonvanishing term in the expansion (6.15) is of the order $n = 3$. When inserted into (6.16) it produces the following term

$$\tilde{Q}_3 = -\frac{i}{24\hbar}[(r\nabla)^3 V(r)]\tilde{C} . \tag{6.19}$$

By the Wigner transformation $\tilde{Q}_3$ is transformed into

$$Q_3 = \frac{1}{24\hbar}[(\nabla_k \cdot \nabla_r)^3 V] f_e . \tag{6.20}$$

If again we introduce the classical momentum $p = \hbar k$, (6.20) becomes

$$Q_3 = \frac{\hbar^2}{24}[(\nabla_p \cdot \nabla_r)^3 V] f_e . \tag{6.21}$$

As the prefactor $\hbar^2$ cannot be absorbed into classical variables, $Q_3$ must be considered as a quantum correction. The order of magnitude of this quantum correction depends on the second derivative of the force scaled with a typical de Broglie wavelength present in $f_e$. The quantum correction therefore becomes important in two cases:

i) if the force $\nabla V$ is inhomogeneous on microscopic distances;
ii) if $f_e$ contains long wavelength contributions. Equivalent to this condition is the existence of long range off-diagonal order (ODLRO) in $C$.

In all other cases the corrections $Q_3$ and higher are negligible and one can use the classical kinetic equation (6.17).

## 6.2 Static Equilibrium

Static equilibrium solutions of the band-edge model are solutions with $Y = 0$ and $\dot{C} = \dot{D} = 0$. The relevant equations are then the two intraband equations

$$i\Omega_{ee}C_{12} = (\dot{C}_{12})_{irr} \tag{6.22a}$$

$$i\Omega_{hh}D_{12} = (\dot{D}_{12})_{irr} \; . \tag{6.22b}$$

Let us concentrate on the discussion of (6.22a), and first neglect the irreversible term. Then (6.22a) reads

$$\frac{\hbar}{2m_e}(\nabla_1^2 - \nabla_2^2)C_{12} + \frac{e}{\hbar}(\Phi_1 - \Phi_2)C_{12} = 0 \; . \tag{6.23}$$

Evidently this equation allows separable solutions in the form

$$C(\boldsymbol{r}_1, \boldsymbol{r}_2) = \psi_1^*(\boldsymbol{r}_1)\psi_2(\boldsymbol{r}_2) \; . \tag{6.24}$$

Inserting (6.24) into (6.23) and separating variables leads to a pair of Schrö-dinger-like equations

$$\left[-\frac{\hbar^2}{2m_e}\nabla_1^2 - e\Phi(\boldsymbol{r}_1)\right]\psi_1(\boldsymbol{r}_1) = E\psi_1(\boldsymbol{r}_1) \tag{6.25a}$$

$$\left[-\frac{\hbar^2}{2m_e}\nabla_2^2 - e\Phi(\boldsymbol{r}_2)\right]\psi_2(\boldsymbol{r}_2) = E\psi_2(\boldsymbol{r}_2) \; . \tag{6.25b}$$

From Hermiticity of $C$ it follows that $\psi_1 = \psi_2$. A general ansatz for $C_{12}$ as a superposition of separable solutions is then of the form

$$C(\boldsymbol{r}_1, \boldsymbol{r}_2) = \sum c_k \psi_k^*(\boldsymbol{r}_1)\psi_k(\boldsymbol{r}_2) \tag{6.26}$$

with $\psi_k$ obeying the Schrödinger equation

$$-\frac{\hbar^2}{2m_e}\nabla^2\psi_k - e\Phi\psi_k = E_k\psi_k\,. \tag{6.27}$$

The coefficients $c_k$ in (6.26) remain indefinite since any superposition of type (6.26) satisfies (6.23). In order to determine the coefficients $c_k$ one must consider the irreversible term in (6.22). When (6.26) is inserted into (6.22a) one has

$$\Omega_{ee}C(\boldsymbol{r}_1,\boldsymbol{r}_2) = 0 \tag{6.28}$$

and therefore one is left with

$$(\dot{C}(\boldsymbol{r}_1,\boldsymbol{r}_2))_{irr} = 0\ . \tag{6.29}$$

Thus the equilibrium distribution is the stationary solution to the irreversible part of the equation of motion. As an example let us consider thermalization by electron-electron collisions. The irreversible term (6.29) in this case is a collision operator operating on the occupation numbers $c_k$ in (6.26):

$$\begin{aligned}(\dot{c}_k)_{irr} = &- \sum_{\ell k'\ell'} [W c_k c_\ell (1-c_{k'})(1-c_{\ell'}) \\ &- W'(1-c_k)(1-c_\ell)c_{k'}c_\ell]\ .\end{aligned} \tag{6.30}$$

The first term in (6.30) belongs to a process $(k,\ell)\rightarrow(k',\ell')$, the second one to the reverse process $(k',\ell')\rightarrow(k,\ell)$; $W$ and $W'$ are the corresponding rate constants. From microscopic time-reversal symmetry it follows that (Landau-Lifschitz, 1981)

$$W = W'\ . \tag{6.31}$$

In addition it is known that in equilibrium one must have detailed balance, as can be for example inferred from Boltzmann's $H$-Theorem (Hittmair-Adam, 1971). Detailed balance means that every term in the summation in (6.30) vanishes separately. The equilibrium distribution therefore obeys the functional equation

$$c_k^0 c_\ell^0(1-c_{k'}^0)(1-c_{\ell'}^0) = c_{k'}^0 c_{\ell'}^0(1-c_k^0)(1-c_\ell^0)\ . \tag{6.32}$$

According to our assumption of nondegenerate energy states the occupation numbers are unique functions of energy ($c_k = c(E_k)$). We therefore write (6.32) (after division by the product $c_k c_\ell c_{k'} c_{\ell'}$) as follows

$$g(E_{k'})g(E_{\ell'}) = g(E_k)g(E_\ell) \tag{6.33}$$

with

$$g(E) = (c^0(E))^{-1} - 1\ . \tag{6.34}$$

In the collision energy is conserved:

$$E_k + E_\ell = E_{k'} + E_{\ell'} \ . \tag{6.35}$$

The functional equation (6.33) together with the subsidiary condition (6.35) is solved by letting

$$\ln g = \beta(E - \mu) \tag{6.39}$$

where $\beta$ and $\mu$ are integration constants. From (6.34) and (6.39) it follows that the occupation numbers $c(E)$ are distributed according to a Fermi distribution

$$c(E) = \frac{1}{1 + \exp\left[\beta(E - \mu)\right]} \tag{6.40}$$

where $\beta$ is related to the inverse temperature

$$\beta = (kT)^{-1} \tag{6.41}$$

and $\mu$ is the chemical potential. This rather well-known text-book argument (Becker and Sauter, 1968) demonstrates that equilibrium properties are among the consequences of the constitutive equations and that no extra assumptions from thermodynamics are needed.

For easy reference the equations determining the static equilibrium of intraband dynamics are summarized. These equations are:
Spectral representation (6.26):

$$C = \sum c_k \psi_k^*(\mathbf{r}_1)\psi_k(\mathbf{r}_2) \ . \tag{6.42a}$$

Wave equation (6.27):

$$-\frac{\hbar^2}{2m_e}\nabla^2\psi_k - e\Phi\psi_k = E_k\psi_k \ . \tag{6.42b}$$

Equilibrium condition (6.29):

$$(\dot{C})_{irr} = 0 \ . \tag{6.42c}$$

Selfconsistency:

$$\Phi = \Phi^{ex} + \Phi^{in} \ ; \quad \varepsilon_0\nabla^2\Phi^{in} = eC(\mathbf{r}, \mathbf{r}) \ . \tag{6.42d}$$

Let us consider two examples:

*a) Unbounded Electron Gas;* $\Phi^{ex} = 0$. From translational invariance then follows:

$$C(\mathbf{r}_1, \mathbf{r}_2) = \overline{C}(\mathbf{r}_1 - \mathbf{r}_2) \ , \tag{6.43}$$

and therefore

$$C(\mathbf{r}, \mathbf{r}) = \text{const.} \ ; \quad \Phi^{in} = 0 \ . \tag{6.44}$$

The wave equation (6.42b) is then solved using plane waves and $C$ becomes according to (6.42a)

$$C(\boldsymbol{r}_1, \boldsymbol{r}_2) = \sum c_{\boldsymbol{k}} e^{-i\boldsymbol{k}\cdot(\boldsymbol{r}_1 - \boldsymbol{r}_2)} \; . \tag{6.45}$$

Replacing summation by integration and using a notation in accordance with the Wigner transformation (6.12b) we write

$$C(\boldsymbol{r}_1, \boldsymbol{r}_2) = (2\pi)^{-3} \int f_e(\boldsymbol{k}) e^{-i\boldsymbol{k}\cdot(\boldsymbol{r}_1 - \boldsymbol{r}_2)} d^3\boldsymbol{k} \tag{6.46}$$

where $f_e(\boldsymbol{k})$ is the Fermi distribution

$$f_e(\boldsymbol{k}) = \frac{2}{1 + \exp\left[\beta(E - \mu)\right]} \; . \tag{6.47}$$

Assuming a parabolic band

$$E(\boldsymbol{k}) = E_c + \frac{\hbar^2}{2m_e}\boldsymbol{k}^2 \tag{6.48}$$

and a not-too-heavily doped semiconductor in the sense that

$$(E_c - \mu)/kT \gg 1 \tag{6.49}$$

we can approximate $f_e(\boldsymbol{k})$ by the Boltzmann distribution

$$f_e(\boldsymbol{k}) \approx 2\exp\left[\left(\mu - E_c - \frac{\hbar^2 \boldsymbol{k}^2}{2m_e}\right)/kT\right] \; . \tag{6.50}$$

Then the integral (6.46) can be evaluated analytically with the result

$$C = 2\left(\frac{m_e kT}{2\pi\hbar^2}\right)^{3/2} \exp\left[-\frac{1}{kT}(E_c - \mu) - \frac{m_e kT}{2\hbar^2}(\boldsymbol{r}_1 - \boldsymbol{r}_2)^2\right] \; . \tag{6.51}$$

It shows that $C(\boldsymbol{r}_1 - \boldsymbol{r}_2)$ is a Gaussian with a width determined by the thermal length

$$\lambda_{th} = \hbar/(m_e kT)^{1/2} \; . \tag{6.52}$$

$\lambda_{th}$ plays the role of the coherence length mentioned in Sect. 2.5. The coherence-length becomes important in the interaction with inhomogeneities, such as e.g. a surface. This is demonstrated by our second example of a static equilibrium situation.

*b) Electron Gas in a Halfspace Semiconductor.* In a bounded medium $C$ has to fulfill boundary conditions. Since in a semiconductor the electron concentration is in general very small compared to the number of states available in the conduction band, the density matrix $C(\boldsymbol{r}_1, \boldsymbol{r}_2)$ is primarily formed from the long wavelength states at the bottom of the conduction band. On the scale defined by the wavelength of these states the exact position and structure of the surface will not be resolved. We therefore may with sufficient accuracy replace the surface by a rigid wall. Then all states contributing to $C$ will have a node on the surface. We are thus led to the boundary condition for $C(\boldsymbol{r}_1, \boldsymbol{r}_2)$ :

$$C(\boldsymbol{r}_1, \boldsymbol{r}_2) = 0 \quad \text{for } \boldsymbol{r}_1 \text{ or } \boldsymbol{r}_2 \text{ on the surface.} \tag{6.53}$$

If $C$ is represented by the spectral decomposition (6.42a) the corresponding boundary condition for the wave functions $\psi_{\boldsymbol{k}}$ is

$$\psi_{\boldsymbol{k}}(\text{surface}) = 0 \ . \tag{6.53a}$$

As a consequence of the boundary condition (6.53) the electronic charge density becomes inhomogeneous near the surface with the result that the induced potential $\Phi^{in}$ is no longer constant. The problem then is to solve the full set of nonlinear equations (6.42). This problem allows no analytical solution. However, since it is a special case of the famous "Surface Quantization" problem in MOSFET-structures a lot of numerical work has been done on it (Baraff and Appelbaum, 1972; Ando, Fowler and Stern, 1982; Ehlers and Mills, 1986). By surface quantization is understood that the wave equation (6.42b) allows bound state solutions in the self-consistent potential $\Phi = \Phi^{ex} + \Phi^{in}$, where $\Phi^{ex}$ is the gate potential (Fig. 6.1).

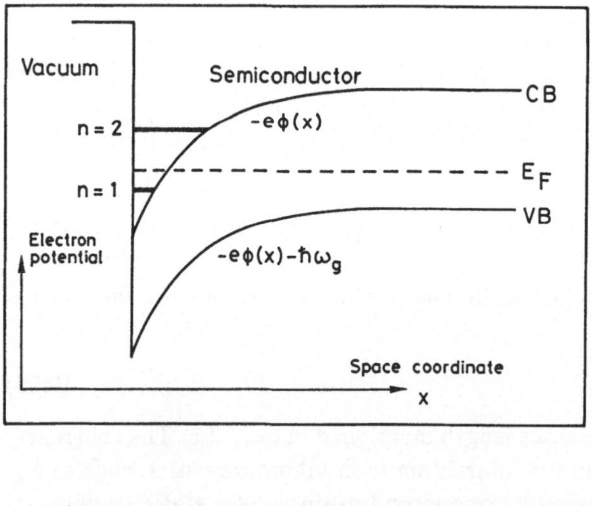


**Fig. 6.1.** Band bending with the formation of quantized surface states ($n = 1, 2$) in the self-consistent potential $\phi(x)$

If the potential $\Phi$ is weak enough to be neglected, the half-space density matrix is represented by an integral over the bulk distribution function $f_e(\boldsymbol{k})$ (Stahl, 1983)

$$C = 4(2\pi)^{-3} \int f_e(\boldsymbol{k}) \exp\left[i\boldsymbol{k}_{\|}(\boldsymbol{r}_1 - \boldsymbol{r}_2)\right] \times \sin(\boldsymbol{k}_{\perp} \cdot \boldsymbol{r}_1) \sin(\boldsymbol{k}_{\perp} \cdot \boldsymbol{r}_2) d^3\boldsymbol{k} \ . \tag{6.54}$$

where $\mathbf{k}_{\parallel}$ and $\mathbf{k}_{\perp}$ are, respectively, wave vectors parallel and normal to the surface. This approximation will be used as the starting point for a treatment of the corresponding dynamical problem in Sect. 6.7.

## 6.3 Stationary Transport Processes

Stationary transport in a system not too far from equilibrium is an important application of intraband dynamics. Examples are electric conduction, diffusive carrier motion, heat conduction etc. Also cross effects such as the Peltier effect fall into this category. The field becomes even richer if, in addition, one considers the influence of magnetic fields on different transport processes. We do not intend to give here a detailed presentation of all these phenomena. The interested reader may consult the literature, e.g. the respective articles in the Handbook on Semiconductors (Moss, 1982). The aim of our discussion of transport processes is only to put the matter into perspective as a special application of the general theory of intraband processes.

As an example let us consider an $n$-type semiconductor which is described by the electronic density matrix $C$. In a stationary state $C$ obeys the equation

$$i\Omega_{ee}C = (\dot{C})_{irr} \ . \tag{6.55}$$

In order to drive transport currents one must apply driving forces such as a temperature gradient $\nabla T$, a longitudinal electric field $-\nabla\Phi$, or a chemical gradient $\nabla\mu$. It is assumed that $T$, $\Phi$, $\mu$ etc. are slowly varying functions on the atomic scale. In this situation it is useful to decompose $C$ into two parts

$$C = C_{loc} + C_1 \tag{6.56}$$

where $C_{loc}$ is the local equilibrium distribution. $C_{loc}$ has the form (6.46) but with spatially dependent parameters in $f = f_{loc}$

$$f_{loc} = 2\{1 + \exp[\beta(\mathbf{r})(E(\mathbf{k}) - \mu(\mathbf{r})]\}^{-1} \ ;$$
$$\mathbf{r} = (\mathbf{r}_1 + \mathbf{r}_2)/2 \ ; \quad \beta^{-1} = kT \ . \tag{6.57}$$

(An expected dependence of $f_{loc}$ on the electric potential $\Phi$ is cancelled by the electric contribution to the electrochemical potential.) $C_1$ has to be chosen in such a way that (6.55) is satisfied. In the linear relaxation time approximation this is achieved by assuming, for the irreversible term, the particular form

$$(\dot{C})_{irr} = -\frac{1}{\tau}(C - C_{loc}) \ . \tag{6.58}$$

With this ansatz (6.55) is solved to first order in the driving forces. In the absence of magnetic fields $\Omega_{ee}C_1$ has no first order contributions and therefore one arrives in this case at the equation

$$C_1 = -i\tau\Omega_{ee}C_{loc} \ . \tag{6.59}$$

The possibility of nonvanishing currents can already be deduced from (6.59) because $C_1$ is imaginary if $C_{loc}$ is real, and currents are carried by the imaginary part of the density matrix (6.7). Usually (6.59) is worked out in terms of distribution functions (Conwell, 1982). As an example we consider the distribution $f_1(\boldsymbol{r},\boldsymbol{k})$ of a system with nonvanishing $\nabla T$ and $\nabla\Phi$ :

$$f_1(\boldsymbol{r},\boldsymbol{k}) = \int \tilde{C}_1(\boldsymbol{r},\boldsymbol{r}')e^{i\boldsymbol{k}\cdot\boldsymbol{r}'}d^3\boldsymbol{r}' \tag{6.60}$$

which in this case becomes

$$f_1 = -\tau\frac{\partial f_{loc}}{\partial\eta}\boldsymbol{v}[(E-\mu)\nabla\beta + e\beta\nabla\Phi] \quad \text{with} \tag{6.61}$$

$$\eta = \beta(E-\mu) \qquad \boldsymbol{v} = \hbar\boldsymbol{k}/m \ .$$

From (6.61) the currents for charge and heat transport respectively can be calculated in the forms [c.f. (6.13)]

$$\boldsymbol{J} = -e\int \boldsymbol{v}f_1 d^3\boldsymbol{v} \qquad \boldsymbol{j}_Q = \int (E-\mu)\boldsymbol{v}f_1 d^3\boldsymbol{v} \ . \tag{6.62}$$

Let us close this section with a comment on the relation of the quantum theoretical formalism based on $C$ and the classical formalism based on $f_e(\boldsymbol{r},\boldsymbol{k})$. In the above example the quantum formulation is pure luxury because the general condition of weak gradients that is characteristic of linear transport theory brings us automatically to the classical limit of the Wigner distribution (compare Sect. 6.1). But this is not always true. A case where quantum effects are nonnegligible even in linear transport theory is that of magnetotransport at low temperatures (Shubnikow de Haas-Effect and Quantum Hall Effect). These effects define a region where the replacement of the quantum mechanical drift operator $\Omega_{ee}$ by the classical Wlassow-Boltzmann expression as described in Sect. 6.1 is inadmissible. The reason lies in the necessity to take into account discrete Landau-levels in a magnetic field. Of particular recent interest is the two-dimensional Quantum Hall Effect which combines Landau-quantization with surface quantization as mentioned in Sect. 6.2 (von Klitzing et al., 1980).

## 6.4 Lindhard's Dielectric Function

In this section we consider the linear response of the mobile carrier system to a wavelike external perturbation

$$\Phi_{ex} = \tilde{\Phi}_{ex}(\boldsymbol{q},\omega)e^{i\boldsymbol{q}\cdot\boldsymbol{r}-i\tilde{\omega}t} \qquad \tilde{\omega} = \omega + i0_+ \ . \tag{6.63}$$

The system is assumed to be unbounded. The equations of motion are the source-free intraband equation (3.1b)

$$\dot{C}_{12} + i\left[\frac{\hbar}{2m_e}(\nabla_1^2 - \nabla_2^2) + \frac{e}{\hbar}(\Phi_1 - \Phi_2)\right]C_{12} = 0 \qquad (6.64)$$

and Poisson's equation for the induced part of the potential

$$\varepsilon_0 \nabla^2 \Phi^{in} = eC(\boldsymbol{r}, \boldsymbol{r}) \ . \qquad (6.65)$$

The linearization is understood to start from thermal equilibrium with $\Phi = 0$ and $C_{equ} = C^0$ taken from Sect. 6.2. The linearized form of (6.64) then becomes

$$\dot{C}_{12} + \frac{i\hbar}{2m_e}(\nabla_1^2 - \nabla_2^2)C_{12} = -\frac{ie}{\hbar}(\Phi_1 - \Phi_2)C_{12}^0 \ . \qquad (6.66)$$

From spatial invariance it also follows that the total self-consistent potential $\Phi = \Phi^{ex} + \Phi^{in}$ is of the form

$$\Phi = \tilde{\Phi}(\boldsymbol{q}, \omega)\mathrm{e}^{i\boldsymbol{q}\cdot\boldsymbol{r} - i\tilde{\omega}t} \ . \qquad (6.67)$$

With

$$C_{12}^0 = (2\pi)^{-3} \int f_e(\boldsymbol{k})\mathrm{e}^{i\boldsymbol{k}\cdot(\boldsymbol{r}_1 - \boldsymbol{r}_2)} d^3\boldsymbol{k}$$

taken from (6.46) and an ansatz for $C_{12}$

$$C(\boldsymbol{r}_1, \boldsymbol{r}_2) = (2\pi)^{-3} \int C_q(\boldsymbol{k}, \omega) \exp\left[i\boldsymbol{q}\cdot\boldsymbol{r}_1 + i\boldsymbol{k}\cdot(\boldsymbol{r}_2 - \boldsymbol{r}_1) - i\tilde{\omega}t\right] d^3\boldsymbol{k} \qquad (6.68)$$

one can solve (6.66) by Fourier transformation. The result is

$$C_q(\boldsymbol{k}, \omega) = -e\tilde{\Phi}(\boldsymbol{q}, \omega)\frac{f_e(\boldsymbol{k}) - f_e(\boldsymbol{k} - \boldsymbol{q})}{E(\boldsymbol{k}) - E(\boldsymbol{k} - \boldsymbol{q}) - \hbar\tilde{\omega}} \qquad (6.69)$$

where $E(\boldsymbol{k}) = (\hbar^2 \boldsymbol{k}^2 / 2m_e)$. From $C_q(\boldsymbol{k}, \omega)$ we can find the charge density response

$$\tilde{\varrho}(\boldsymbol{q}, \omega) = -e(2\pi)^{-3} \int C_q(\boldsymbol{k}, \omega) d^3\boldsymbol{k} \ . \qquad (6.70)$$

Evidently $\tilde{\varrho}(\boldsymbol{q}, \omega)$ is the prefactor of the wavelike charge density fluctuation

$$\varrho(\boldsymbol{r}, t) = -eC(\boldsymbol{r}, \boldsymbol{r}, t) = \tilde{\varrho}(\boldsymbol{q}, \omega)\mathrm{e}^{i\boldsymbol{q}\boldsymbol{r} - i\tilde{\omega}t} \qquad (6.71)$$

being produced by the disturbing potential.

From (6.70) we find

$$\tilde{\varrho}(\boldsymbol{q}, \omega) = e^2(2\pi)^{-3}\tilde{\Phi}(\boldsymbol{q}, \omega) \int \frac{f_e(\boldsymbol{k}) - f_e(\boldsymbol{k} - \boldsymbol{q})}{E(\boldsymbol{k}) - E(\boldsymbol{k} - \boldsymbol{q}) - \hbar\tilde{\omega}} d^3\boldsymbol{k} \ . \qquad (6.72)$$

The implicit dependence of $\tilde{\Phi}(\boldsymbol{q},\omega)$ on the density via the induced part $\Phi^{in}$ can be eliminated if we apply Poisson's equation (6.65) in Fourier transform

$$\tilde{\varrho}(\boldsymbol{q},\omega) = \varepsilon_0 \boldsymbol{q}^2 \tilde{\Phi}^{in}(\boldsymbol{q},\omega) = \varepsilon_0 q^2 (\tilde{\Phi}(\boldsymbol{q},\omega) - \tilde{\Phi}^{ex}(\boldsymbol{q},\omega)) \ . \tag{6.73}$$

From (6.72,73) one can derive either a formula giving the charge density response to an external potential, or more conveniently, the dielectric function $\varepsilon(\boldsymbol{q},\omega)$ of the free carrier system which is defined as the screening ratio between $\Phi^{ex}$ and $\Phi$.

$$\tilde{\Phi}^{ex}(\boldsymbol{q},\omega) \equiv \varepsilon(\boldsymbol{q},\omega)\tilde{\Phi}(\boldsymbol{q},\omega) \ . \tag{6.74}$$

Combining (6.72) and (6.73) we find for $\varepsilon(\boldsymbol{q},\omega)$ Lindhard's famous result (Lindhard, 1954)

$$\varepsilon(\boldsymbol{q},\omega) = 1 - \frac{e^2}{\varepsilon_0 q^2}(2\pi)^{-3} \int \frac{f_e(\boldsymbol{k}) - f_e(\boldsymbol{k}-\boldsymbol{q})}{E(\boldsymbol{k}) - E(\boldsymbol{k}-\boldsymbol{q}) - \hbar\tilde{\omega}} d^3\boldsymbol{k} \ . \tag{6.75}$$

The properties of $\varepsilon(\boldsymbol{q},\omega)$ are discussed in many textbooks (Ziman, 1964; Kittel, 1963). The important features come out most clearly if one considers the extreme case where $f(\boldsymbol{k})$ is a Fermi distribution at $T = 0$. Then the integral in (6.75) can be done analytically with the following result (Landau-Lifschitz, 1981)

$$\varepsilon(\boldsymbol{q},\omega) = 1 + \frac{3\omega_p}{4q^2 v_F^2}\left[1 - g\left(\omega + \frac{\hbar q^2}{2m_e}\right) + g\left(\omega - \frac{\hbar q^2}{2m_e}\right)\right]$$

with

$$g(x) = \frac{m_e[x^2 - (qv_F)^2]}{2\hbar v_F q^3} \ln\left|\frac{x + qv_F}{x - qv_F}\right| \ . \tag{6.76}$$

$v_F = \hbar k_F/m_e$ is the Fermi-velocity, $\omega_p = (ne^2/m_e\varepsilon_0)^{1/2}$ is the plasma frequency. If one uses the convention

$$\ln u = \ln|u| - i\pi\Theta(-u) \tag{6.77}$$

one can see that $\mathrm{Im}\{\varepsilon\} \neq 0$ is possible. Two important features are extracted from (6.76).

1. The curve in the $\boldsymbol{q}$-$\omega$-plane where wavelike excitations are possible; these excitations which are known as plasmons follow a dispersion curve defined through

$$\varepsilon(\boldsymbol{q},\omega) = 0 \ . \tag{6.78}$$

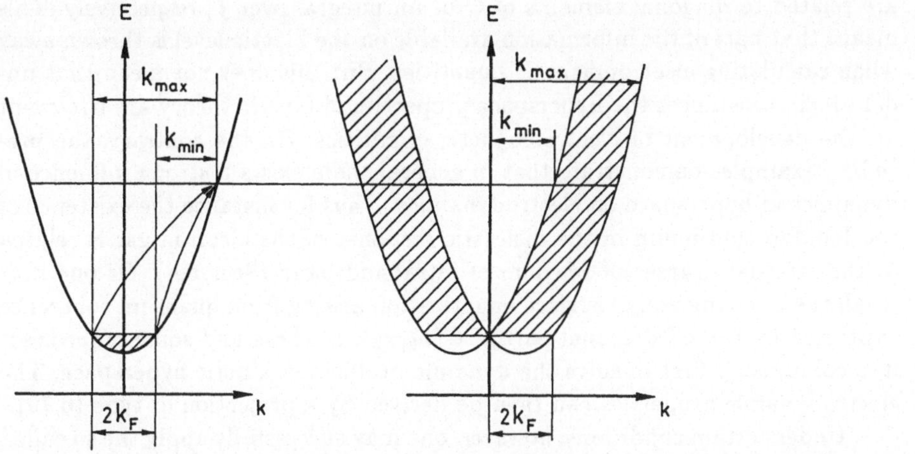


**Fig. 6.2.** Kinematics of intraband pair excitations; the *shaded region* is the Landau continuum of asymptotically free real particles. The plasmon mode (Fig. 6.3) is outside the Landau region and therefore consists of virtual pair excitations

2. the region of coherent Landau-damping: It is characterized by $\mathrm{Im}\{\varepsilon\} \neq 0$. The contour of the Landau continuum is due to a kinematic restriction. This restriction means that the $(\boldsymbol{q}, \omega)$-value of the excitation is produced by transfer of a particle from an occupied state inside the Fermi-sphere to an empty state outside this sphere (Fig. 6.2). Since these states are real particle states the particle and the hole in the Fermi sphere can fly apart.

On the plasmon dispersion curve these kinematic requirements for the production of real pairs are not fulfilled. Hence the plasmon must be interpreted as a superposition of virtual pairs.

## 6.5 Hydrodynamic Approximation

A description of intraband processes based on the density matrix $C(\boldsymbol{r}_1, \boldsymbol{r}_2)$ or the corresponding distribution function $f_e(\boldsymbol{r}, \boldsymbol{k})$ is called a kinetic theory. Typical property of a kinetic theory is its use of a six dimensional hyperspace. Electromagnetic quantities such as the charge density

$$\varrho(\boldsymbol{r}) = -eC(\boldsymbol{r}, \boldsymbol{r}) = -e(2\pi)^{-3} \int f_e(\boldsymbol{r}, \boldsymbol{k}) d^3\boldsymbol{k} \tag{6.79a}$$

or the conduction current density

$$\boldsymbol{J}(\boldsymbol{r}) = -\frac{ie\hbar}{2m_e}(\nabla_1 - \nabla_2)C|_{\boldsymbol{r}_1=\boldsymbol{r}_2=\boldsymbol{r}} = -e(2\pi)^{-3} \int \frac{\hbar\boldsymbol{k}}{m_e} f_e(\boldsymbol{r}, \boldsymbol{k}) d^3\boldsymbol{k} \tag{6.79b}$$

are related to diagonal elements of $C$ or an integral over $f$, respectively. This means that part of the information available on the kinetic level is thrown away when calculating electromagnetic quantities. But this does not mean that under all circumstances the hyperspace properties of kinetic theory are irrelevant for the development of electromagnetic quantities. On the contrary, the preceding examples demonstrate that in general there exists a strong influence of dynamics in hyperspace on electrodynamics. Thus for instance the existence of the Landau continuum in the dielectric response of the electron gas is related to the internal degrees of freedom of intraband pairs (Sect. 6.4). Or one may recall the phenomenon of surface quantization arising from quantum coherence expressed by the off diagonal part of $C(\boldsymbol{r}_1, \boldsymbol{r}_2)$. In these and some other cases it is compulsory first to solve the dynamic problem in kinetic hyperspace. The electrodynamic properties can then be derived by a projection of type (6.79).

Under certain conditions, however, one may successfully apply the so called hydrodynamic approximation. In this approximation the projection onto electrodynamic subspace (conventionally called hydrodynamic space) is made before the dynamic problem is solved. There exists a certain analogy between the hydrodynamic treatment of intraband processes and the Hopfield-Thomas (H.T.) wave equation for polarization waves (4.95). The related feature of the H.T.-theory is that the bilocal wave equation (4.83) for the exciton amplitude $Y(\boldsymbol{r}_e, \boldsymbol{r}_h)$ is replaced by the real space wave equation for the electromagnetic quantity $\boldsymbol{P}(\boldsymbol{r}, t)$. Because of this simplification the H.T.-theory is restricted to a single resonance. In particular it does not describe the continuum of *e-h* pair production. An analogous restriction will occur in hydrodynamic plasma theory: it does not account for the continuum of Landau-damping.

From the intraband constitutive equation (6.64)

$$\dot{C}_{12} + i\left[\frac{\hbar}{2m_e}(\nabla_1^2 - \nabla_2^2) + \frac{e}{\hbar}(\Phi_1 - \Phi_2)\right]C_{12} = 0 \tag{6.80}$$

one derives balance equations for $\varrho$ and $\boldsymbol{J}$ as defined in (6.79). Assuming in (6.80) $\boldsymbol{r}_1 = \boldsymbol{r}_2 = \boldsymbol{r}$ one finds the continuity equation

$$\dot{\varrho} + \nabla\cdot\boldsymbol{J} = 0 \ . \tag{6.81}$$

The equation for the current becomes

$$\frac{\partial \boldsymbol{J}}{\partial t} + \frac{e\hbar^2}{4m_e}(\nabla_1 + \nabla_2)(\nabla_1 - \nabla_2) \otimes (\nabla_1 - \nabla_2)C\big|_{\boldsymbol{r}_1=\boldsymbol{r}_2=\boldsymbol{r}} = \frac{e}{m_e}\varrho\nabla\Phi \ . \tag{6.82}$$

We write (6.82) as follows

$$\frac{\partial \boldsymbol{J}}{\partial t} + \nabla\cdot\underset{\sim}{\Sigma} = -\frac{e}{m_e}\varrho\boldsymbol{E} \tag{6.83}$$

with

$$\boldsymbol{E} = -\nabla\Phi \ . \tag{6.83a}$$

The tensor $\underset{\sim}{\Sigma}$ has the elements

$$\Sigma_{ik} = \frac{e\hbar^2}{4m_e^2}\left(\frac{\partial}{\partial r_{1i}} - \frac{\partial}{\partial r_{2i}}\right)\left(\frac{\partial}{\partial r_{1k}} - \frac{\partial}{\partial r_{2k}}\right)C(\boldsymbol{r}_1, \boldsymbol{r}_2)\Big|_{\boldsymbol{r}_2=\boldsymbol{r}_2=\boldsymbol{r}} \quad . \tag{6.83b}$$

Equation (6.83) is still exact if the term div $\underset{\sim}{\Sigma}$ is calculated from the full density matrix. In the hydrodynamic approximation one tries to express div $\underset{\sim}{\Sigma}$ in a simple phenomenological ansatz by "hydrodynamic" quantities such as $\varrho$ and $\boldsymbol{J}$. Most widely used is the form (Forstmann and Gerhards, 1982)

$$\nabla\cdot\underset{\sim}{\Sigma} = \beta\nabla\varrho \ . \tag{6.84}$$

The ansatz (6.84) can be motivated from a solution of the Boltzmann equation with a result for the coefficient $\beta$ in the degenerate case

$$\beta = \tfrac{3}{5}v_F^2 \ . \tag{6.85}$$

We note that (6.85) is an approximation relevant for $\omega \gg kv_F$. In the low frequency limit $\omega\to 0$, (6.85) has to be replaced by (Becker and Sauter, 1968)

$$\beta_0 = \tfrac{1}{3}v_F^2 \ . \tag{6.85a}$$

When (6.84) is inserted into (6.83) one obtains a material equation for the intraband current in hydrodynamic approximation

$$\frac{\partial \boldsymbol{J}}{\partial t} + \beta\nabla\varrho = -\frac{e}{m}\varrho\boldsymbol{E} \ . \tag{6.86}$$

A solution describing plasma waves is found from (6.86) after the linearization

$$\varrho\boldsymbol{E} \approx -en_0\boldsymbol{E} \tag{6.87}$$

with $n_0$ being the equilibrium density of electrons. The field $\boldsymbol{E}$ has to be derived self-consistently from $\varrho$

$$\varepsilon_0\nabla\cdot\boldsymbol{E} = \varrho \ . \tag{6.88}$$

In this way one is led to the following wave equation for $\varrho$ [after applying div to (6.86)]:

$$\ddot{\varrho} + \omega_p^2\varrho - \beta\nabla^2\varrho = 0 \ . \tag{6.89}$$

A solution to (6.89) are plasma waves

$$\varrho = \varrho_0 e^{i\boldsymbol{q}\boldsymbol{r}-i\omega t} \tag{6.90}$$

with a dispersion relation

$$\omega^2 = \omega_p^2\beta q^2 \ . \tag{6.91}$$

The hydrodynamic material equation (6.86) was derived from the reversible part of the intraband equation of motion (6.80). It therefore yields an undamped plasmon-dispersion. But if necessary, damping can be incorporated by a Drude-like phenomenological relaxation term. Then the material equation (6.86) is generalized to

$$\frac{\partial}{\partial t}\boldsymbol{J} + \beta\nabla\varrho + \frac{1}{\tau}\boldsymbol{J} = -\frac{e}{m_e}\varrho\boldsymbol{E} \ . \tag{6.92}$$

Equation (6.92) also covers the field of low frequency conduction phenomena where it coincides with the Drude-Lorentz Model (Drude, 1900; Lorentz, 1909).

## 6.6 Plasmon LO-Phonon Modes

As an application of hydrodynamic plasma theory we discuss the coupled modes which are formed when a plasma and an optically active phonon interact via the common electric field. These modes can for example be analyzed with the help of Raman scattering (Mooradian, 1969; Nowak, Richter and Sachs, 1981).

The plasma we describe by (6.92) applying the linearisation (6.87). The LO-phonon is assumed to be a nondispersive oscillator with displacement field $\boldsymbol{w}$. The longitudinal field obeys Maxwell's div-equation. We therefore have the following set of model equations

$$\frac{\partial}{\partial t}\boldsymbol{J} + \beta\nabla\varrho + \frac{1}{\tau}\boldsymbol{J} = \frac{en_0}{m}\boldsymbol{E} \tag{6.93a}$$

$$\ddot{\boldsymbol{w}} + \omega_0^2\boldsymbol{w} = \frac{\tilde{q}}{\mu\Omega^{1/2}}\boldsymbol{E} \tag{6.93b}$$

$$\varepsilon_0\nabla\cdot\boldsymbol{E} = \varrho \tag{6.93c}$$

$\mu$ is the ionic reduced mass; $\tilde{q}$ the effective charge of the phonon oscillator; $\Omega$ is the volume of a lattice cell [c.f. (3.39)]. Together with the continuity equation

$$\nabla\cdot\boldsymbol{J} = -\varrho_e$$

one can derive from (6.93a) an equation for the electronic charge density $\varrho_e$

$$\beta\nabla^2\varrho_e - \ddot{\varrho}_e - \frac{1}{\tau}\dot{\varrho}_e = \omega_p^2\varrho \ . \tag{6.94}$$

Similarly from (6.93b) there follows an equation for the polarization charge of the phonon

$$\varrho_p = -\frac{\tilde{q}}{\Omega^{1/2}}\nabla\cdot\boldsymbol{w} \ . \tag{6.95}$$

$\varrho_p$ obeys the equation

$$\ddot{\varrho}_p + \omega_0^2 \varrho_p = \frac{\tilde{q}^2}{\Omega \mu \varepsilon_0} \varrho \; . \tag{6.96}$$

Using the relation

$$\varrho = \varrho_p + \varrho_e$$

and the abbreviations

$$\omega_L^2 = \omega_0^2 + \omega_c^2 \qquad \omega_c^2 = \frac{\tilde{q}^2}{\Omega \mu \varepsilon_0} \tag{6.97}$$

the coupled mode equations (6.94) and (6.96) take on the form

$$\ddot{\varrho}_e + \frac{1}{\tau} \dot{\varrho}_e - \beta \nabla^2 \varrho_e + \omega_p^2 (\varrho_e + \varrho_p) = 0 \tag{6.98a}$$

$$\ddot{\varrho}_p + \omega_L^2 \varrho_p + \omega_c^2 \varrho_e = 0 \; . \tag{6.98b}$$

$\omega_L$ is the longitudinal phonon frequency; the coupling $\omega_c$ determines the LT-splitting $\omega_L - \omega_0$. The coupled plasmon-phonon modes are solutions of (6.98) with a space-time structure

$$\varrho_e \propto \varrho_p \propto e^{i\boldsymbol{q}\boldsymbol{r} - i\omega t} \; . \tag{6.99}$$

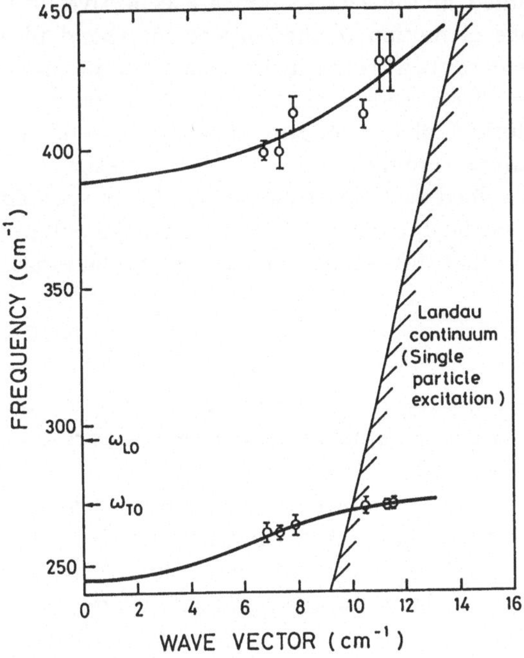


**Fig. 6.3.** Plasmon-phonon modes in GaAs; the bars around experimental points denote the spectral width; measurements were made by Nowak, Richter, Sachs (1981)

The solubility condition yields the dispersion equation

$$(\omega_p^2 + \beta q^2 - i\frac{\omega}{\tau} - \omega^2)(\omega_L^2 - \omega^2) - \omega_p^2\omega_c^2 = 0 \ . \tag{6.100}$$

Depending on the experimental situation (6.100) is to be solved either for $\omega$ real with $q$ becoming complex or with $q$ real and $\omega$ complex. The former choice corresponds to a situation as in IR reflection spectroscopy where the real frequency is imposed by the incident wave. The system response is then a transmitted wave that is damped in its direction of propagation. The second case where $q$ is real and $\omega$ complex corresponds to a Raman experiment because in this experiment $q$ is the real momentum transfer between incident and scattered wave which is selected by the scattering angle. The complex $\omega$ is then observed as the broadened scattering line.

In the limit of weak damping the solutions of (6.100) can be drawn in a real $q$-$\omega$ plane. As an example in Fig. 6.3 are shown theoretical and experimental results for a doped GaAs sample.

If one goes beyond the hydrodynamic approximation one can study the influence of coherent Landau damping on the plasmon-phonon modes (Lemmens and Devreese, 1974).

## 6.7 The Polaron Model

The plasmon-phonon modes studied in the last section are collective excitations of the linearized equations of motion of the coupled intraband plus phonon system. The modes were described in the hydrodynamic approximation which means that coherence effects were neglected. We shall now focus on a rather different type of solutions of the coupled intraband-LO-phonon dynamics known as Fröhlich polarons (Fröhlich et al., 1950). In quasiparticle language the polaron model is described as an electron surrounded by a cloud of virtual phonons. In the context of the dynamical band-edge model a polaron solution is generated if one assumes that the density matrix $C(\boldsymbol{r}_1, \boldsymbol{r}_2)$ factorizes

$$C(\boldsymbol{r}_1, \boldsymbol{r}_2) = \psi^*(\boldsymbol{r}_1)\psi(\boldsymbol{r}_2) \ . \tag{6.101}$$

This means that $C$ represents a pure state in the sense of quantum ensemble theory. The polaron solution as a case of maximum coherence is in this respect complementary to the solution in Sect. 6.6.

The interest in polaron solutions to the coupled electron-phonon dynamics lies in the strong interaction between electrons and LO-phonons. To set up the polaron model we have to diagonalize the strongest interaction first. It is then assumed that the polaron interacts with the rest of the world only via relatively weak forces.

The starting point for the construction of a polaron model are (6.80) for the intraband density matrix $C$ and an equation of type (3.39) for the displacement field $\boldsymbol{w}$ of a phonon oscillator

$$\dot{C}_{12} + i\left[\frac{\hbar}{2m_e}(\nabla_1^2 - \nabla_2^2) + \frac{e}{\hbar}(\Phi_1 - \Phi_2)\right]C_{12} = 0 \tag{6.102a}$$

$$\ddot{\boldsymbol{w}} + \omega_L^2 \boldsymbol{w} = \frac{\tilde{q}}{\mu\Omega^{1/2}}\boldsymbol{E} \; . \tag{6.102b}$$

$\boldsymbol{E}$ is the field produced by the electronic charge and $\Phi$ the potential field due to the phonon polarization. Note that the use of an eigenfrequency $\omega_L$ amounts to assuming that the self-interaction of the LO-mode via its own field is already included [compare also (6.97)].

Assuming the factorization (6.101) the electronic equation (6.102a) yields a pair of equations for $\psi$ and $\psi^*$

$$i\hbar\dot{\psi}^* - \frac{\hbar^2}{2m_e}\nabla^2\psi^* - e\Phi\psi^* = \eta\psi^* \tag{6.103a}$$

$$-i\hbar\dot{\psi} - \frac{\hbar^2}{2m_e}\nabla^2\psi - e\Phi\psi = \eta\psi \tag{6.103b}$$

$\eta$ being a separation constant. We consider (6.103) together with (6.102b) as the equations defining the polaron model if $\Phi$ and $\boldsymbol{E}$ are assumed as follows

$$\Phi(\boldsymbol{r}) = \frac{1}{4\pi\varepsilon_0\Omega^{1/2}}\int \frac{1}{|\boldsymbol{r}-\boldsymbol{r}'|}\nabla'(-\tilde{q}\cdot\boldsymbol{w}(\boldsymbol{r}'))d^3\boldsymbol{r}' \; , \tag{6.104}$$

$$\boldsymbol{E} = -\nabla\frac{1}{4\pi\varepsilon_0}\int \frac{-e|\psi(\boldsymbol{r}')|^2}{|\boldsymbol{r}-\boldsymbol{r}'|}d^3\boldsymbol{r}' \; . \tag{6.105}$$

An attractive way to characterize the polaron model is by an action principle for the Lagrangian $L(\psi, \psi^*, \boldsymbol{w})$ :

$$L = L_e + L_{ph} + L_{int} \tag{6.106}$$

$$L_e = \int \left[\frac{\hbar}{2i}(\psi\dot{\psi}^* - \psi^*\dot{\psi}) - \frac{\hbar^2}{2m_e}|\nabla\psi|^2\right]d^3\boldsymbol{r} \tag{6.106a}$$

$$L_{ph} = \int \frac{\mu}{2}(\dot{\boldsymbol{w}}^2 - \omega_L^2\boldsymbol{w}^2)d^3\boldsymbol{r} \tag{6.106b}$$

$$L_{int} = -\frac{e\tilde{q}}{4\pi\varepsilon_0\Omega^{1/2}}\iint \frac{|\psi(\boldsymbol{r}')|^2(\nabla\boldsymbol{w})}{|\boldsymbol{r}-\boldsymbol{r}'|}d^3\boldsymbol{r}'d^3\boldsymbol{r} \; . \tag{6.106c}$$

In formulating the action principle a constraint has to be imposed, viz, that $\psi$ is normalized

$$\int |\psi(\boldsymbol{r}')|^2 d^3\boldsymbol{r}' = 1 \ . \tag{6.107}$$

This can be done by the method of Lagrange multiplyers. The action principle under a constraint becomes

$$\delta \int (L + \eta \int |\psi|^2 d^3\boldsymbol{r}')dt = 0 \ . \tag{6.108}$$

By the Euler-Lagrange procedure one then is led back to the equations (6.103) and (6.102b).

In the literature it is customary to start the polaron theory with Fröhlich's Hamiltonian. Given the Lagrangian (6.106) this Hamiltonian is found by the usual procedure: One first introduces canonical "momenta" as functional derivatives of $L$ :

$$p_\psi \equiv \frac{\delta L}{\delta \dot{\psi}} = -\frac{\hbar}{2i}\psi^* \tag{6.109a}$$

$$p_{\psi^*} = \frac{\delta L}{\delta \dot{\psi}^*} = \frac{\hbar}{2i}\psi \tag{6.109b}$$

$$p_w = \frac{\delta L}{\delta \dot{\boldsymbol{w}}} = \mu \dot{\boldsymbol{w}} \ . \tag{6.109c}$$

The Hamiltonian then is defined in the usual way as

$$H = \int (p_\psi \dot{\psi} + p_{\psi^*}\dot{\psi}^* + p_w \dot{\boldsymbol{w}})d^3r - L \tag{6.110}$$

with the result

$$H = H_e + H_{ph} + H_{int} \ ,$$

$$H_e = \frac{\hbar^2}{2m_e}\int |\psi|^2 d^3\boldsymbol{r} \ , \tag{6.111a}$$

$$H_{ph} = \int \left(\frac{1}{2\mu}p_w^2 + \frac{\mu}{2}\omega_L^2 \boldsymbol{w}^2\right) d^3\boldsymbol{r} \ , \tag{6.111b}$$

$$H_{int} = \frac{e\tilde{q}}{4\pi\varepsilon_0 \Omega^{1/2}} \iint \frac{|\psi(\boldsymbol{r}')|^2 \nabla \boldsymbol{w}(\boldsymbol{r})}{|\boldsymbol{r}-\boldsymbol{r}'|} d^3\boldsymbol{r}\, d^3\boldsymbol{r}' \ . \tag{6.111c}$$

The polaron is defined as an eigenstate of the Hamiltonian (6.111). Owing to the nonlinear interaction term the eigenvalue problem can only be solved in an approximate sense. A number of different approximation methods have been developed. For a review the reader may consult the book by Devreese (1972) and herein especially the article by Evrard (1972). On a textbook level the polaron problem has been treated, among others, by Haken (1973).

An important result of the polaron theory is a renormalization of the electronic energy due to the phonon cloud. By the method described by Haken (1973), originally developed by Lee, Low and Pines (1953, 1955), one finds for the shifted band edge

$$E_g' = E_g - \hbar\omega_L\alpha \tag{6.112}$$

and a renormalized effective mass

$$m_{pol} = m_e\left(1 + \frac{\alpha}{6}\right) \tag{6.113}$$

where $\alpha$ is the so called polaron coupling constant. In terms of the parameters in the Hamiltonian (6.111) $\alpha$ becomes

$$\alpha = \frac{e^2}{4\pi\varepsilon_0}\frac{\tilde{q}}{2\varepsilon_0}\frac{1}{\hbar\mu\omega_L\Omega}\left(\frac{2m_e}{\hbar\omega_L}\right)^{1/2} . \tag{6.114}$$

## 6.8 Plasma Dynamics in a Bounded Semiconductor

A typical situation in which plasma oscillations in a semiconductor are excited, is by electron scattering at the surface (Lüth, 1981). In the analysis of such experiments it is important to know the influence of the surface on the dynamics of the plasma. This is a nontrivial problem since the response of the semiconductor plasma is nonlocal in two respects. Firstly, one has spatial dispersion on the hydrodynamic scale as can, for instance, be seen from (6.91). In addition a nonnegligible influence of bilocal coherence can occur because the typical coherence length in a semiconductor plasma is large compared to the atomic scale. The bounded plasma therefore poses a problem which in many respects is similar to that encountered in the theory of exciton reflection at a boundary (compare Chap. 5). Again one must treat the boundary value problem in a higher space and then calculate the electromagnetic quantities by means of a projection. The result will be (as in the excitonic case) the formation of a transition layer with a reduced electromagnetic response, roughly approximated by a "dead layer". The depth of the transition layer is determined by the coherence length. It becomes noticeable if the coherence length exceeds the relevant length for electromagnetic screening. In the nondegenerate case the latter length is the Debye length. Ritz and Lüth (1984) have demonstrated the existence of the transition layer phenomenon in electron energy loss spectra of InSb (Fig. 6.4).

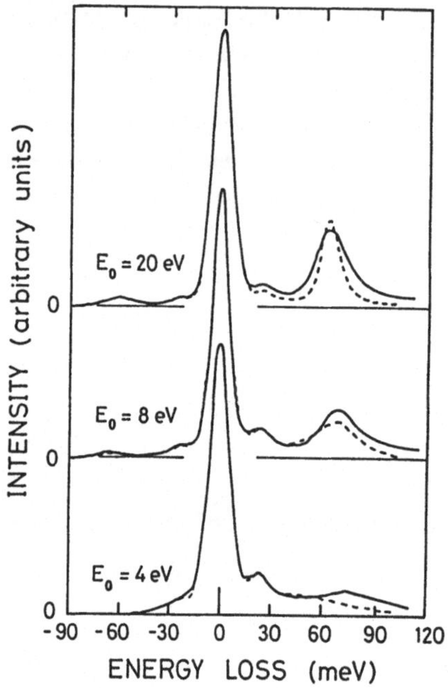


**Fig. 6.4.** Electron energy loss spectrum of InSb; the two peaks at $\Delta E \approx 25\,\mathrm{meV}$ and $\Delta E \approx 60\,\mathrm{meV}$ are assigned to phonon and plasmon excitation, respectively; depending on the primary energy the relative strength of the plasmon and the phonon peak varies drastically; this phenomenon is attributed by Ritz and Lüth (1984) to surface quenching of the plasmon

It should be stressed that the transition layer phenomenon is typical for semiconductors with their relatively large coherence length. In metals where the coherence length is typically of the order of the lattice constant the situation is different. For metals an acceptable treatment of the boundary value problem is then either within the hydrodynamic approximation (Das Sarma and Quinn, 1979; Forstmann and Gerhards, 1982; Schaich, 1982) or in a kinetic theory using the Boltzmann distribution function (Wagner, 1966; Harris, 1972). If these approximations are not reliable enough one must use a fully microscopic theory (Feibelman, 1980).

Our treatment in this section will be in the spirit of Lindhard's theory (Sect. 6.4). This means that we consider the half-space boundary value problem of the linearized intraband equation (6.66)

$$\dot{C}_{12} + \frac{i\hbar}{2m_e}(\nabla_1^2 - \nabla_2^2)C_{12} + \frac{ie}{\hbar}(\Phi_1 - \Phi_2)C^0 = 0 \; . \qquad (6.115)$$

Let the semiconductor fill the half-space $z>0$. Then we assume as boundary condition

$$C(\boldsymbol{r}_1, \boldsymbol{r}_2) = 0 \quad \text{for } \boldsymbol{r}_1 \text{ or } \boldsymbol{r}_2 \text{ on the surface.} \qquad (6.116)$$

Of course, the static distribution $C^0$ must not be identified with the bulk equilibrium distribution (6.46) because $C^0$ must also satisfy the boundary condition (6.116). In Sect. 6.2 we have already formulated the static boundary value problem. Unfortunately there is no rigorous analytical solution taking self-consistent fields into account. We shall therefore content ourselves with the approximation (6.54). This approximation is determined by the following conditions:

i) differential equation: $(\nabla_1^2 - \nabla_2^2)C_{12}^0 = 0$,
ii) boundary condition: (6.116),
iii) asymptotic condition: $\lim_{z_1,z_2\to\infty} C^0 = C^0_{bulk}$.

The use of (6.54) can be considered as a first step of an iteration. The second step would be to calculate a self-consistent potential from (6.54) and to determine an improved $C^0$ as a distribution in this potential. But this possibility will not be pursued here.

We are interested in a solution to (6.115) that has plane wave structure parallel to the surface. $C^0$ we insert from (6.54); the induced potential is determined self-consistently by

$$\varepsilon_0 \nabla^2 \Phi^{in}(\boldsymbol{r}) = eC(\boldsymbol{r},\boldsymbol{r}) \ . \tag{6.117}$$

Because translational invariance in the $z$ direction is broken by the presence of the surface a complete solution by Fourier transformation as in Lindhard's bulk case is no longer possible. But at least parallel to the surface the Fourier method still works. We try therefore the following ansatz:

$$C(\boldsymbol{r}_1,\boldsymbol{r}_2) = (2\pi)^{-2}\int C_q(\boldsymbol{k}_{||};z_1,z_2) \times\exp\left[i\boldsymbol{q}_{||}\cdot\boldsymbol{r}_1 + i\boldsymbol{k}_{||}\cdot(\boldsymbol{r}_2-\boldsymbol{r}_1) - i\omega t\right]d^3\boldsymbol{k} \tag{6.118}$$

$$\Phi(\boldsymbol{r},t) = \Phi_q(z)\exp\left(i\boldsymbol{q}_{||}\cdot\boldsymbol{r} - i\omega t\right) \tag{6.119}$$

$$\varrho(\boldsymbol{r},t) = \varrho_q(z)\exp\left(i\boldsymbol{q}_{||}\cdot\boldsymbol{r} - i\omega t\right) \ . \tag{6.120}$$

From $\varrho(\boldsymbol{r}) = -eC(\boldsymbol{r},\boldsymbol{r})$ we get

$$\varrho_q(z) = -e(2\pi)^{-2}\int C_q(\boldsymbol{k}_{||};z,z)d^2\boldsymbol{k}_{||} \ . \tag{6.121}$$

Note the analogy to Lindhard's bulk solution: (6.118) corresponds to (6.68), (6.121) to (6.70).

With the help of (6.118) and (6.119), (6.115) can be transformed into the integro-differential equation

$$\left(\frac{\partial^2}{\partial z_1^2} - \frac{\partial^2}{\partial z_2^2} - \kappa^2\right)C_q = \frac{4em_e}{\pi\hbar^2}\int_0^\infty \sin(k_\perp z_1)\sin(k_\perp z_2) \times\left[\Phi_q(z_1)f(\boldsymbol{k}) - \Phi_q(z_2)f(\boldsymbol{k}-\boldsymbol{q}_{||})\right]dk_\perp \tag{6.122}$$

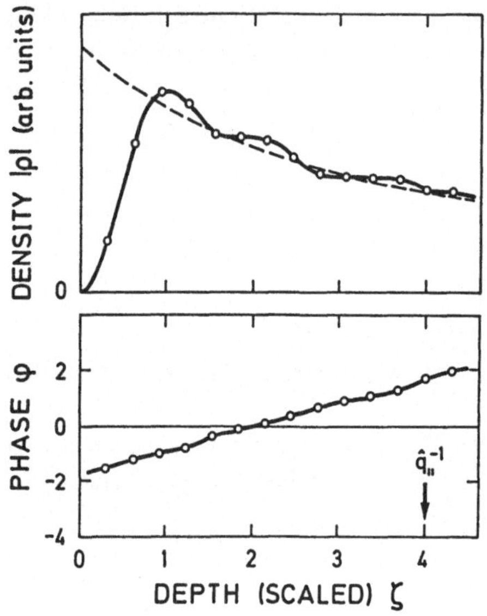


**Fig. 6.5.** Calculated profile of the dynamical density response near a semiconductor surface; the coordinate $\varsigma$ is scaled by the thermal length $\lambda_{th} = \hbar/(m_e kT)^{1/2}$

with

$$\kappa^2 = \frac{2m_e\omega}{\hbar} - k_{\parallel}^2 + (\boldsymbol{k}_{\parallel} - \boldsymbol{q}_{\parallel})^2 \ .$$

Since, in semiconductors, one is usually in the nondegenerate case, one can insert a Boltzmann-distribution for $f(\boldsymbol{k})$.

Equation (6.122) must be solved with boundary conditions (6.116). This has been done numerically by Stahl (1983) and Lederhofer (1985). From the results we show in Fig. 6.5 the density response $\varrho_q(z)$ as a function of the distance $z$ from the surface. It can be seen that in a thin layer under the surface the oscillating part of the density goes to zero. (In fact also the static part $C^0(\boldsymbol{r}, \boldsymbol{r})$ has a similar property.) From the calculations it follows that the dominating length determining the depth of the transition layer in the plasma is the thermal length (6.52). This result once more underlines the analogy to the excitonic boundary value problem (Chap. 5) where it also was found that the transition layer depends on the relevant coherence length.

# 7. Parametric and Nonlinear Processes

## 7.1 Introductory Remarks

Despite the success of linearized solutions presented in preceeding chapters, the band-edge dynamics developed in Chaps. 2, 3 is a nonlinear theory. In the present chapter we shall focus on nonlinear behaviour of the band edge. The constitutive equations (3.1,2) of the band edge contain a number of terms listed below, which can cause nonlinear effects.

i) Nonlinear source terms $\sim EC$ and $\sim ED$ in the interband equation, or $\sim EY$ and $\sim EY^*$ in the intraband equations (3.1); these terms are responsible for saturation effects and for stimulated emission.

ii) Nonlinearities caused by self consistent potentials via the terms $\sim \Phi Y$ or $\sim \Phi C$ in the edge equations. The self-interaction leading to surface quantization is related to this nonlinearity (Sect. 6.2).

iii) The electron-phonon interaction is represented by the nonlinear term $\sim \xi Y$ in the interband equation and by a nonlinear source $\sim |Y|^2$ in the phonon equation (3.39). A consequence of this nonlinearity are Raman-type optical processes.

iv) The nonlinear coupling between phonons and intraband densities is the origin of polaron effects (Sect. 6.7).

v) Nonlinearities can also occur in the terms describing the irreversible change of $Y$, $C$, $D$. An example is furnished by the electron-electron collision operator (3.76). Among the simplest consequences of this nonlinearity is the determination of the static intraband solution. About its impact on dynamics, very little is known.

The potentially nonlinear terms in the constitutive equations are also used to describe the class of parametric processes. In these processes one of the factors in a nonlinear term acts as a given external perturbation. One example is that of spontaneous scattering by thermally activated phonons via the term $\sim \xi Y$ in the interband equation. This process will be treated in Sect. 7.2.

A wide class of nonlinear optical effects is described in terms of nonlinear susceptibilities. The underlying perturbation expansion in powers of the electric field is particularly well suited for the treatment of steady state phenomena. Examples will be presented in Sects. 7.3, 4.

A fascinating and relatively new field is that of optical transient effects. In this field the above mentioned perturbation expansion is of little use and must be replaced by new methods aiming at a more direct solution of the equations of motion. In atomic physics, where the relevant equations are the optical Bloch equations (2.16), a successful method is the "slowly-varying envelope" approximation. This approximation exploits a separation of time scales between the fast oscillation with a frequency $\omega \approx \omega_g$ and a slow envelope motion dominated by the Rabi frequency $\omega_F$ (compare Sect. 7.5). The third scale present in the equations of motion is the scale of irreversible change. In the simplest case this scale is represented by relaxation terms with time constants $T_1$ and $T_2$, respectively. In atomic systems relaxation processes are often much slower than the above mentioned Rabi time scale. Under these conditions some very spectacular optical effects are observable as e.g. optical nutation, photon echos or self induced transparency (SIT). For the phenomenology of the effects and a review of the pertinent theory we refer to the literature (Abella, 1969; Lamb 1971; Poluektov et al., 1975; Brewer, 1977).

While the dynamics of a two level transition in isolated atoms is fairly well understood the same is not true for the transient behaviour of the semiconductor band edge under high excitation conditions. Compared to the atomic system the semiconductor offers several additional complications. All these complications are related more or less to mobility in the initial and final states of the transition. The problems caused by mobility in optical transients fall roughly into two groups: Effects related to coherence in higher space, and effects related to fast irreversibility due to electron-electron scattering. The prospects of treating the coherence effects by a suitable extension of the slow envelope solution known from atomic physics are reasonable as will be shown in the SIT example discussed in Sect. 7.5. But unfortunately all the nice coherence effects are, in practice, usually blurred by fast irreversibility. Qualitatively this is immediately seen by looking at the edge equations (3.1). According to these equations the nonlinear coupling between interband and intraband processes depends on intraband coherence represented by the off-diagonal elements of $C$ and $D$. But this intraband coherence will be destroyed by electron-electron scattering events on a time scale of 1 ps or even less. We therefore may expect that optical transients in semiconductors can only be understood by considering in detail the coupling between coherent and incoherent dynamics. Remarkable recent experiments on the dynamics of excitonic screening (Ulbrich, 1985; Knox et al., 1985) form a major challenge to invest some labor into that problem.

While so little is known about the short-time behaviour of a highly excited semiconductor, the long-time behaviour dominated by a quasi-equilibrium state of the electron-hole plasma is much better understood. A brief discussion of this relaxed high excitation state will be given in Sect. 7.7; compare also the review by Klingshirn and Haug (1981).

The vector character of the electric field $\boldsymbol{E}$, the dipole density $\boldsymbol{M}$, and related quantities will be suppressed in the notation of Sects. 7.2–6.

## 7.2 Spontaneous Raman Scattering

When a light wave traversing a crystal is deflected and shifted in frequency by the influence of an internal excitation of the crystal this process is called a Raman-type process. Excitations causing Raman processes may be phonons, plasmons, spin-waves etc. We shall restrict our discussion to Raman processes induced by phonons or plasmons.

Let us consider propagation of light through a semiconductor. If the frequency is near the gap frequency and the intensity not too high, the relevant equations are the linearized polariton equations discussed extensively in Chap. 4. These equations may be used to study phonon-induced Raman processes if the phonon dependent drift operator of Sect. 3.3 is introduced. To be specific, let us consider a deformation potential coupling acting on the conduction band edge. Then the relevant operator $\Omega_{eh}(\xi)$ is (3.44a) with $d_c \neq 0$ and $d_v = 0$. Using a simplified model where all influences of crystal anisotropy are suppressed we may write the basic equations as follows:

$$i\dot{Y} - \left(\omega_g - \frac{\hbar}{2m}\nabla_R^2 - \frac{\hbar}{2\mu}\nabla_r^2 + \frac{V_{eh}}{\hbar}\right)Y + \frac{M_0}{\hbar}\delta(\boldsymbol{r})E(\boldsymbol{R})$$
$$= \frac{d_c}{\hbar}\xi(\boldsymbol{r}_e)Y \tag{7.1a}$$

$$\tilde{\varepsilon}\ddot{E} - c^2\nabla_R^2 E + \frac{2M_0}{\varepsilon_0}\mathrm{Re}\{\ddot{Y}(\boldsymbol{R},0)\} = 0 \ . \tag{7.1b}$$

Note that the coordinates refer to the center of mass basis of the electron-hole pair; $\boldsymbol{r}_e$ is accordingly to be understood as

$$\boldsymbol{r}_e = \boldsymbol{R} + \frac{m_h}{m}\boldsymbol{r} \ . \tag{7.2}$$

In principle $\xi(\boldsymbol{r}_e)$ consists of two terms

$$\xi(\boldsymbol{r}_e) = \xi_0(\boldsymbol{r}_e) + \xi_{in}(\boldsymbol{r}_e) \tag{7.3}$$

where $\xi_0$ represents the "spontaneous" phonon amplitude having its origin in thermal fluctuations and in zero-point quantum noise; $\xi_{in}$ is the induced phonon amplitude. The necessity to introduce an induced phonon field has been explained in Sect. 3.3. In the special case of a deformation potential coupling as in (7.1) the equation for $\xi_{in}$ is according to (3.45)

$$\mu(\ddot{\xi}_{in}(\boldsymbol{r}) + \omega_0^2\xi_{in}(\boldsymbol{r})) = -d_c\int\left|Y\left(\frac{m_h}{m}\boldsymbol{r}' + \frac{m_e}{m}\boldsymbol{r}, \boldsymbol{r}' - \boldsymbol{r}\right)\right|^2 d^3\boldsymbol{r}' \ . \tag{7.4}$$

At low intensities the spontaneous effect due to $\xi_0$ will be dominant because the induced part $\xi_{in}$ is of second order in $Y$. In the present section we shall concentrate on this case. This means that $\xi = \xi_0$ is treated as a given function. Then the r.h.s. of (7.1a) describes a parametric influence on the polariton via its excitonic component $Y$. A convenient method of handling such a problem is the Born iteration known from quantum-mechanical scattering theory. Let the incident polariton $(E_i, Y_i)$ be a solution of the homogeneous pair of equations (written in the compact notation of Sect. 3.2):

$$i\dot{Y}_i - \Omega_{eh} Y_i + \frac{M_0}{\hbar}\delta(\boldsymbol{r}) E_i(\boldsymbol{R}) = 0 \tag{7.5a}$$

$$\tilde{\varepsilon}\ddot{E} - c^2 \nabla_R^2 E + \frac{2M_0}{\varepsilon_0}\mathrm{Re}\{\ddot{Y}_i(\boldsymbol{R},0)\} = 0 \ . \tag{7.5b}$$

The relevant solutions of (7.5) are discussed in Chap. 4. The first-order scattered wave according to the Born approximation scheme is found by replacing $Y$ on the r.h.s. of (7.1a) by the incident solution $Y_i$

$$i\dot{Y}_s - \Omega_{eh} Y_s + \frac{M_0}{\hbar}\delta(\boldsymbol{r}) E_s(\boldsymbol{R}) = \frac{d_c}{\hbar}\xi(\boldsymbol{r}_e) Y_i \tag{7.6a}$$

$$\tilde{\varepsilon}\ddot{E}_s - c^2 \nabla_R^2 E_s + \frac{2M_0}{\varepsilon_0}\mathrm{Re}\{Y_s(\boldsymbol{R},0)\} = 0 \ . \tag{7.6b}$$

Solving (7.6) amounts to finding the polariton Green's function. It is a 2×2 matrix function defined by the equations

$$i\dot{G}_{YY} - \Omega_{eh} G_{YY} + \frac{M_0}{\hbar}\delta(\boldsymbol{r}) G_{EY} = \delta(\boldsymbol{r}-\boldsymbol{r}')\delta(\boldsymbol{R}-\boldsymbol{R}')\delta(t-t') \tag{7.7a}$$

$$\tilde{\varepsilon}\ddot{G}_{EY} - c^2 \nabla_R^2 G_{EY} + \frac{2M_0}{\varepsilon_0}\mathrm{Re}\{G_{YY}\}|_{\boldsymbol{r}=0} = 0 \tag{7.7b}$$

$$iG_{YE} - \Omega_{eh} G_{YE} + \frac{M_0}{\hbar}\delta(\boldsymbol{r}) G_{EE} = 0 \tag{7.7c}$$

$$\tilde{\varepsilon}\ddot{G}_{EE} - c^2 \nabla_R^2 G_{EE} + \frac{2M_0}{\varepsilon_0}\mathrm{Re}\{G_{YE}\}|_{\boldsymbol{r}=0} = \delta(\boldsymbol{R}-\boldsymbol{R}')\delta(t-t') \ . \tag{7.7d}$$

The soluton of (7.7a,b) is given in Appendix E. The scattered field $E_s$ is a convolution of $G_{EY}$ with the source in (7.6a)

$$E_s(\boldsymbol{R},t) = \frac{d_c}{\hbar}\int G_{EY}(\boldsymbol{R},t,\boldsymbol{r}',\boldsymbol{R}',t')\xi\left(\boldsymbol{R}' + \frac{m_h}{m}\boldsymbol{r}', t\right)$$

$$\times Y_i(\boldsymbol{R}',\boldsymbol{r}',t') d^3\boldsymbol{r}' d^3\boldsymbol{R}' \, dt' \ . \tag{7.8}$$

The functions forming the integrand in (7.8) are known analytically from methods developed in Chap. 4 and Appendix E, but unfortunately there still exists no means of calculating the integral itself by other than numerical methods (Dörpelkus, 1985). Once $E_s$ is known, the scattering cross-section and other interesting quantities are accessible by methods to be found in the pertinent literature (Cardona, 1982; Pinczuk and Burstein, 1975) (compare also Appendix G). A spectrum derived from (7.8) is displayed in Fig. 7.1.

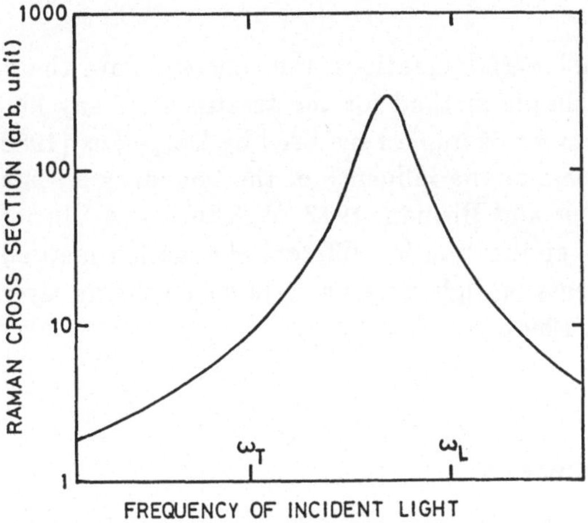


**Fig. 7.1.** Resonance enhancement of the Raman scattering cross section. The curve shows the contribution of the lower branch polariton in case of backward scattering. The parameters used are typical for GaAs. $\hbar/T_2$ is 0.01 meV (Dörpelkus, 1986)

So far we have restricted the discussion to deformation-potential scattering by phonons. The extension of the treatment to Fröhlich-type interactions poses no difficulty; one must only replace the r.h.s. of (7.6a) by the appropriate Fröhlich term taken from (3.31); then (7.6a) goes over into

$$i\dot{Y}_s - \Omega_{eh}Y_s + \frac{M_0}{\hbar}\delta(\boldsymbol{r})E_s(\boldsymbol{R}) = \frac{e}{\hbar}(\Phi_\xi(\boldsymbol{r}_h) - \Phi_\xi(\boldsymbol{r}_e))Y_i \ . \tag{7.6c}$$

In a similar way also scattering by plasmons via the electrooptic interaction can be treated. Then $\Phi_\xi$ is interpreted as the self consistent potential of the plasmon (compare Sect. 6.4). It is interesting to note that, besides the electrooptic coupling, the plasmon can also influence the polariton via the saturating terms on the r.h.s. of the interband equation (3.1a). Whether this is a new scattering channel hitherto unknown or only a new way of looking at known results has not yet been explored.

Although the technical handling of (7.8) surely needs further exploration, some attractive features of the coherent polariton approach to the theory of light scattering are already noticeable at the present stage:

i) The fluctuation aspect of classical scattering theory (Einstein, 1910; Placzek, 1934) is retained and combined with the correct resonance properties typical of the microscopic scattering theory (Loudon, 1963). In this respect the coherent-wave approach matches the program initiated by Barker and Loudon (1972) with their coupled oscillator model.

ii) The necessity of accounting for the electron-hole interaction which is a major problem in standard theory (Mills and Burstein, 1969) is included in the coherent-wave method without great difficulty.

iii) Polariton effects, the importance of which has been stressed a number of times (Hopfield, 1969; Bendow, 1970), are incorporated by the use of the polariton Green's function $G_{EY}$.

iv) As a theory based on differential equations, the coherent-wave theory offers a conceptually clear and simple method for the treatment of any kind of spatial inhomogeneity. This has for example been used by Dörpelkus (1986) to settle the long-standing problem of the influence of the boundary on light scattering spectra (Brenig, Zeyher and Birman, 1972; Weisbuch and Ulbrich, 1982). An attractive new field of application for differential equation methods is opened up by recent experiments on light scattering in space charge layers and heterojunctions (Abstreiter, 1984).

## 7.3 Induced Raman Processes

In the treatment of spontaneous scattering in the last section the induced phonon amplitude in (7.3) has been neglected. Now, we shall retain it and assume that it is so strong that the spontaneous fluctuations $\xi_0$ become negligible. Then a number of coherent Raman type processes can take place (Bloembergen, 1967; Vogt, 1982). As examples we shall consider so called "Stimulated Raman Scattering" (SRS) and "Coherent Antistokes Raman Scattering" (CARS). The relevant equations have already been given in Sect. 7.2., but for reference we repeat them here. The complete set of equations consists of two material equations for $Y$ and $\xi$

$$i\dot{Y} - \Omega_{eh}Y - \frac{d_e}{\hbar}Y\,\xi(\boldsymbol{r}_e) = -\frac{M_0}{\hbar}\delta(\boldsymbol{r})E(\boldsymbol{R}) \tag{7.9a}$$

$$\ddot{\xi} + 2\gamma\dot{\xi} + \omega_0^2\xi + \frac{d_c}{\mu}\int\left|Y\left(\frac{m_h}{m}\boldsymbol{r}' + \frac{m_e}{m}\boldsymbol{r}_e, \boldsymbol{r}' - \boldsymbol{r}_e\right)\right|^2 d^3\boldsymbol{r}' \ , \tag{7.9b}$$

and the field equation for the transverse field

$$\tilde{\varepsilon}\ddot{E} - c^2\nabla_R^2 E + \frac{2M_0}{\varepsilon_0}\mathrm{Re}\{\ddot{Y}(\boldsymbol{R},0)\} = 0 \ . \tag{7.9c}$$

As in Sect. 7.2, the influences of crystal anisotropy have been suppressed in order to keep the model as simple as possible. In (7.9b) a phenomenological damping term has been added.

In SRS as well as in CARS experiments one uses two laser beams of different frequencies $\omega_1$ and $\omega_2$. To be definite, let us assume $\omega_2<\omega_2$. The two beams are superimposed in the sample and the beat signal is used to drive the phonon oscillator $\xi$. So far there is no difference between SRS and CARS. The difference is introduced by the selection of the frequency to be detected. In SRS one selects the output at one of the pump frequencies, while in CARS one is interested in the output at the frequency $\omega_a = 2\omega_1 - \omega_2$. If the phonon is at resonance with the beat, $\omega_a$ coincides with the Antistokes frequency $\omega_1 + \omega_0$ (Fig. 7.2).

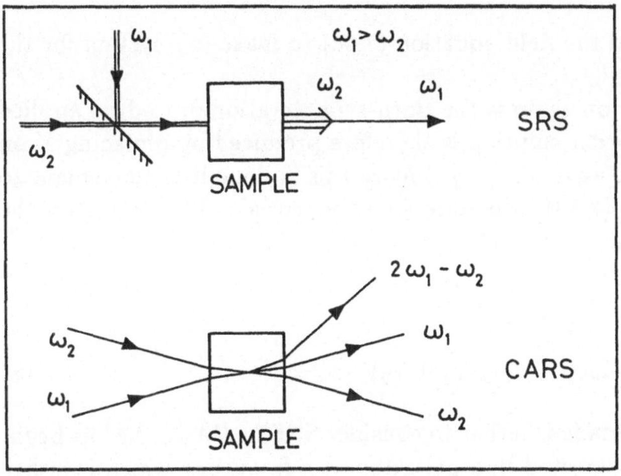


**Fig. 7.2.** Generation of a nonlinear SRS or CARS signal, respectively, by coherent mixing of two polariton waves

From the experimental conditions it is almost obvious how one has to handle equations (7.9) in order to set up a theoretical model for SRS and CARS respectively.

i) In a first step one solves the "free" polariton equations (7.9a,c) without phonon terms. Let $(E_1, Y_1)$ and $(E_2, Y_2)$ be the corresponding solutions at two frequencies $\omega_1$ and $\omega_2$ respectively. The free solutions are discussed in Chap. 4.

ii) In the next step $Y_1$ and $Y_2$ are superimposed and the result is inserted into the source term of the phonon oscillator equation (7.9b)

$$\ddot{\xi} + 2\gamma\dot{\xi} + \omega_0^2\xi = \frac{d_c}{\mu}\int(|Y_1|^2 + |Y_2|^2 + 2\mathrm{Re}\,\{Y_1^* Y_2\})d^3\mathbf{r}' \ . \tag{7.10}$$

The interesting term is the interference term which oscillates with a frequency $\omega_{12} = \omega_1 - \omega_2$. It will drive the phonon oscillator into a forced oscillation with the same frequency. The general solution of (7.10) in the limit $\gamma \to 0$ has the form

$$\xi = A\cos(\omega_{12}t - \delta) + \xi_0 \ . \qquad (7.11)$$

Resonance occurs if

$$\omega_{12} \approx \omega_0 \ . \qquad (7.12)$$

iii) In the third step the induced phonon amplitude (7.11) is inserted into the interband equation (7.9). The resulting equation becomes

$$i\dot{Y} - \Omega_{eh}Y + \frac{M_0}{\hbar}\delta(\boldsymbol{r})E(\boldsymbol{R}) = \frac{d_c}{\hbar}A\cos(\omega_{12}t - \delta)Y \ . \qquad (7.13)$$

To this equation is added the field equation (7.9c) to make it a system for the secondary polariton.

The method of solution again is the Born-type iteration procedure applied in Sect. 7.1. The lowest order solution is therefore produced by replacing $Y$ on the r.h.s. by the incident wave $Y_1 + Y_2$. Before this is done it is convenient to decompose the cosine in (7.13) into complex exponentials. The source on the r.h.s. of (7.13) then reads

$$s = \frac{Ad_c}{2\hbar}\{\exp[i(\omega_1 - \omega_2)t - i\delta] + \exp[-i(\omega_1 - \omega_2)t + i\delta]\}(Y_1 + Y_2) \ . \qquad (7.14)$$

Now the decision can be made whether to consider SRS or CARS. Let us begin with SRS. Here we are interested in a polariton at a frequency which is either $\omega_1$ or $\omega_2$. We therefore must select from the source (7.14) terms driving a polariton at one of those frequencies. The appropriate sources are

$$s_1(\omega_1) = \frac{Ad_c}{2\hbar}Y_2\exp[-i(\omega_1 - \omega_2)t + i\delta] \qquad (7.15a)$$

$$s_2(\omega_2) = \frac{Ad_c}{2\hbar}Y_1\exp[i(\omega_1 - \omega_2)t - i\delta] \ . \qquad (7.15b)$$

The SRS-signals $(E_{NL}, Y_{NL})$ then are found by solving the polariton equations with the corresponding source terms

$$i\dot{Y}_{NL} - \Omega_{eh}Y_{NL} + \frac{M_0}{\hbar}\delta(\boldsymbol{r})E_{NL}(\boldsymbol{R}) = s_j(\omega_j) \quad j = 1,2 \ , \qquad (7.16a)$$

$$\tilde{\varepsilon}\ddot{E}_{NL} - c^2\nabla_R^2 E_{NL} + \frac{2M_0}{\varepsilon_0}\mathrm{Re}\{\ddot{Y}_{NL}(\boldsymbol{R},0)\} = 0 \ . \qquad (7.16b)$$

If instead of the SRS signal the CARS signal is wanted one must select from (7.14) the source term with the Antistokes frequency $\omega_a = 2\omega_1 - \omega_2$

$$s_a(\omega_a) = \frac{Ad_c}{2\hbar}Y_1\exp[-i(\omega_1 - \omega_2)t + i\delta] \ . \qquad (7.17)$$

If $s_a$ is inserted instead of $s_j$ on the r.h.s. of (7.16) one obtains the polariton equations for the nonlinear CARS signal.

A method by which (7.16) may be solved has been already described in Sect. 7.2. It involves the polariton Green's function $G_{EY}$ or $G_{YY}$ respectively. By this method the nonlinear response is represented as a convolution of the source with the Green's function; in shorthand notation as

$$E_{NL} = G_{EY} * s \ , \tag{7.18a}$$

$$Y_{NL} = G_{YY} * s \ . \tag{7.18b}$$

Depending on the geometry, part of the convolution may be resolved by Fourier transformation applying the convolution theorem.

When the general relation (3.11) is applied to $Y_{NL}$ it yields a nonlinear polarization

$$P_{NL} = 2M_0 \mathrm{Re}\{Y_{NL}(\boldsymbol{r},\boldsymbol{r})\} \ . \tag{7.19}$$

Tracing back powers of the incident field, i.e. the $E$ component of the incident polariton, one easily identifies $P_{NL}$ in (7.19) as being of order $E^3$ ($\xi$ being $\xi \sim E^2$ and $Y_{NL} \sim \xi Y_i \sim \xi E_i$). In a formal theory of nonlinear material response SRS and CARS is therefore implied in the third-order material equation. For example the CARS response is covered by the formal relation

$$P_{CARS}(\omega_a) = \chi^{(3)}(-\omega_a, \omega_1, \omega_1, -\omega_2) E(\omega_1) E(\omega_1) E^*(\omega_2) \ . \tag{7.20}$$

The dynamic approach to induced Raman scattering described in this section could therefore be used to calculate a particular contribution to $\chi^{(3)}$. But for reasons similar to those given at the end of Sect. 7.2 the potential of the dynamic method is even greater: As a polariton-based approach it yields a single solution for both the material dynamics and the field propagation problem, and since it is a differential equation method it is applicable to various kinds of geometry.

## 7.4 Nonlinear Refraction by Saturation

As a further example of the use of band-edge dynamics in nonlinear optics we shall discuss the intensity dependence of the index of refraction due to the saturation of an interband transition. This phenomenon has attracted some recent attention because of its possible application in nonlinear optical devices (Miller, Miller and Smith, 1981). From the point of view of coherent band-edge dynamics the effect is of particular interest because it probes the nonlinear source terms in the basic equations (3.1) which are a peculiarity of that theory.

In this section we give a sketchy derivation of the effect. For details the reader is referred to the paper by Frank and Stahl (1984). The discussion is restricted to the onset of nonlinear refraction if one approaches the band edge from below. Then the dependence $n(I)$ of the refractive index on intensity is small enough to try an expansion

$$n(I) = n_0 + n_2 I \ . \tag{7.21}$$

Since the effect is observed at a single frequency the relevant expansion of the polarization in powers of the field is

$$P(\omega) = \varepsilon_0(\chi^{(1)}(\omega) + \chi^{(3)}(\omega, -\omega, \omega)|E(\omega)|^2 E(\omega) + \dots) \ . \tag{7.22}$$

The nonlinear index $n_2$ is simply related to the susceptibility $\chi^{(3)}$: From the definition of intensity

$$I = \frac{c}{2} n_0 \varepsilon_0 |E|^2 \quad \text{and} \tag{7.23}$$

$$n_0^2 = 1 + \chi^{(1)}$$

one finds

$$n_2 = \frac{\mathrm{Re}\{\chi^{(3)}\}}{c\varepsilon_0 n_0^2} \ . \tag{7.24}$$

$\chi^{(3)}$ is calculated iteratively from the dynamic equations (3.1). Excitonic effects will be neglected. For the problem at hand it is important to incorporate appropriate irreversible relaxation terms into the basic equations because the band edge resonance of $\chi^{(3)}$ is due to concerted action of coherent and incoherent processes, a fact that has first been recognized by Wherrett and Higgins (1982).

The first step in the iteration consists of solving the equation

$$\dot{Y}_{12}^{(1)} + i\Omega_{eh} Y_{12}^{(1)} + T_2^{-1} Y_{12}^{(1)} = \frac{iM_0}{\hbar} \delta_{12} E \ . \tag{7.25}$$

The solution yields the first order wave-function $Y^{(1)}$ for $E$ given. The problem has been treated in Chap. 4.

Then $Y^{(1)}$ is inserted into the source terms of the intraband equations (3.1b,c). The resulting inhomogeneous equations for the density matrices $C^{(2)}$ and $D^{(2)}$ are enriched by appropriate irreversible terms (compare Sect. 3.7); the superscript indicates the order of $C^{(2)}$ and $D^{(2)}$ with respect to $E$.

$$\dot{C}_{12}^{(2)} + i\Omega_{ee} C_{12}^{(2)} - (\dot{C}_{12})_{irr} = -\frac{iM_0}{\hbar}(E_1 Y_{12}^{(1)} - E_2 Y_{21}^{(1)*}) \tag{7.26a}$$

$$\dot{D}_{12}^{(2)} + i\Omega_{hh} D_{12}^{(2)} - (\dot{D}_{12})_{irr} = -\frac{iM_0}{\hbar}(E_1 Y_{21}^{(1)} - E_2 Y_{12}^{(1)*}) \ . \tag{7.26b}$$

In order to keep things as simple as possible a relaxation-time approximation is assumed for the irreversible terms. From general conservation laws one is forced to distinguish between the relaxation of the symmetric and antisymmetric parts of $C$ and $D$.

Take as example the decomposition of $C^{(2)}$

$$C^{(2)} = C_s + C_a \ . \tag{7.27}$$

Going over to a Wigner representation according to (6.12) we introduce the distribution functions

$$f_{s/a}(\boldsymbol{r}, \boldsymbol{q}) = \int C_{s/a}(\boldsymbol{r} + \tfrac{1}{2}\boldsymbol{r}', \boldsymbol{r} - \tfrac{1}{2}\boldsymbol{r}') \exp(i\boldsymbol{q}\cdot\boldsymbol{r}') d^3\boldsymbol{r}' \ . \tag{7.28}$$

Then the relaxation time ansatz for the irreversible terms is assumed as follows

$$\begin{aligned}(\dot{f}_s)_{irr} = & -\tau^{-1}[f_s(\boldsymbol{r}, \boldsymbol{q}) - f_0(\boldsymbol{q}) \int f_s(\boldsymbol{r}, \boldsymbol{p}) d^3\boldsymbol{p}] \\ & - T_1^{-1} \delta(\boldsymbol{q}) \int f_s(\boldsymbol{r}, \boldsymbol{q}) d^3\boldsymbol{p}\end{aligned} \tag{7.29}$$

$$(\dot{f}_a)_{irr} = -\tau^{-1} f_a(\boldsymbol{r}, \boldsymbol{q}) \ . \tag{7.30}$$

The first term in (7.29) describes intraband relaxation to the thermal equilibrium distribution $f_0$; the second term accounts for a loss of electron density through recombination processes. The latter must therefore have a counterpart in $(\dot{D})_{irr}$. The relaxation ansatz (7.30) states that collective motions die out.

For $(\dot{D})_{irr}$ one sets up a similar pair of equations as (7.29,30).

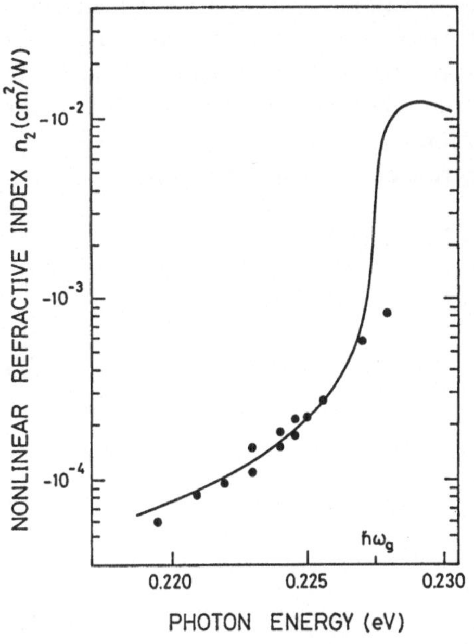


**Fig. 7.3.** Band edge resonance of intensity dependent index of refraction in InSb; calculated curve by Frank and Stahl (1984), experimental points by Miller et al. (1981)

The simple relaxation time ansatz allows an analytical solution of (7.26). The result then is inserted into the saturating terms of the interband equation (3.1a). The resulting equation for the third order electron-hole amplitude $Y^{(3)}$ reads

$$\dot{Y}_{12}^{(3)} + T_2^{-1} Y_{12}^{(3)} + i\Omega_{eh} Y_{12}^{(3)} = -\frac{iM_0}{\hbar}(E_1 C_{12}^{(2)} + E_2 D_{21}^{(2)}) \; . \tag{7.31}$$

Equation (7.31) has the same structure as (4.5, 7.25) and is therefore solved by methods developed in Chap. 4. The resulting nonlinear wavefunction $Y^{(3)}$ is used to calculate the nonlinear susceptibility $\chi^{(3)}(\omega, -\omega, \omega)$ and the intensity-dependent index of refraction $n_2$. In Fig. 7.3 the nonlinear index $n_2$ of InSb which has been calculated in this way is compared with experimental results by Miller, Seaton, Prise and Smith (1981). Note that $n_2$ is negative, as expected from the fact that saturation tends to reduce the oscillator strength. It is not surprising that the theory fails when the frequency lies too close to the band-edge resonance. In this region the higher-order terms in the expansion (7.21) should be considered; they may in fact invalidate the expansion when they become too large.

## 7.5 Optical Transients in Two-Level Materials and Semiconductors

In the relatively new field of optical transient spectroscopy one studies fast amplitude variations of an optical carrier signal. For reasons already mentioned in Sect. 7.1 there exists a significant difference between the transient behaviour of a two-level resonance and that of a two-band semiconductor. In order to make this difference explicit we begin our discussion with an outline of the envelope dynamics in a two level transition.

We start from (2.10) rewritten in terms of the real quantities

$$s' = \mathrm{Re}\{s\} \; ; \quad s'' = \mathrm{Im}\{s\} \; ; \quad I = (n + p - N)/2 \; .$$

$I$ is called the "inversion". The real Bloch equations become

$$\dot{s}' = \omega_g s'' \; ; \quad \dot{s}'' = -\omega_g s' - 2M_0\hbar^{-1}EI \; ; \quad \dot{I} = 2M_0\hbar^{-1}Es'' \tag{7.32}$$

or, after elimination of $s''$

$$\ddot{s}' + \omega_g^2 s' = -2M_0\hbar^{-1}\omega_g EI \; ; \quad \dot{I} = 2M_0\hbar^{-1}\omega_g^{-1}E\dot{s}' \; . \tag{7.33}$$

Let us assume that the electric field is of the form

$$E = E_0 \cos \omega t \tag{7.34}$$

with $\omega \approx \omega_g$. Inserting (7.34) into (7.33) we get

$$\ddot{s}' + \omega_g^2 s' = -2\omega_g F I \cos \omega t \;\; ; \quad \dot{I} = 2\omega_g^{-1} F \dot{s}' \cos \omega t \;\; . \qquad (7.35)$$

As an abbreviation we have introduced the Rabi flopping frequency

$$F = M_0 E_0 \hbar^{-1} \;\; . \qquad (7.36)$$

For any reasonable field strength it turns out that

$$F \ll \omega_g \approx \omega \;\; . \qquad (7.37)$$

This inequality is the basis of the "slowly-varying-envelope-approximation" (SVEA), also known as "rotating wave approximation". In this approximation one considers an ansatz

$$s' = A(t) \sin \omega t + B(t) \cos \omega t \;\; . \qquad (7.38)$$

Under the assumption that $A$ and $B$ are slowly varying on the scale of the fast oscillation the amplitudes $A$ and $B$ are shown to obey approximately the equations (McCall and Hahn, 1969)

$$\dot{A} - \Delta\omega B + F I = 0 \;\; ; \quad \dot{B} + \Delta\omega A = 0 \;\; ; \quad \Delta\omega = \omega - \omega_g \;\; . \qquad (7.39a)$$

For $I$, which is slowly varying anyhow, one obtains the equation

$$\dot{I} = F A \;\; . \qquad (7.39b)$$

A solution to (7.39) with initial values

$I(0) = -N/2$ ; $A(0) = 0$ is

$$A = \frac{N\omega_1}{2F} \sin \omega_1 t \;\; ; \quad B = \frac{N\Delta\omega}{2F} \cos \omega_1 t \;\; ; \quad I = -\frac{N}{2} \cos \omega_1 t \qquad (7.40)$$

where

$$\omega_1^2 = F^2 + (\Delta\omega)^2 \;\; . \qquad (7.41)$$

In case of exact resonance, i.e. $\Delta\omega = 0$, from (7.40) one finds for $s = s' + is''$ and $I$ the solution

$$s = \tfrac{1}{2} N \sin (Ft) e^{i(\pi/2 - \omega t)} \qquad (7.42a)$$

$$I = -\tfrac{1}{2} N \cos (Ft) \;\; . \qquad (7.42b)$$

The solution describes the phenomenon of optical nutation. It consists of an oscillation with frequency $F$ between the ground state ($I = -N/2$) and the

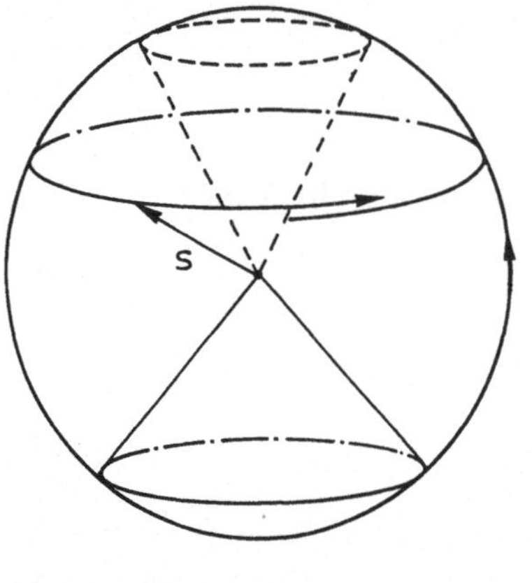


**Fig. 7.4.** Schematic picture showing the nutation of the optical Bloch vector $\boldsymbol{S} = 2(s', s'', I)/N$; the lower cone corresponds to a noninverted situation while the *dashed cone* refers to an inverted situation

fully inverted state ($I = +N/2$). The oscillation of the inversion is accompanied by an amplitude modulation of the optical response $s$. The solution (7.42) may be visualized as a motion of the optical Bloch vector $\boldsymbol{S} = 2(s', s'', I)/N$ on a unit sphere (Fig. 7.4).

So far $E$ has been treated as a prescribed external field. At a sufficiently high density of two-level atoms this is no longer justified because the polarization $P = 2M_0 s'$ acts as a source of an induced field. The induced field obeys the wave equation

$$\ddot{\boldsymbol{E}} + c^2 \nabla \times \nabla \times \boldsymbol{E} = -\frac{2M_0}{\varepsilon_0} \ddot{s}' \ . \tag{7.43}$$

One can now ask for self-consistent solutions to the closed system of equations (7.32,43) (Maxwell-Bloch (MB) equations). External sources then only act via boundary conditions.

Since according to (7.43) solutions will develop as waves in space and time the fast carrier of (7.34) is now assumed wavelike and of the form

$$E = E_0 \cos{(\omega t - kx - \delta)} \ . \tag{7.44}$$

In a general SVEA approximation $E_0$ and $\delta$ may be assumed to be slowly varying functions of $x$ and $t$. Let us specialize to the case of a fixed phase ($\delta = 0$). Then the SVEA applied to (7.43) yields the dispersion relation

$$\omega = ck$$

and an equation for the Rabi flopping frequency (7.36) which now has become a slowly varying function of $x, t$:

$$c\frac{\partial F}{\partial x} + \frac{\partial F}{\partial t} = -\alpha A \qquad \alpha = \frac{M_0^2 \omega}{\hbar \varepsilon_0} \ . \tag{7.45a}$$

Assuming exact resonance ($\omega = \omega_g$) the system of envelope equations is completed by two equations taken from (7.39), namely

$$\dot{A} = -FI \ ; \quad \dot{I} = FA \ . \tag{7.45b}$$

Let us define the area function $\vartheta$ by

$$F = \dot{\vartheta} \qquad \vartheta = \int_{-\infty}^{t} F(t')dt' \ . \tag{7.46}$$

With the help of (7.46) one can transform (7.45) into

$$A = \frac{1}{2} N \sin \vartheta \qquad I = -\frac{1}{2} N \cos \vartheta \tag{7.47a}$$

$$c \frac{\partial^2 \vartheta}{\partial x \partial t} + \frac{\partial^2 \vartheta}{\partial t^2} = -\frac{1}{2} N \alpha \sin \vartheta \ . \tag{7.47b}$$

Assuming that (7.47b) has solutions in the form of a distortionless travelling pulse

$$\vartheta = \vartheta(\tau) \qquad \tau = t - \frac{x}{v} \tag{7.48}$$

one finds that the pulse area $\vartheta$ has to obey the equation

$$\ddot{\vartheta} - \Omega^2 \sin \vartheta = 0 \ , \tag{7.49}$$

$$\Omega^2 = \frac{1}{2} N \alpha \left( \frac{c}{v} - 1 \right)^{-1} > 0 \ . \tag{7.49a}$$

Equation (7.49) is the inverted-pendulum equation with $\vartheta = 0$ corresponding to the upper (unstable) equilibrium position. According to (7.47a) $\vartheta = 0$ describes a noninverted state with polarization amplitude $A = 0$. The famous $2\pi$ pulse (McCall and Hahn, 1969) corresponds to a single sweep of the pendulum between

$$\vartheta(\tau \to -\infty) = 0 \qquad \vartheta(\tau \to \infty) = 2\pi \ . \tag{7.50}$$

A first integral of (7.49) obeying the condition (7.50) is derived from energy conservation

$$\frac{d\vartheta}{d\tau} = 2\Omega \sin \frac{1}{2} \vartheta \ . \tag{7.51}$$

This equation is satisfied if we let

$$\cos \tfrac{1}{2} \vartheta = -\tanh \Omega \tau \ . \tag{7.52}$$

From this it then follows that the flopping frequency $F = \dot{\vartheta}$ has a pulse form

$$F = 2\Omega \operatorname{sech} \Omega \tau \ . \tag{7.53}$$

Since, by (7.36), $F$ is related to the electric field amplitude, then (7.53) also gives the shape of the field envelope. The field strength reaches its maximum for $\tau = 0$

$$E_{\max} = 2\hbar M_0^{-1}\Omega = 2\hbar M_0^{-1}\left\{\frac{N\alpha}{2}\left(\frac{c}{v}-1\right)^{-1}\right\}^{1/2} . \tag{7.54}$$

Hence the yet undefined pulse velocity $v$ is a function of the pulse height.

The complete phenomenon of a $2\pi$ pulse travelling through a two-level material is described as follows: In the raising stage ($\tau<0$) the medium is pumped from the initial state $I = -N/2$ to the inverted state $I = N/2$ while the field strength is growing; in the second half of the pulse the medium returns by stimulated emission to its ground state. Since, at least in our idealized theory, the pulse migration is loss-free, although the light is at resonance with the transition, the phenomenon has been named "self induced transparency" (SIT).

Let us briefly consider how the damping terms of (2.16) influence the envelope dynamics. By a straightforward calculation it is found that equations (7.39) are replaced by

$$\dot{A} + \gamma_2 A - \Delta B + FI = 0 \tag{7.55a}$$

$$\dot{B} + \gamma_2 B + \Delta\omega A = 0 \tag{7.55b}$$

$$\dot{I} + \gamma_1(I - I_0) = FA \tag{7.55c}$$

$$\gamma_1 = T_1^{-1} \ ; \quad \gamma_2 = (2T_2)^{-1} .$$

The optical-nutation solution is now a superposition of a steady state and a transient. In case of strict resonance ($\Delta\omega = 0$) it becomes

$$A = a\cos(\overline{\omega}t + \alpha)\exp(-\lambda t) + \overline{A} \tag{7.56a}$$

$$B = b\exp(-\gamma_2 t) \tag{7.56b}$$

$$I = c\cos(\overline{\omega}t + \beta)\exp(-\lambda t) + \overline{I} \tag{7.56c}$$

$$\overline{\omega} = F[1 - \tfrac{1}{4}F^{-2}(\gamma_1 - \gamma_2)^2]^{1/2} \tag{7.56d}$$

$$\lambda = \tfrac{1}{2}\gamma_1 + \tfrac{1}{2}\gamma_2 \tag{7.56e}$$

$$\overline{I} = \frac{\gamma_1\gamma_2 I_0}{\gamma_1\gamma_2 + F^2} \qquad \overline{A} = \frac{\gamma_1 F I_0}{\gamma_1\gamma_2 + F^2} . \tag{7.56f}$$

The amplitudes $a$, $b$, $c$ and the phases $\alpha$, $\beta$ depend on the initial conditions. In the weak-field limit the steady state part of (7.56) becomes the linear-response solution.

Let us now turn to the case of a two-band semiconductor. Remembering the point of view adopted in Chap. 2 according to which a semiconductor is an array of two level systems with intralevel mobility we are facing mainly two problems: The first problem concerns the influence of mobility on coherent dynamics while the second problem refers to intraband dissipation.

### 7.5.1 Coherent Mobility Effects

If the electron-hole interaction is neglected the extension of the two-level model to a two-band model can be achieved in several ways. A simple method is suggested by the theory of inhomogeneous broadening. In this theory the two-band system is considered as an ensemble of two-level systems, each representing a transition between a pair of Bloch states $\varphi_v(\boldsymbol{k}) \rightarrow \varphi_c(\boldsymbol{k})$. The $\boldsymbol{k}$ vector is assumed to be conserved in the transition. The transition amplitude to a fixed value of $\boldsymbol{k}$ is denoted as $s_{\boldsymbol{k}}$. Equation (7.35) is then replaced by

$$\ddot{S}'_{\boldsymbol{k}} + \omega_{\boldsymbol{k}}^2 s' = -2\omega_{\boldsymbol{k}} F I_{\boldsymbol{k}} \cos \omega t \tag{7.57}$$

$$\dot{I}_{\boldsymbol{k}} = \frac{2}{\omega_{\boldsymbol{k}}} F \dot{s}'_{\boldsymbol{k}} \cos \omega t \; .$$

The source in the field equation is a sum of the transition dipoles with different values of $k$.

$$\ddot{\boldsymbol{E}} + c^2 \nabla \times \nabla \times \boldsymbol{E} = -\frac{2M_0}{\varepsilon_0} \sum_{\boldsymbol{k}} \ddot{s}'_{\boldsymbol{k}} \; . \tag{7.58}$$

The envelope equations resulting from (7.57,58) have been analyzed by Poluektov et al. (1975). Their result is that this model under certain conditions allows solutions corresponding to SIT. The fact that the effect has not yet been clearly verified experimentally may be attributed to several influences not accounted for by the model, in particular the neglect of electron-hole interactions and intraband scattering.

One situation at least where scattering might be less important is that of SIT on an excitonic resonance. This problem has been studied by Goll and Haken (1978) in a manner based on a model proposed by Haken and Schenzle (1973). A peculiarity of the Haken-Schenzle model is the assumption of an exciton wave-function which is "rigid" in the sense that it is not influenced by the coherent pulse propagation. In a treatment of SIT based on the band-edge model in its one-dimensional simplification, (Sect. 5.4), Huhn (1985) has solved the pulse propagation problem in electron-hole configuration space without projection to a rigid wave function. As is to be expected from the nonlocal character of the source terms in the edge equations (3.1), the qualitative properties of the solution strongly depend on a proper treatment of coherence between the wave function in relative space and the carrier wave. In Fig. 7.5a are displayed results from Huhn's paper concerning the dependence of the internal wave function on the carrier frequency. The predicted dispersion law is shown in Fig. 7.5b.

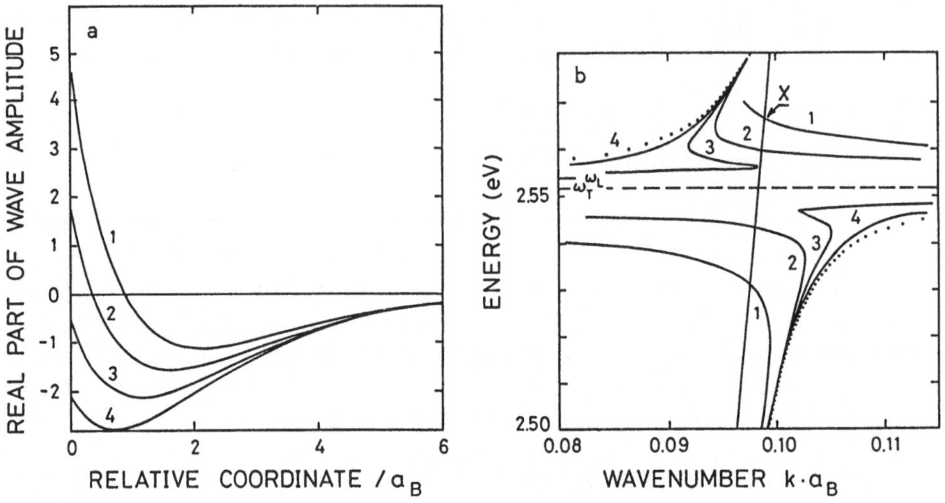


**Fig. 7.5.(a)** Energy dependence of the exciton wave function in a pulsed polariton: The four curves correspond to constant pulse duration and intensity, but different values of the carrier frequency; the frequency variation is chosen to include an intersection point of the pulse dispersion curve with the free light cone (point $X$ in Fig. 7.5b); *(1,2)* correspond to an energy slightly below the intersection, *(3,4)* to an energy above the intersection; energetic distance between *(1)* and *(4)* $\approx 2\omega_{LT}$. Note that at the intersection the polarization $P = \mathrm{Re}\{Y(z_{rel} = 0)\}$ undergoes a change of sign (Huhn, 1986); **(b)** Calculated dispersion curves for pulses of different duration: *(1)* $\tau = 0.1\,\mathrm{ps}$, *(2)* $\tau = 0.5\,\mathrm{ps}$, *(3)* $\tau = 1.0\,\mathrm{ps}$, *(4)* $\tau = 2.0\,\mathrm{ps}$; the pulse area is normalized, so $\tau$ is also a measure of intensity; *dotted lines* refer to linear polariton dispersion; absolute values correspond to the $A$ exciton in CdS; for meaning of point $X$ compare Fig. 7.5a (Huhn, 1986)

### 7.5.2 Incoherent Mobility Effects

The mobile states in a two-band model allow scattering processes which are not present in the two-level model. So we know that the fast intraband thermalization process is due to the electron-electron scattering term (3.76) in the intraband equations. A number of questions can be raised concerning the impact of scattering processes on transient dynamics of the edge, as for instance:

i) Is the phenomenological $T_2$-damping in the interband equation a sufficient approximation in case of transient dynamics of the edge?

ii) What is the indirect influence of $(\dot{C})_{irr}$ on the irreversible behaviour of $Y$ when the coupling via the edge equations is considered?

iii) Can there be found any reasonable linearization procedure for the nonlinear scattering term or is the nonlinearity essential for an understanding of transient dynamics of the edge?

iv) Is the semiconductor plasma immediately after its generation by a short light pulse in a state with "Off-diagonal long range order" with respect to $C$ and $D$?

No answer is known to any of these questions. Some progress in this field could perhaps also lead to a better understanding of certain transient phenomena which have been reported from recent experiments on plasma expansion (Forchel et al., 1983) and on the dynamics of screening (Ulbrich, 1985; Knox et al., 1985).

## 7.6 Biexcitons

While a Wannier exciton is considered as an analog of the hydrogen atom, the biexciton, being a two-exciton state, is the analog of the hydrogen molecule. The first evidence for the existence of a biexciton state was found in the luminescence spectrum of CuCl (Mysyrowicz et al., 1968). In the meantime the biexciton state has been verified in many other substances.

The existence of the biexciton as a long-living bound state provides a strong resonance enhancement to a number of nonlinear optical processes as for instance coherent four wave mixing (Maruani et al., 1978), hyper-Raman scattering (Bechstedt and Henneberger, 1977), two-photon absorption (Vu Duy Phach et al., 1977), two-photon reabsorption (Hvam et al., 1973), and induced absorption (Gogolin and Rashba, 1973; Bivas et al., 1977). An elaborate theory of biexciton formation in which details of the four-particle wave function are considered is rather complicated (Quattropani and Forney, 1977). On the other hand it has been shown by Bigot and Hönerlage (1984) that typical features of the dynamics in the exciton-biexciton system are reproduced by a relatively simple model. In this model one considers three levels, the ground state $|0\rangle$, the exciton state $|x\rangle$ and the biexciton state $|b\rangle$. Mobility effects in the sense of Sect. 2.2 are neglected and the quasi-scalar notation of Sect. 2.1 is adopted. A subset of relevant variables then can be chosen by analogy with that of Sect. 2.1:

$$\begin{aligned}
\hat{n}_0 &= |0\rangle\langle 0| && \text{number of systems in ground state}\\
\hat{n}_x &= |x\rangle\langle x| && \text{number of systems in exciton state}\\
\hat{s} &= |0\rangle\langle x| && \text{exciton annihilator}\\
\hat{s}^\dagger &= |x\rangle\langle 0| && \text{exciton creator.}
\end{aligned}$$

In addition one considers the following variables:

$$\begin{aligned}
\hat{n}_b &= |b\rangle\langle b| && \text{number of biexcitons}\\
\hat{u} &= |x\rangle\langle b| && \text{transition from } |b\rangle \text{ to } |x\rangle\\
\hat{u}^\dagger &= |b\rangle\langle x| && \text{transition from } |x\rangle \text{ to } |b\rangle\\
\hat{v} &= |0\rangle\langle b| && \text{biexciton annihilator}\\
\hat{v}^\dagger &= |b\rangle\langle 0| && \text{biexciton creator.}
\end{aligned} \tag{7.59}$$

The undisturbed Hamiltonian is

$$H_0 = \hbar\omega_{0x}\hat{n}_x + \hbar\omega_{0b}\hat{n}_b \; . \tag{7.60}$$

Dipolar transitions are allowed between $|0\rangle$ and $|x\rangle$ with transition moment $M_{0x}$, and between $|x\rangle$ and $|b\rangle$ with transition moment $M_{xb}$. The Hamiltonian describing interactions of these dipoles with an electric field $E$ is

$$H_{EM} = -M_{0x}(\hat{s} + \hat{s}^{\dagger})E - M_{xb}(\hat{u} + \hat{u}^{\dagger}) \ . \tag{7.61}$$

In the next step, Heisenberg's equations of motion are set up for the operators listed in (7.59). When this has been done the operators are replaced by their expectation values as described in Sect. 2.1. In this way one arrives at the following set of constitutive equations for the exciton-biexciton system

$$\dot{s} + i\omega_{0x}s = i\hbar^{-1}M_{x0}(n_0 - n_x)E + i\hbar^{-1}M_{xb}vE \tag{7.62a}$$

$$\dot{u} + i\omega_{xb}u = i\hbar^{-1}M_{xb}(n_x - n_b)E - i\hbar^{-1}M_{0x}vE \tag{7.62b}$$

$$\dot{v} + i\omega_{0b}v = i\hbar^{-1}M_{xb}sE - i\hbar^{-1}M_{0x}uE \tag{7.62c}$$

$$\dot{n}_0 = i\hbar^{-1}M_{0x}(s - s^*)E \tag{7.62d}$$

$$\dot{n}_x = -i\hbar^{-1}M_{0x}(s - s^*)E + i\hbar^{-1}M_{xb}(u - u^*)E \tag{7.62e}$$

$$\dot{n}_b = -i\hbar^{-1}M_{xb}(u - u^*)E \tag{7.62f}$$

where $\hbar\omega_{xb} = \hbar\omega_{0b} - \hbar\omega_{0x}$ is the difference between the biexciton energy $\hbar\omega_{0b}$ and the exciton energy $\hbar\omega_{0x}$.

Evidently (7.62) is a generalization of the constitutive equations introduced in (2.1) for two-level transitions. In principle it is therefore clear how they could be further generalized to account for mobility effects and for internal wave function coherence. This would amount to repeating procedures described in Chaps. 2,3, but so far this program has not yet been pursued.

Here a remark on the relation to the work by Hõnerlage and Bigot (1984) might be appropriate. These authors have treated the same model, but in the language of density matrices. Hence their constitutive equations, though superficially rather different, are equivalent to the set (7.62).

Because of the parametric dependence on $E$, the solutions to (7.62) describe in general a nonlinear response. In the spirit of an expansion in powers of $E$ the simplest approximation is linear response theory. In this theory one considers from the set (7.62) only (7.62a) with $n_x$ and $v$ neglected and $n_0 = \overline{n}_0$ fixed

$$\dot{s}_1 + i\omega_{0x}s_1 = i\hbar^{-1}M_{0x}\overline{n}_0E \ . \tag{7.63}$$

Considering also the field equation for $E$

$$-\varepsilon_0 c^2 \nabla\times\nabla\times \boldsymbol{E} - \varepsilon_0\tilde{\varepsilon}\ddot{\boldsymbol{E}} = 2M_{0x}\mathrm{Re}\{\tilde{s}_1\} \tag{7.64}$$

one has a model describing the subset of linear exciton polaritons. The polariton solutions generated by (7.63,64) are a crude approximation to the solutions discussed extensively in Chap. 4.

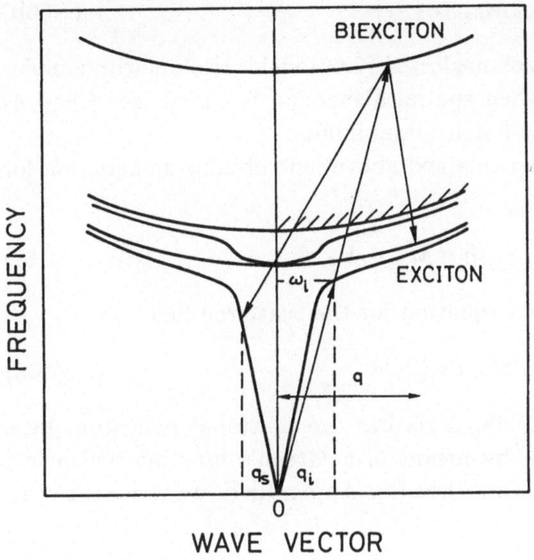


**Fig. 7.6.** Transition scheme in a hyper-Raman process mediated by the biexciton level: Two incident polaritons with frequency $\omega_i$ and wave vector $q_i$ drive a biexciton; the latter emits spontaneously an exciton $(\omega, q)$, in the figure assumed as longitudinal. The scattered polariton in the picture is observed in backward configuration with wave vector $q_s$

Going on in the expansion with respect to $E$ we can find solutions describing nonlinear optical processes. As an example let us consider hyper-Raman scattering. In the hyper-Raman process incident light at frequency $\omega_i$ is tuned to drive the biexciton at $\omega_{0b} \approx 2\omega_i$. The biexciton then undergoes a spontaneous scattering process in which an exciton polariton on the lower branch is formed. The residual energy is left in the crystal either as a longitudinal exciton or as another polariton (see Fig. 7.6).

The iterative solution of (7.62) corresponding to the above process goes through the following steps:

1) (7.63) is solved with $E_i = E_{i0}\exp(i\boldsymbol{q}_i\cdot\boldsymbol{r} - i\omega_i t) + \text{c.c.}$

2) The solution $s_1$ of the first step is inserted together with $E_i$ on the r.h.s. into (7.62c) to give the second-order equation for the biexciton amplitude $v$:

$$\dot{v}_2 + i\omega_{0b}v_2 = i\hbar^{-1}M_{xb}s_1E_i \ . \tag{7.65}$$

Since $s_1$ oscillates with the incident frequency $\omega_i$, the biexciton is driven at resonance for $\omega_{0b} = 2\omega_i$.

3) In the next step one goes back to (7.62a), inserting $v_2$ on the r.h.s.; as the second factor in the product $vE$ one must not use the incident field $E_i$, but the field of a spontaneously formed polariton. In order to make contact with Appendix G and the theory of scattering by phonons we denote the spontaneous field by $E_\xi$. The complete mode equations of the spontaneous mode are (7.63,64) which in the $\xi$ notation are written as

$$\dot{s}_\xi + i\omega_{0x}s_\xi = i\hbar^{-1}M_{0x}\bar{n}_0E_\xi \tag{7.66a}$$

$$-\varepsilon_0 c^2 \nabla \times \nabla \times \boldsymbol{E}_\xi - \varepsilon_0 \tilde{\varepsilon} \ddot{\boldsymbol{E}}_\xi = 2\boldsymbol{M}_{0x} \mathrm{Re}\,\{\ddot{s}_\xi\} \ . \tag{7.66b}$$

In our model with spatial dispersion neglected (7.66) yields two polariton modes (longitudinal and transverse); when spatial dispersion is added (see Chap. 4) one finds two transverse plus one longitudinal mode.

Inserting $E_\xi$ into (7.62a), as explained above, one obtains an equation for the scattered exciton $s_s$

$$\dot{s}_s + i\omega_{0x} s_s - i\hbar^{-1} M_{0x} \bar{n}_0 E_s = i\hbar^{-1} M_{xb} v_2 E_\xi \ . \tag{7.67}$$

To this equation is added the field equation for the scattered field

$$-\varepsilon_0 c^2 \nabla \times \nabla \times \boldsymbol{E}_s - \varepsilon_0 \tilde{\varepsilon} \ddot{\boldsymbol{E}}_s = 2\boldsymbol{M}_{0x} \mathrm{Re}\,\{\ddot{s}_s\} \ . \tag{7.68}$$

Equation (7.67) together with (7.68) describes the scattered polariton. Integrating formally as in Sect. 7.2 by means of a Green's function which is a simplified version of the function considered in Appendix E, we write the scattered polariton

$$s_s = G_{ss} * E_\xi \ ; \quad E_s = G_{Es} * E_\xi \ . \tag{7.69}$$

The scattered intensity then is found in the way as explained in Appendix G, the only difference being that the incident amplitude $Y_i$ is replaced by $v_2$. Since the condition $kT \ll \hbar\omega_{0x}$ is certainly fulfilled, only the Stokes signal from the ground-state fluctuation will contribute. Assuming that two transverse and one longitudinal mode are present, the theory predicts scattering signals at the following $(\boldsymbol{q}, \omega)$-values

$$\begin{aligned} \omega_{s1} &= 2\omega_i - \omega_{t1} \qquad & \boldsymbol{q}_{s1} &= 2\boldsymbol{q}_i - \boldsymbol{q}_{t1} \\ \omega_{s2} &= 2\omega_i - \omega_{t1} \qquad & \boldsymbol{q}_{s1} &= 2\boldsymbol{q}_i - \boldsymbol{q}_{t2} \\ \omega_{s3} &= 2\omega_i - \omega_\ell \qquad & \boldsymbol{q}_{s3} &= 2\boldsymbol{q}_i - \boldsymbol{q}_\ell \ . \end{aligned} \tag{7.70}$$

The resonance enhancement of the scattering is given by the appropriate polariton Green's functions used in solving (7.62a) in the first step, (7.65) in the second step, and (7.67, 68) in the last step.

Experimentally hyper-Raman scattering has been observed in various substances. It has become a valuable tool for $\boldsymbol{q}$-space spectroscopy (Hõnerlage et al., 1985).

Besides the incoherent hyper-Raman effect with the source polarization

$$P_s \propto E_i^2 E_\xi \tag{7.71}$$

there exists also a corresponding coherent effect (Maruani et al., 1978; Chemla et al., 1979). A theory of the coherent hyper-Raman effect is almost the same as above, only in (7.67) the random field $E_\xi$ is replaced by the field $E_t$ of an externally driven test beam.

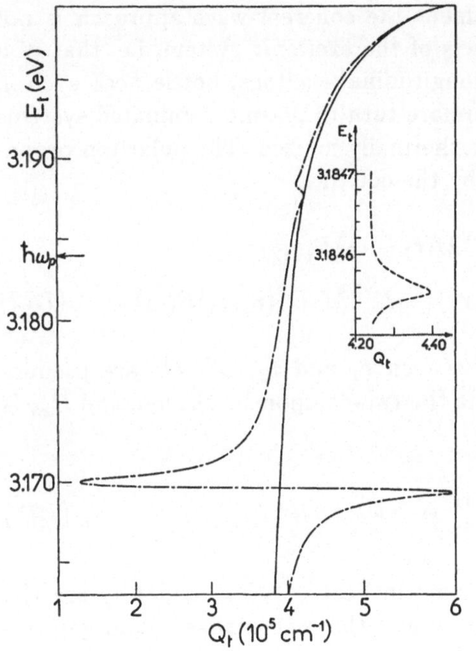


**Fig. 7.7.** Biexciton anomaly in the polariton dispersion of CuCl. (After Hönerlage and Bigot, 1984)

Further information on nonlinear processes in the exciton-biexciton system can be obtained from the model equations (7.62) if one considers nonpertubative solutions (Hönerlage and Bigot, 1984). Using a method of coupled Fourier amplitudes these authors developed a theory of so-called "renormalization effects". An example is an observed anomaly in the dispersion of the exciton polariton near the frequency $\omega = \omega_{0b}/2$ (Fig. 7.7).

In conclusion it can be said that the density matrix approach or the equivalent coherent-wave treatment of the dynamics of the exciton-biexciton system provides valuable insight even if, as a crude approximation, mobility effects are neglected completely.

## 7.7 Processes in a Dense Electron-Hole Plasma

The irreversible processes following photoexcitation near and above the band edge give rise to a large variety of phenomena. At low excitation densities the informative effects are photoconductivity and photoluminescence.

In the case of higher intensity of the primary illumination, the pronounced pile-up of excitations alters the dielectric response significantly. As reviewed thoroughly by Klingshirn and Haug (1981) the high excitation effects fall into two categories, namely those of a dense exciton-biexciton gas and those of a dense electron-hole plasma.

In the present stage of development the coherent-wave approach is not suited for treating the electrodynamcis of the excitonic system, i.e. that of a thermalized gas of triplet excitons, longitudinal excitons, bottle neck exciton polaritons and/or biexcitons. We therefore turn to plasma dominated systems in which excitons are screened out or thermally ionized. The polariton propagation in such a system is described by the equation

$$\dot{Y}(\boldsymbol{r}_1,\boldsymbol{r}_2) + i\overline{\Omega}_{eh}Y(\boldsymbol{r}_1,\boldsymbol{r}_2) = i\hbar^{-1}\boldsymbol{M}(\boldsymbol{r}_1-\boldsymbol{r}_2)\boldsymbol{E}(\boldsymbol{r}_{12}) - i\hbar^{-1}\boldsymbol{M}_0C^0(\boldsymbol{r}_1,\boldsymbol{r}_2)\boldsymbol{E}(\boldsymbol{r}_1) - i\hbar^{-1}\boldsymbol{M}_0D^0(\boldsymbol{r}_2,\boldsymbol{r}_1)\boldsymbol{E}(\boldsymbol{r}_2) \tag{7.72}$$

where $\boldsymbol{r}_{12}$ is a position somewhere between $\boldsymbol{r}_1$ and $\boldsymbol{r}_2$, $C^0$, $D^0$ are pseudo-equilibrium density matrices describing the two-component plasma, and $\overline{\Omega}_{eh}$ is a modified drift operator

$$\overline{\Omega}_{eh} = -\frac{\hbar}{2m_h}\nabla_1^2 - \frac{\hbar}{2m_e}\nabla_2^2 + \hbar^{-1}\overline{V}_{eh} + \overline{\omega}_g + \Omega_x \tag{7.73}$$

where $\Omega_x$ is the exchange term discussed in Sect. 3.2; $\overline{V}_{eh}$ and $\hbar\overline{\omega}_g$ are the plasma-screened electron-hole interaction and the renormalized band gap, respectively. Both quantities have been subject to extensive studies (Zimmermann et al., 1978; Schweizer et al., 1983; Rösler et al., 1984; Banyai and Koch, 1986). A very simple way of obtaining closed formulae for $\overline{\omega}_g$ and $\overline{V}_{eh}$ is to extend the two-test charge model of Banyai and Koch (1986) to include a Lindhard-type response. Details of the derivation are given in Appendix H. The results can be expressed as

$$\hbar(\overline{\omega}_g - \omega_g) = \frac{e^2}{(2\pi)^3\varepsilon_s\varepsilon_0}\int\frac{R}{q^2(\varepsilon_s\varepsilon_0q^2 - R)}d^3\boldsymbol{q} \tag{7.74}$$

$$\overline{V}_{eh}(r) = -\frac{e^2}{(2\pi)^3\varepsilon_s\varepsilon_0}\int\frac{\cos(\boldsymbol{q}\cdot\boldsymbol{r})}{q^2[1 - R/(\varepsilon_s\varepsilon_0q^2)]}d^3\boldsymbol{q} \tag{7.75}$$

where

$$R = \frac{e^2}{(2\pi)^3}\int\left(\frac{f_e(\boldsymbol{k}) - f_e(\boldsymbol{k}-\boldsymbol{q})}{E_e(\boldsymbol{k}) - E_e(\boldsymbol{k}-\boldsymbol{q})} + \frac{f_h(\boldsymbol{k}) - f_h(\boldsymbol{k}-\boldsymbol{q})}{E_h(\boldsymbol{k}) - E_h(\boldsymbol{k}-\boldsymbol{q})}\right)d^3\boldsymbol{k}\,. \tag{7.76}$$

In this expression for $R$, $f_e$ and $f_h$ are distribution functions of electrons and holes [cf. (6.72)].

Neglecting the $\boldsymbol{q}$-dependence of $R$ the integrals (7.74,75) can be evaluated analytically. The results are

$$\hbar(\overline{\omega}_g - \omega_g) = \frac{-e^2}{4\pi\varepsilon_0\varepsilon_s}\kappa_0 \tag{7.77}$$

$$\overline{V}_{eh}(r) = \frac{-e^2}{4\pi\varepsilon_0\varepsilon_s r}e^{-\kappa_0 r} \tag{7.78}$$

$$\kappa_0 = (-R/\varepsilon_0\varepsilon_s)^{1/2} . \tag{7.79}$$

The parameter $\kappa_0$ is the inverse Thomas-Fermi or Debye-Hückel screening length, and the Mott criterion (Mott, 1974) for the absence of exciton states is $\kappa_0 a_B \gtrsim 1$.

The linear optical response controlled by (7.72,73) can be calculated by very simple methods in the common case of long waves in center-of-mass space. Then, $\boldsymbol{E}$ varies little on the scale of the coherence length of $C^0$ and $D^0$. By analogy with the arguments leading to (4.24), the susceptibility becomes

$$\chi_\pm = (\varepsilon_0\hbar)^{-1} \iint M(\boldsymbol{r}')G_\pm(\boldsymbol{r},\boldsymbol{r}')[M(\boldsymbol{r}) - M_0(\tilde{C}^0(0,\boldsymbol{r}) + \tilde{D}^0(0,\boldsymbol{r})]d^3\boldsymbol{r}\,d^3\boldsymbol{r}' . \tag{7.80}$$

Here, $\tilde{C}^0$ and $\tilde{D}^0$ are density matrices expressed as functions of the midpoint and the relative coordinates [cf. (6.3,4)]. As we consider a homogeneous plasma, $\tilde{C}^0$ and $\tilde{D}^0$ are independent of the midpoint coordinate.

The evaluation of (7.80) is particularly simple when $\overline{V}_{eh}$ and $\Omega_x$ are neglected and the shell-source approximation for $M(\boldsymbol{r})$ (Sect. 4.3) is applied. As Green's function we use (4.41), but we replace the variable $\kappa$ by $k = i\kappa$ :

$$k = \left[\frac{2\mu}{\hbar}(\omega + i\eta - \overline{\omega}_g)\right]^{1/2} . \tag{7.81}$$

Then,

$$g_0(r,\varrho) = \frac{2i\mu}{4\pi\hbar k r\varrho}\sin(kr)e^{ik\varrho} \qquad r<\varrho \tag{7.82a}$$

$$g_0(r,\varrho) = \frac{2i\mu}{4\pi\hbar k r\varrho}\sin(k\varrho)e^{ikr} \qquad \varrho<r . \tag{7.82b}$$

Denoting the shell radius by $\varrho_0$ and inserting (7.82) into (7.80) one finds by straightforward algebra:

$$\begin{aligned}
\chi_- &= (\varepsilon_0\hbar)^{-1}M_0^2[g_0(\varrho_0,\varrho_0) - \int g_0(r,\varrho_0)(\tilde{C}^0(0,r) + \tilde{D}^0(0,r))4\pi r^2 dr] \\
&= \frac{2\mu M_0^2}{4\pi\varepsilon_0\hbar^2}\left\{\frac{i[1-\exp 2ik\varrho_0]}{2k\varrho_0^2}\right. \\
&\quad - i\varrho_0^{-1}\sin(k\varrho_0)\int_0^\infty (kr)^{-1}(\tilde{C}^0(0,r) + \tilde{D}^0(0,r))\sin(kr)4\pi r^2 dr \\
&\quad - \varrho_0^{-1}\cos(k\varrho_0)\int_0^{\varrho_0} (kr)^{-1}(\tilde{C}^0(0,r) + \tilde{D}^0(0,r))\sin(kr)4\pi r^2 dr \\
&\quad \left. - \varrho_0^{-1}\sin(k\varrho_0)\int_{\varrho_0}^\infty (kr)^{-1}(\tilde{C}^0(0,r) + \tilde{D}^0(0,r))\cos(kr)4\pi r^2 dr\right\} .
\end{aligned} \tag{7.83}$$

In equilibrium $\tilde{C}^0$ and $\tilde{D}^0$ are real. It then follows from (7.80) and (7.82) that $\chi_-$ is real below the (renormalized) gap frequency. Above the gap the last two terms in the square bracket of (7.83) are real, whereas the second term is imaginary. One can relate this second term very simply to the $k$-space distribution functions $f_e(k)$ and $f_h(k)$ of electrons and holes, respectively. Using the Fourier transformation (6.12b) the imaginary part of $\chi$ becomes

$$\chi'' = \frac{2M_0^2\mu \sin(k\varrho_0)}{4\pi\hbar^2\varepsilon_0\varrho_0}\left(\frac{\sin(k\varrho_0)}{k\varrho_0} - f_e(k) - f_h(k)\right) \; ; \quad \omega > \overline{\omega}_g \; . \tag{7.84}$$

To first order in $k\varrho_0$, corresponding to the validity range of the shell source approximation, one finds the text-book result

$$\chi'' = \chi''_{dark}(1 - f_e(k) - f_h(k)) \tag{7.85}$$

where $\chi_{dark}$ is the absorptive part of $\chi$ in the absence of the electron-hole plasma. At zero temperature

$$f_e(k) = f_h(k) = \theta(k_F - k) \; .$$

$k_F$ is the Fermi wave vector. Then there is optical gain between $\overline{\omega}_g$ and $\omega_F$, $\hbar\omega_F$ being the pair chemical potential.

This gain in highly excited semiconductors has been studied by several experimental groups (Göbel, 1974; Leheny and Shah, 1976; Hildebrand and Göbel, 1976; Bohnert et al., 1978; Majumder et al., 1985). The theories in which both electron-hole interaction and band filling effects are included are complicated. Several groups have treated this problem by means of many-body Green's function techniques in a $\boldsymbol{k}$-space representation (Schmitt-Rink et al., 1980; Arya and Hanke, 1980; Klingshirn and Haug, 1981; Haug and Schmitt-Rink, 1984). These theories can reproduce the essential features of the experiments.

When applying the coherent wave theory it is relatively easy to include the directly screened interaction $\overline{V}_{eh}$, because $\overline{\Omega}_{eh}$ retains its structure as a differential operator. Simple numerical methods can be employed for calculating real space Green's functions $G(\boldsymbol{r}_1, \boldsymbol{r}_2)$ in this case. But when the exchange term $\Omega_x$ in (7.73) is included then the calculation of a Green's function becomes rather complicated because then the pertinent operator has become an integro-differential operator. As shown in Sect. 3.2, $\Omega_x Y$ contains terms of the form

$$\hbar^{-1}\int \overline{V}_{eh}(\boldsymbol{r}_1 - \boldsymbol{r})C(\boldsymbol{r}_2, \boldsymbol{r})Y(\boldsymbol{r}_1, \boldsymbol{r})d^3\boldsymbol{r} \; .$$

It remains to be seen if the numerical procedure in which such Green's functions are found and inserted in (7.80) offers any advantage compared to previously published methods, for instance with respect to speed and accuracy. Besides the exchange screening $\Omega_x$ having its counterpart in the Green's function approach (Haug and Schmitt-Rink, 1984), the coherent wave theory yields a variety of other nonlinear terms, some of which are of the source type. At present very little is known about the relevance of these terms.

It should finally be noted, that all the considerations in this section can be applied equally well to the case of doped (rather than photoexcited) semiconductors. Using (7.76–80) with $f_h = D = 0$ one can explore the influence of the electron-hole interaction on the Burstein-shifted absorption edge in degenerate $n$-type semiconductors.

## 7.8 Spontaneous Band-to-Band Recombination

In the previous section the stimulated processes in a dense plasma were studied. In contrast to this, the present section deals with the spontaneous radiative recombination of electron-hole pairs. This process is nonlinear in the sense that the emission can be calculated from the expectation value of $\hat{Y}^\dagger \hat{Y}$.

We consider as the source for the band-to-band luminescence a distribution of electrons and holes described by the density matrices $C(\boldsymbol{r}_e, \boldsymbol{r}'_e)$ and $D(\boldsymbol{r}_h, \boldsymbol{r}'_h)$. If radiative damping is neglected, $C$ and $D$ can be for all time identified with their initial values at time zero. The densities $n_e = C(\boldsymbol{r}, \boldsymbol{r}, 0)$ and $n_h = D(\boldsymbol{r}, \boldsymbol{r}, 0)$ do not necessarily have an appreciable overlap. In junctions, for example, the electrons and holes are often separated in space (Yuan et al., 1985; Döhler, 1983). Then recombination will still occur if there exists a sizeable overlap of the wave functions tailing in the forbidden gap. For the density matrices $C$ and $D$ this amounts to an overlap of the off-diagonal parts. In this sense the tunneling luminescence is a coherence effect.

A calculation of the emitted intensity of radiation involves the following steps:

a) The expectation value $\hat{Y}^\dagger \hat{Y}$ at time zero is rearranged into a product of density matrices:

$$\begin{aligned}\langle \hat{Y}^\dagger_{12} \hat{Y}_{34} \rangle &= \langle c^\dagger_2 d^\dagger_1 d_3 c_4 \rangle = \langle c^\dagger_2 c_4 d^\dagger_1 d_3 \rangle \\ &= \langle c^\dagger_2 c_4 \rangle \langle d^\dagger_1 d_3 \rangle = C_{24} D_{13} \ .\end{aligned} \tag{7.86}$$

In this step it is assumed that initially electrons and holes are uncorrelated.

b) The interband edge equations

$$\left( \frac{\partial}{\partial t} + i\Omega_{eh} \right) \hat{Y} = 0 \tag{7.87a}$$

$$\left( \frac{\partial}{\partial t} - i\Omega_{eh} \right) \hat{Y}^\dagger = 0 \tag{7.87b}$$

are used to relate $\hat{Y}$ and $\hat{Y}^\dagger$ at a time $t$ with the inital value considered in (7.86). We use the source-free equations because we neglect stimulated processes.

c) Knowing $\hat{Y}^{\dagger}(\boldsymbol{r}_h, \boldsymbol{r}_e, t)$ and $\hat{Y}(\boldsymbol{r}_h, \boldsymbol{r}_e, t)$ one can find the operator-valued polarization

$$\hat{P}(\boldsymbol{r}, t) = M_0(\hat{Y}(\boldsymbol{r}, \boldsymbol{r}, t) + \hat{Y}^{\dagger}(\boldsymbol{r}, \boldsymbol{r}, t)) \ . \tag{7.88}$$

$\hat{P}(\boldsymbol{r}, t)$ acts as a source in the field equation

$$c^2 \nabla^2 \hat{E} - \frac{\partial^2}{\partial t^2}\hat{E} = \frac{1}{\varepsilon_0}\frac{\partial^2}{\partial t^2}\hat{P} \ . \tag{7.89}$$

d) Finally the intensity spectrum at a given point is related to $\hat{E}(\boldsymbol{r}, t)$.

Let us perform the above steps in a reverse succession and in more detail. Beginning with step (d) we use the result of Appendix K (K.8)

$$\frac{dI}{d\omega} = \frac{1}{2\pi}\int \mathrm{e}^{-i\omega\tau}\langle \hat{E}^{\dagger}(\boldsymbol{r}, t+\tau)\hat{E}(\boldsymbol{r}, t)\rangle d\tau \ . \tag{7.90}$$

The field equation (7.89) together with (7.88) is solved by means of the retarded Green's function

$$\hat{E}(\boldsymbol{r}, t) = M_0 \int \frac{\partial^2}{\partial t^2}\frac{\hat{Y}(\boldsymbol{r}', \boldsymbol{r}', t - |\boldsymbol{r} - \boldsymbol{r}'|c^{-1})}{4\pi\varepsilon_0|\boldsymbol{r} - \boldsymbol{r}'|}d^3\boldsymbol{r}' \ . \tag{7.91}$$

Note that in principle we should add to (7.91) a free-field solution. But if we assume that the field is initially in the vacuum state, this free-field part will not contribute to the expectation value. Hence it has been omitted. A further simplification is possible in (7.91) if $\boldsymbol{r}$ is in the far-field region and the emitting volume, i.e. the overlap region of the two density matrices, is small compared to the wavelength. Then $|\boldsymbol{r} - \boldsymbol{r}'|$ in (7.91) can be replaced by $r = |\boldsymbol{r}|$ everywhere and we get

$$\hat{E}(\boldsymbol{r}, t) = \frac{M_0}{4\pi\varepsilon_0 r}\int \frac{\partial^2}{\partial t^2}\hat{Y}[\boldsymbol{r}', \boldsymbol{r}', t - (r/c)]d^3\boldsymbol{r}' \ . \tag{7.92}$$

Together with the analogous formula for $\hat{E}^{\dagger}$, (7.92) has to be inserted into (7.90)

$$\frac{dI}{d\omega} = \frac{M_0^2\omega^4}{8\pi^2\varepsilon_0 r^2}\int \mathrm{e}^{i\omega\tau}\langle \hat{Y}^{\dagger}(\boldsymbol{r}'', \boldsymbol{r}'', 0)\hat{Y}(\boldsymbol{r}', \boldsymbol{r}', \tau)\rangle d^3\boldsymbol{r}' d^3\boldsymbol{r}'' d\tau \ . \tag{7.93}$$

Note that the time derivatives of $\hat{Y}^{\dagger}$ and $\hat{Y}$ have been replaced by the factor $\omega^4$ in front of the integral. In the next step (7.87a) is integrated in order to trace $\hat{Y}(\boldsymbol{r}', \boldsymbol{r}', \tau)$ back to its initial value. The result is expressed in terms of the time development operator $U(\boldsymbol{r}_h, \boldsymbol{r}_e, \boldsymbol{r}'_h, \boldsymbol{r}'_e, t)$ solving the initial value problem of (7.85a)

$$\hat{Y}(\boldsymbol{r}_h, \boldsymbol{r}_e, t) = \int U(\boldsymbol{r}_h, \boldsymbol{r}_e, \boldsymbol{r}'_h, \boldsymbol{r}'_e, t)\hat{Y}(\boldsymbol{r}'_h, \boldsymbol{r}'_e, 0)d^3\boldsymbol{r}'_h d^3\boldsymbol{r}'_e \ . \tag{7.94}$$

$U$ obeys (7.87a) and the initial condition

$$U(\boldsymbol{r}_h, \boldsymbol{r}_e, \boldsymbol{r}'_h, \boldsymbol{r}'_e, 0) = \delta(\boldsymbol{r}_h - \boldsymbol{r}'_h)\delta(\boldsymbol{r}_e - \boldsymbol{r}'_e) \ . \tag{7.95}$$

Applying (7.94) and (7.93) one is led to an expression with an equal-time product $\hat{Y}^\dagger \hat{Y}$ under the expectation value. Hence we can make use of (7.86) and replace this product by a product of density matrices. In this way we get

$$\frac{dI}{d\omega} = \frac{M_0^2 \omega^4}{8\pi^2 \varepsilon_0 r^2} \int e^{i\omega\tau} U(\boldsymbol{r}', \boldsymbol{r}', \boldsymbol{r}'_h, \boldsymbol{r}'_e, \tau) C(\boldsymbol{r}'', \boldsymbol{r}'_e) D(\boldsymbol{r}'', \boldsymbol{r}'_h) d\tau\, d^4\Omega \tag{7.96}$$

where $d^4\Omega$ denotes integration over all spatial variables.

The final formula (7.96) is still rather cumbersome because of all the spatial integrations which must be performed. A considerable simplification is achieved if the electron-hole interaction is neglected in the calculation of the time development operator $U$. Then $U$ can be represented in terms of single-particle stationary states which are found by a separation of variables in (7.87)

$$U(\boldsymbol{r}_h, \boldsymbol{r}_e, \boldsymbol{r}'_h, \boldsymbol{r}'_e, t) = \sum_{m,n} \varphi_n(\boldsymbol{r}_h) \varphi_n^*(\boldsymbol{r}'_h) \times \psi_m(\boldsymbol{r}_e) \psi_m^*(\boldsymbol{r}'_e) e^{-i(\omega_g + \tilde{\omega}_n + \omega_m)t} \ . \tag{7.97}$$

The stationary-state wave-functions $\varphi_m$, $\varphi_n$ (for electrons and holes, respectively) obey the equations

$$\left(-\frac{\hbar^2}{2m_e}\nabla^2 - e\Phi - \hbar\omega_m\right)\psi_m = 0 \tag{7.98a}$$

$$\left(-\frac{\hbar^2}{2m_h}\nabla^2 + e\Phi - \hbar\tilde{\omega}_n\right)\varphi_n = 0 \ . \tag{7.98b}$$

The same basis may be used in a spectral representation (6.42a) of the density matrices

$$C(\boldsymbol{r}''_e, \boldsymbol{r}'_e) = \sum_m f_{em} \psi_m^*(\boldsymbol{r}''_e) \psi_m(\boldsymbol{r}'_e) \tag{7.99a}$$

$$D(\boldsymbol{r}''_h, \boldsymbol{r}'_h) = \sum_n f_{hn} \varphi_n^*(\boldsymbol{r}''_h) \varphi_n(\boldsymbol{r}'_h) \ . \tag{7.99b}$$

Inserting (7.97) and (7.99) into (7.96) one can perform part of the spatial integrations and the temporal integration. Thus one arrives at

$$\frac{dI}{d\omega} = \frac{M_0^2 \omega^4}{4\pi \varepsilon_0 r^2} \int \sum_{mn} f_{em} \psi_m^*(\boldsymbol{r}'') \psi_m(\boldsymbol{r}') f_{hn} \varphi_n^*(\boldsymbol{r}'') \varphi_n(\boldsymbol{r}') \times \delta(\omega_g + \tilde{\omega}_n + \omega_m - \omega) d^2\Omega \tag{7.100}$$

where $d^2\Omega = d^3\mathbf{r}'d^3\mathbf{r}''$. More realistic than the artificial discretization is the assumption of a continuous manifold of states parameterized by the variables $\omega_e, \omega_h$. Then (7.100) goes over into

$$\frac{dI}{d\omega} = \frac{M_0^2\omega^4}{4\pi\varepsilon_0 r^2} \int g_e f_e g_h f_h \psi_e^*(\mathbf{r}'')\psi_e(\mathbf{r}')\varphi_h^*(\mathbf{r}'')\varphi_h(\mathbf{r}')$$
$$\delta(\omega_g + \omega_e + \omega_h - \omega) d^2\Omega \, d\omega_e d\omega_h \qquad (7.101)$$

$g_e(\omega_e)$ and $g_h(\omega_h)$ denoting densities of states for electrons and holes, respectively.

The intuitively simple result (7.101) shows that the spectral density of the emitted light is given by that part of the trace product

$$\int C(\mathbf{r}'', \mathbf{r}')D^*(\mathbf{r}', \mathbf{r}'')d^2\Omega$$

which is selected by the spectral window. The selection is achieved by the $\delta$-function $\delta(\omega_g + \omega_e + \omega_h - \omega)$. How much this typical result of a single particle approximation will be modified if in (7.96) the electron-hole interaction is taken into account remains to be investigated.

# Appendix

## A. Band-Limited Sampling

Suppose we have a lattice with position vectors $\boldsymbol{R}_j$. To each lattice site we assign a number $a_j$. The set of all numbers $a_j$ is denoted by $\{a_j\}$. Our aim is to construct a function $a(\boldsymbol{r})$ on the continuum of space points $\boldsymbol{r}$ which interpolates between the discrete set $\{a_j\}$ in such a way that $a(\boldsymbol{r})$ contains only macroscopic structures. This task is given a precise meaning if we use the concept of band limitation which in a crystal is realized by Fourier analysis restricted to the first Brillouin zone. We therefore define the Fourier representation $A(\boldsymbol{k})$ of the interpolating function $a(\boldsymbol{r})$ as follows:

$$A(\boldsymbol{k}) = \sum_j a_j \mathrm{e}^{i\boldsymbol{k}\boldsymbol{R}_j} \tag{A.1}$$

where $\boldsymbol{k}$ is confined to the first Brillouin zone. From $A(\boldsymbol{k})$ we derive $a(\boldsymbol{r})$ :

$$a(\boldsymbol{r}) = \Omega^{1/2}(2\pi)^{-3} \int_{BZ} \mathrm{e}^{i\boldsymbol{k}\boldsymbol{r}} A(\boldsymbol{k}) d^3\boldsymbol{k} \tag{A.2}$$

with $\Omega$ being the volume of a crystal cell. This gives

$$a(\boldsymbol{r}) = \sum_j a_j \Delta(\boldsymbol{r} - \boldsymbol{R}_j) \tag{A.3}$$

with the sampling function $\Delta$ being

$$\Delta(\boldsymbol{r} - \boldsymbol{R}_j) = \Omega^{1/2}(2\pi)^{-3} \int_{BZ} \mathrm{e}^{i\boldsymbol{k}(\boldsymbol{r}-\boldsymbol{R}_j)} d^3\boldsymbol{k} \ . \tag{A.4}$$

The prefactor is chosen in such a way that the sampling functions form an orthonormalized set

$$\int \Delta(\boldsymbol{r} - \boldsymbol{R}_j)\Delta(\boldsymbol{r} - \boldsymbol{R}_k) d^3\boldsymbol{r} = \delta_{jk} \ . \tag{A.5}$$

On the other hand it has the effect that the "interpolation" provides a factor $\Omega^{-1/2}$ :

$$a(\boldsymbol{r} = \boldsymbol{R}_j) = \Omega^{-1/2} a_j \ . \tag{A.6}$$

The set of functions $\Delta(\boldsymbol{r}-\boldsymbol{R}_j)$ is also complete in the subspace of band-limited functions, i.e. if

$$\varphi(\boldsymbol{r})=\sum_j \varphi_j \Delta(\boldsymbol{r}-\boldsymbol{R}_j) \tag{A.7}$$

then

$$\int \varphi(\boldsymbol{r}') \sum_\ell \Delta(\boldsymbol{r}'-\boldsymbol{R}_\ell)\Delta(\boldsymbol{r}-\boldsymbol{R}_\ell) d^3\boldsymbol{r}'=\varphi(\boldsymbol{r}) \; . \tag{A.8}$$

This expresses in the subspace of band-limited functions the typical property of a $\delta$-function and thus leads to the concept of a band limited $\delta$-function

$$\delta_B(\boldsymbol{r}'-\boldsymbol{r})=\sum_\ell \Delta(\boldsymbol{r}'-\boldsymbol{R}_\ell)\Delta(\boldsymbol{r}-\boldsymbol{R}_\ell) \; . \tag{A.9}$$

$\delta_B$ is the band-limited continuum representative of the discrete Kronecker symbol:

$$\delta_B(\boldsymbol{r}_1-\boldsymbol{r}_2)=\sum_{ij} \delta_{ij} \Delta(\boldsymbol{r}_1-\boldsymbol{R}_i)\Delta(\boldsymbol{r}_2-\boldsymbol{R}_j) \; . \tag{A.10}$$

# B. Local Gauge in the Interband Hamiltonian

The standard gauge in solid-state theory is the global Coulomb gauge (Jackson, 1975; Power, 1964). In this gauge, transverse fields are derived from a vector potential $\boldsymbol{A}$, and longitudinal fields are derived from a scalar potential $\Phi$

$$\boldsymbol{E}_t=-\dot{\boldsymbol{A}} \; ; \quad \nabla\cdot\boldsymbol{A}=0 \; ; \quad \boldsymbol{E}_\ell=-\nabla\Phi \; . \tag{B.1}$$

The interaction of electrons with transverse fields is incorporated in the canonical form of the kinetic energy

$$H_{kin}=\frac{1}{2m_0}\int \psi^*(i\hbar\nabla-e\boldsymbol{A})^2\psi \, d^3\boldsymbol{x} \tag{B.2}$$

while the interaction with the scalar potential is rerresented by the expression known from electrostatics

$$H^\ell=-e\int \psi^*\psi\Phi \, d^3\boldsymbol{x} \; . \tag{B.3}$$

In atomic physics it is often useful to apply a gauge transformation by which electric multipole contributions are removed from $H_{kin}$ and accounted for by a term $H^t$ formally similar to (B.3)

$$H^t=-\int \boldsymbol{P}\cdot\boldsymbol{E}_t d^3\boldsymbol{r} \; . \tag{B.4}$$

$\boldsymbol{P}$ is the polarization (Loudon, 1973). Such a transformation can be found if the field interacts with localized charge-current distributions. The necessary localization means that $H_{kin}$ is a sum of terms referring to individual atoms

$$H_{kin} = \frac{1}{2m_0} \sum_i \int_i \psi^* (i\hbar\nabla - e\boldsymbol{A})^2 \psi \, d^3\boldsymbol{x} \ . \tag{B.5}$$

The important property of (B.5) is that in each term $\boldsymbol{A}$ does not vary too much within the range of integration. If this is the case then a local gauge transformation applied to each term in the sum $\sum_i$ leads to the desired result (Loudon, 1973).

In a two-band semiconductor the field $\boldsymbol{A}$ in general interacts with localized interband currents and with delocalized intraband currents. Then a local gauge transformation in the sense of atomic physics can only be performed in the interband part of the Hamiltonian. This Hamiltonian we write in a Wannier representation

$$H^t = \sum_{ij} H^t_{ij} \ ,$$

$$H^t_{ij} = d_i c_j \int w_v(\boldsymbol{x} - \boldsymbol{R}_i) \frac{1}{2m_0} (i\hbar\nabla - e\boldsymbol{A})^2 w_c(\boldsymbol{x} - \boldsymbol{R}_j) d^3\boldsymbol{x} + \text{h.c.} \tag{B.6}$$

From the fact that a Wannier function $w(\boldsymbol{x} - \boldsymbol{R}_j)$ is localized to the neighbourhood of the lattice point $\boldsymbol{R}_i$ it follows that the integration in $H_{ij}$ comprises at most a few lattice cells.

A gauge transformation leaving the electromagnetic fields and the field-matter interaction unchanged has the general form

$$\boldsymbol{A} \to \boldsymbol{A} - \nabla\chi \qquad \Phi \to \Phi + \dot{\chi}$$

$$\psi \to \psi \ \exp(ie\chi/\hbar) \tag{B.7}$$

with an arbitrary gauge function $\chi(\boldsymbol{x}, t)$. If $\chi$ in addition obeys Laplace's equation

$$\nabla^2 \chi = 0 \tag{B.8}$$

then the gauge transformation even preserves the Coulomb constraint

$$\nabla \cdot \boldsymbol{A} = 0 \ . \tag{B.9}$$

For the local gauge transformation applied to $H_{ij}$ the gauge function $\chi_{ij}$ is chosen as an expansion

$$\chi_{ij}(\boldsymbol{x}) = \left[1 + \frac{1}{2}\boldsymbol{x}_{ij}\nabla' + \frac{1}{3!}(\boldsymbol{x}_{ij}\nabla')^2 + \dots\right](\boldsymbol{x}_{ij}\boldsymbol{A}(\boldsymbol{x}')|_{\boldsymbol{x}'=\boldsymbol{R}_{ij}} \tag{B.10a}$$

$$\boldsymbol{x}_{ij} = \boldsymbol{x} - \boldsymbol{R}_{ij} \ . \tag{B.10b}$$

The reference point $\boldsymbol{R}_{ij}$ can in principle be chosen arbitrarily. For fast convergence of the multipole expansion $\boldsymbol{R}_{ij}$ must be a point in the neighbourhood of $\boldsymbol{R}_i$ and $\boldsymbol{R}_j$. A convenient choice is sometimes to take $\boldsymbol{R}_{ij}$ as the center of mass of the electron hole pair

$$\boldsymbol{R}_{ij} \approx (m_h \boldsymbol{R}_i - m_e \boldsymbol{R}_j)/m \tag{B.11}$$

where $\approx$ means lattice vector closest to the r.h.s. Note that in (B.11) there are used effective masses, whereas in (B.2), $m_0$ is the bare electron mass.

From (B.7) it is found that the transformed scalar potential is

$$\Phi_{ij} = \Phi - (1 + \tfrac{1}{2}\boldsymbol{x}_{ij}\nabla' + \ldots)(\boldsymbol{x}_{ij}\boldsymbol{E}_t(\boldsymbol{x}'))|_{\boldsymbol{x}'=\boldsymbol{R}_{ij}} \ . \tag{B.12}$$

The extra term in $H^\ell$ that is generated from (B.12) becomes

$$\begin{aligned} \Delta H = ed_i c_j \int w_v(\boldsymbol{x} - \boldsymbol{R}_i)(1 + \tfrac{1}{2}\boldsymbol{x}_{ij}\nabla' + \ldots) \\ \times \boldsymbol{x}_{ij}\boldsymbol{E}_t(\boldsymbol{x}')w_c(\boldsymbol{x} - \boldsymbol{R}_j)d^3\boldsymbol{x} + h.c. \end{aligned} \tag{B.13}$$

Let us consider the lowest-order term

$$\begin{aligned} H_{ij}^D &= -(\hat{Y}_{ij} + \hat{Y}_{ij}^\dagger)\boldsymbol{m}_{ij}\boldsymbol{E}_t(\boldsymbol{R}_{ij}) \\ \boldsymbol{m}_{ij} &= \int w_v^*(\boldsymbol{x} - \boldsymbol{R}_i)(-e\boldsymbol{x}_{ij})w_c(\boldsymbol{x} - \boldsymbol{R}_j)d^3\boldsymbol{x} \ . \end{aligned} \tag{B.14}$$

$H_{ij}^D$ is a dipolar interaction between the field at $\boldsymbol{R}_{ij}$ and the dipole moment produced in the transition $w_v(\boldsymbol{x} - \boldsymbol{R}_i) \rightarrow w_c(\boldsymbol{x} - \boldsymbol{R}_j)$.

In a similar way the higher-order terms in (B.13) are related to higher-order electric multipoles. In particular the term $O(x_{ij}^2)$ describes electric quadrupole interaction

$$\begin{aligned} H_{ij}^Q = -(\hat{Y}_{ij} + \hat{Y}_{ij}^\dagger) \int w_v^*(\boldsymbol{x} - \boldsymbol{R}_i)\left(-\tfrac{1}{2}e\boldsymbol{x}_{ij} \otimes \boldsymbol{x}_{ij}\right) \\ \times w_c(\boldsymbol{x} - \boldsymbol{R}_j)d^3\boldsymbol{x}[\nabla \otimes \boldsymbol{E}_t]_{\boldsymbol{R}_{ij}} \ . \end{aligned} \tag{B.15}$$

It is clear from general properties of gauge invariance that the electric multipole energies which in a local gauge are accounted for in (B.13), by the transformation (B.7) are removed from $H_{kin}$. What is left are the magnetic multipole energies. As an example we consider the lowest-order contribution related to a magnetic dipole transition. It is derived from the transformed vector potential up to order $O(x^2)$. This vector potential is, according to (B.7) and (B.10),

$$A'_\nu(\boldsymbol{x}) = A_\nu(\boldsymbol{x}) - \partial_\nu(x_{ij\kappa}A_\kappa(\boldsymbol{R}_{ij}) + \tfrac{1}{2}x_{ij\kappa}x_{ij\lambda}\partial'_\kappa A_\lambda(\boldsymbol{R}_{ij}))$$
$$= A_\nu(\boldsymbol{x}) - A_\nu(\boldsymbol{R}_{ij}) - \tfrac{1}{2}(x_{ij\lambda}\partial'_\nu A_\lambda(\boldsymbol{R}_{ij})$$
$$+ \tfrac{1}{2}x_{ij\kappa}\partial'_\kappa A_\nu(\boldsymbol{R}_{ij}))\ . \tag{B.16}$$

Note that a summation convention has been applied, and the abbreviation is used

$$\partial'_\nu \boldsymbol{A}(\boldsymbol{R}_{ij}) = \frac{\partial}{\partial x'_\nu}\boldsymbol{A}(\boldsymbol{x}')|_{\boldsymbol{x}'=\boldsymbol{R}_{ij}}\ .$$

Expanding $\boldsymbol{A}$ in a Taylor series up to the order considered, we find

$$A_\nu(\boldsymbol{x}) = A_\nu(\boldsymbol{R}_{ij}) + x_{ij\kappa}\partial'_\kappa A_\nu(\boldsymbol{R}_{ij})\ . \tag{B.17}$$

Inserting (B.17) into (B.16) one obtains

$$A'_\nu(\boldsymbol{x}) = \tfrac{1}{2}(x_{ij\kappa}\partial'_\kappa A_\nu(\boldsymbol{R}_{ij}) - x_{ij\kappa}\partial'_\nu A_\kappa(\boldsymbol{R}_{ij}))\ . \tag{B.18}$$

This result must be inserted into the interaction term derived from (B.6). The lowest-order contribution is then a term linear in $\boldsymbol{A}$. This term becomes

$$H_{ij}^{M1} = d_i c_j \int w_v^*(\boldsymbol{x}-\boldsymbol{R}_i)\frac{e\hbar}{im_0}A'_\nu\partial_\nu w_c(\boldsymbol{x}-\boldsymbol{R}_i)d^3\boldsymbol{x} + \text{h.c.} \tag{B.19}$$
$$= d_i c_j \frac{e\hbar}{im_0}\int w_v^*(\boldsymbol{x}-\boldsymbol{R}_i)(x_{ij\kappa}\partial_\nu - x_{ij\nu}\partial_\kappa)$$
$$\times\ w_c(\boldsymbol{x}-\boldsymbol{R}_j)d^3\boldsymbol{x}\{\partial'_\kappa A_\nu - \partial'_\nu A_\nu\} + \text{h.c.} \tag{B.20}$$

The content of the curly bracket is identified as the magnetic field at the point $\boldsymbol{R}_{ij}$:

$$\{\partial_\kappa A_\nu(\boldsymbol{R}_{ij}) - \partial'_\nu A_\nu(\boldsymbol{R}_{ij})\} = e_{\kappa\nu\mu}B_\mu(\boldsymbol{R}_{ij}) \tag{B.21}$$

where $e_{\kappa\nu\mu}$ is the antisymmetric unit tensor. The quantity in the matrix element is a component of $\boldsymbol{x}_{ij}\times\boldsymbol{J}$; after integration it becomes the magnetic dipole moment. This verifies our assertion that the lowest order interaction term contained in $H'_{kin}$ is the magnetic dipole interaction.

Extrapolating the results found explicitely for $E1$, $E2$ and $M1$-transitions we arrive at the following conclusion: The local gauge generated by the gauge function (B.10) accounts for electric multipole transitions by a Hamiltonian of the form (B.4) where $\boldsymbol{P}$ is composed of electric multipoles according to the scheme

$$\boldsymbol{P} = \boldsymbol{D} - \nabla \underset{\sim}{\boldsymbol{Q}} + \dots \tag{B.22}$$

($\boldsymbol{D}$ =dipole density, $\underset{\sim}{Q}$ =quadrupole density etc.). The transformed vector potential $\boldsymbol{A}'$ left in the kinetic energy couples in the local gauge only to magnetic multipoles.

The electric multipole interaction with the longitudinal field is described by the interband part of the Hamiltonian $H^{\ell}$ (B.3)

$$H^{\ell} = -e\int(\psi_v^*\psi_c + \psi_c^*\psi_v)\Phi\, d^3\boldsymbol{x} \ . \tag{B.23}$$

Expanding $\psi_v$ and $\psi_c$ in the Wannier basis one obtains

$$H^{\ell} = -e\sum_{ij}(d_i c_j + c_j^{\dagger}d_i^{\dagger})\int w_v^*(\boldsymbol{x}-\boldsymbol{R}_i)w_c(\boldsymbol{x}-\boldsymbol{R}_j)\Phi(\boldsymbol{x})d^3\boldsymbol{x} \ . \tag{B.24}$$

Let us consider a single term of this sum

$$H_{ij}^{\ell} = -e(\hat{Y}_{ij} + \hat{Y}_{ij}^{\dagger})\int w_v^*(\boldsymbol{x}-\boldsymbol{R}_i)w_c(\boldsymbol{x}-\boldsymbol{R}_j)\Phi(\boldsymbol{x})d^3\boldsymbol{x} \tag{B.25}$$

and expand $\Phi$ around a reference point $\boldsymbol{R}_{ij}$ in a power series

$$\begin{aligned} H_{ij}^{\ell} = (\hat{Y}_{ij} + \hat{Y}_{ij}^{\dagger})\sum_n \frac{-e}{n!}\int w_v^*(\boldsymbol{x}-\boldsymbol{R}_i) \\ \times w_c(\boldsymbol{x}-\boldsymbol{R}_j)(\boldsymbol{x}_{ij}\nabla')^n\Phi(\boldsymbol{x}')|_{\boldsymbol{R}_{ij}}d^3\boldsymbol{x} \ . \end{aligned} \tag{B.26}$$

Evidently this has the form of a multipolar interaction; in particular the lowest-order term is the dipole interaction

$$H_{ij}^{D} = -(\hat{Y}_{ij} + \hat{Y}_{ij}^{\dagger})\boldsymbol{E}_{\ell}(\boldsymbol{R}_{ij})\int w_v^*(\boldsymbol{x}-\boldsymbol{R}_i)(-e\boldsymbol{x}_{ij})w_c(\boldsymbol{x}-\boldsymbol{R}_j)d^3\boldsymbol{x} \ . \tag{B.27}$$

Combining the results obtained above we may write the Hamiltonian $H_{ij}^{EM}$ for the total electromagnetic interaction

$$H_{ij}^{EM} = H_{ij}^{D} + H_{ij}^{Q} + \ldots + H_{ij}^{M1} + \ldots \tag{B.28}$$

where electric multipole contributions $D, Q, \ldots$ (dipolar, quadrupolar, ...) can be expressed in terms of the electric field irrespective of the longitudinal or transverse character [compare (B.27) with (B.16)], while the magnetic multipole contributions are derived from the kinetic energy Hamiltonian only.

It is of interest to express the polarization operator by the Wannier overlap integral $\boldsymbol{m}_{ij}$ (B.14b) and the pair operators $\hat{Y}_{ij}$, $\hat{Y}_{ij}^{\dagger}$. Dropping the subscript $t$, we can express the dipole interaction Hamiltonian (B.4) as

$$H^{D} = -\int \hat{\boldsymbol{P}}\cdot\boldsymbol{E}\, d^3\boldsymbol{r} = -\sum_{\boldsymbol{R}_{ij}}\hat{\boldsymbol{P}}(\boldsymbol{R}_{ij})\boldsymbol{E}(\boldsymbol{R}_{ij}) \tag{B.29}$$

where $\hat{P}$ is the band limited interpolation of $\hat{P}(R_{ij})$. On the other hand, from (B.14a) and (B.27)

$$H^D = \sum_{ij} H_{ij}^D = -\sum_{R_i}\sum_{R_j}(\hat{Y}_{ij} + \hat{Y}_{ij}^\dagger)m_{ij}E(R_{ij}) \ . \tag{B.30}$$

Rearranging the double sum in (B.30) we finally get

$$\hat{P}(R_{ij}) = \sum_{R=R_i-R_j} (\hat{Y}_{ij} + \hat{Y}_{ij}^\dagger)m_{ij} \ . \tag{B.31}$$

We close with a comment on the proper choice of the reference point $R_{ij}$. In order to achieve fast convergence of the expansion (B.17), $R_{ij}$ should be in the neighbourhood of the overlap region of $w_v(x - R_i)$ and $w_c(x - R_j)$. As long as this condition is satisfied one has a certain freedom to choose $R_{ij}$. Besides the center of mass (B.11), a convenient choice would be the midpoint between $R_i$ and $R_j$. A change of the reference point does not influence the electric dipole interaction (B.16) because of orthogonality of the Wannier functions. In higher order multipoles such as (B.15) or (B.20) a change of the reference point leads to extra terms, but these will be negligible if the field varies sufficiently slowly.

# C. Lagrange Formulation of Interband Dynamics

In this appendix we shall formulate an action principle for the constitutive equations of an interband transition. The coupling to phonons as discussed in Sect. 3.3 is included. Explicitely excluded is the interaction with intraband densities $C$ and $D$. This means that saturation effects are neglected. By omitting the coupling to $C$ and $D$ we elude typical formal difficulties related to the non-holonomous spin-like structure of complete band edge dynamics.

Without phonons the linearized interband equation (3.1a) reads

$$\frac{\hbar}{i}\dot{Y}_{12} + \left[E_g - \frac{\hbar^2}{2m_e}\nabla_2^2 + e(\Phi_1 - \Phi_2) + V_{eh}\right]Y_{12} = M_0\delta_{12}E \ . \tag{C.1}$$

As a complex equation it must be considered as two equations for the two independent functions

$$Y' = \mathrm{Re}\{Y\} \ ; \quad Y'' = \mathrm{Im}\{Y\} \ .$$

Another equivalent possibility is to consider besides (C.1) the equation

$$-\frac{\hbar}{i}\dot{Y}_{12}^* + \left[E_g - \frac{\hbar^2}{2m_h}\nabla_1^2 - \frac{\hbar^2}{2m_e}\nabla_2^2 + e(\Phi_1 - \Phi_2) + V_{eh}\right]Y_{12}^*$$
$$= M_0\delta_{12}E \tag{C.2}$$

and to interprete $Y$ and $Y^*$ as two independent functions. We shall follow the latter possiblity.

Then our problem is stated as follows: Can there be found a Lagrange-density $\Lambda_e$ in configuration space such that (C.1,2) follow from the action principle

$$\delta \int \Lambda_e d^3\boldsymbol{r}_1 d^3\boldsymbol{r}_2 dt = 0 \ ? \tag{C.3}$$

Let us show that the following function has the desired property:

$$\begin{aligned}\Lambda_e = {} & \frac{\hbar}{2i}(Y\dot{Y}^* - Y^*\dot{Y}) - [E_g + e(\Phi_1 - \Phi_2) + V_{eh}]|Y|^2 - \frac{\hbar^2}{2m_h}|\nabla_1 Y|^2 \\ & - \frac{\hbar^2}{2m_e}|\nabla_2 Y|^2 + \boldsymbol{M}_0\delta_{12}\boldsymbol{E}_1(Y + Y^*) \ .\end{aligned} \tag{C.4}$$

$Y$, as always, is a function of two position variables $\boldsymbol{r}_1, \boldsymbol{r}_2$.

The Euler-Lagrange equations derived from the action principle (C.3) are

$$\frac{d}{dt}\frac{\partial \Lambda_e}{\partial \dot{Y}^*} + \sum_\mu \frac{\partial}{\partial x_\mu}\frac{\partial \Lambda_e}{\partial(\partial Y^*/\partial x_\mu)} - \frac{\partial \Lambda_e}{\partial Y^*} = 0 \qquad \text{and} \tag{C.5a}$$

$$\frac{d}{dt}\frac{\partial \Lambda_e}{\partial \dot{Y}} + \sum_\mu \frac{\partial}{\partial x_\mu}\frac{\partial \Lambda_e}{\partial(\partial Y/\partial x_\mu)} - \frac{\partial \Lambda_e}{\partial Y} = 0 \ . \tag{C.5b}$$

($x_\mu$ is to be summed among the six components of $\boldsymbol{r}_1$ and $\boldsymbol{r}_2$). As can be immediately seen, insertion of (C.4) into (C.5) leads to (C.1) and (C.2) respectively.

The Lagrangian (C.4) allows a generalization in various respects:

a) Other coordinate systems in configuration space may be chosen, e.g. relative displacement $\boldsymbol{r}$, and center of mass $\boldsymbol{R}$, of the electron-hole pair. These coordinates are particularly useful if $\Phi = 0$. Then the Lagrangian in coordinates $\boldsymbol{r}$, $\boldsymbol{R}$ reads

$$\begin{aligned}\Lambda_e = {} & \frac{\hbar}{2i}(Y\dot{Y}^* - Y^*\dot{Y}) - [E_g + V_{eh}(\boldsymbol{r})]|Y|^2 - \frac{\hbar^2}{2\mu}|\nabla_r Y|^2 \\ & - \frac{\hbar^2}{2m}|\nabla_R Y|^2 + \boldsymbol{M}_0\delta(\boldsymbol{r})\boldsymbol{E}(\boldsymbol{R})(Y + Y^*) \ .\end{aligned} \tag{C.6}$$

b) The "smeared-out" dipolar coupling $\boldsymbol{M}(\boldsymbol{r})$ as explained in Sect. 3.6 may be introduced. In this case one has to make the replacement in (C.6)

$$\boldsymbol{M}_0\delta(\boldsymbol{r})\boldsymbol{E}(\boldsymbol{R})(Y + Y^*) \rightarrow \boldsymbol{M}(\boldsymbol{r})\boldsymbol{E}(\boldsymbol{R})(Y + Y^*) \ . \tag{C.7}$$

c) Coupling to phonons is introduced by considering also a phonon Lagrangian density $\Lambda_\xi$ and an interaction Lagrangian density $\Lambda_{int}$ as described in Sect. 3.3. Note that $\Lambda_e$ and $\Lambda_{int}$ are densities in configuration space while $\Lambda_\xi$ is a density in real space. The action principle therefore is of the form

$$\delta \int dt \left[ \int \Lambda_\xi d^3\boldsymbol{r} + \int\int \Lambda_e d^3\boldsymbol{r}_1 d^3\boldsymbol{r}_2 + \int\int\int \Lambda_{int} d^3\boldsymbol{r}\, d^3\boldsymbol{r}_1 d^3\boldsymbol{r}_2 \right] . \tag{C.8}$$

How this type of action principle is to be handled has been shown in Sect. 3.3.

d) In principle the radiation field $\boldsymbol{E}$ can also be incorporated into the Lagrangian. Any attempts to do this will be confronted with the problem of the freedom of gauge and the corresponding auxiliary conditions (Power, 1964). A possible choice for the free field Lagrangian to be added is

$$\Lambda_F(\boldsymbol{A}, \dot{\boldsymbol{A}}) = \frac{\varepsilon_0}{2} \left[ \frac{1}{c^2} \dot{\boldsymbol{A}}^2 - (\nabla \times \boldsymbol{A})^2 \right] \tag{C.9}$$

where $\boldsymbol{A}$ is the vector potential. Note that (C.9) refers to Coulomb gauge. Therefore $\boldsymbol{A}$ is related to the transverse radiation fields as follows

$$\boldsymbol{E}_t = -\dot{\boldsymbol{A}} \qquad \boldsymbol{B} = \nabla \times \boldsymbol{A} . \tag{C.10}$$

If the field is treated in terms of a Lagrangian (C.9), one should replace the field $\boldsymbol{E}$ in (C.7) by $\dot{\boldsymbol{A}}$.

# D. Bloch and Wannier Representations of Transition Matrix Elements

Consider a crystal with $N$ unit cells with lattice vectors $\boldsymbol{R}_i$ and volume $\Omega$. A Bloch state with energy $E_n(k)$ has the wave function

$$\varphi_n(\boldsymbol{k}, \boldsymbol{x}) = e^{i\boldsymbol{k}\cdot\boldsymbol{x}} u_n(\boldsymbol{k}, \boldsymbol{x}) \tag{D.1}$$

where $n$ is a band index and $u_n(\boldsymbol{k}, \boldsymbol{x})$ is periodic in the (real) $\boldsymbol{x}$ space and the reciprocal $\boldsymbol{k}$ space. In the first Brillouin zone there are $N$ discrete values of $\boldsymbol{k}$ which are evenly distributed. Bloch states are orthogonal:

$$\int \varphi_m^*(\boldsymbol{k}_i, \boldsymbol{x}) \varphi_n(\boldsymbol{k}_j, \boldsymbol{x}) d^3\boldsymbol{x} = \delta_{ij}\delta_{mn} . \tag{D.2}$$

Wannier functions are nonstationary states. The function derived from band $n$ and localized at $\boldsymbol{R}_i$ is given by (Knox, 1963)

$$w_n(\boldsymbol{x} - \boldsymbol{R}_i) = N^{-1/2} \sum_j e^{i\boldsymbol{k}_j\cdot\boldsymbol{R}_i} \varphi_n(\boldsymbol{k}_j, \boldsymbol{x}) \tag{D.3}$$

where the sum runs over allowed values of $\boldsymbol{k}$ in the first Brillouin zone.

In the coherent-wave theory it is of interest to calculate the transition dipole

$$m_{ij} = -e \int w_v^*(\boldsymbol{x} - \boldsymbol{R}_i)(\boldsymbol{x} - \boldsymbol{R}_{ij}) w_c(\boldsymbol{x} - \boldsymbol{R}_j) d^3\boldsymbol{x} \tag{D.4}$$

where $\boldsymbol{R}_{ij}$ is somewhere between $\boldsymbol{R}_i$ and $\boldsymbol{R}_j$. When (D.1,2) is inserted in (D.4) one obtains

$$m_{ij} = \frac{e}{N} \sum_{\ell m} e^{i(\boldsymbol{k}_\ell \cdot \boldsymbol{R}_j - \boldsymbol{k}_m \cdot \boldsymbol{R}_i)} \int \varphi_v^*(\boldsymbol{k}_m, \boldsymbol{x})(\boldsymbol{x} - \boldsymbol{R}_{ij}) \varphi_c(\boldsymbol{k}_\ell, \boldsymbol{x}) d^3\boldsymbol{x} \ . \tag{D.5}$$

According to the orthogonality (D.2) we have for any lattice vector $\boldsymbol{R}$

$$\int \varphi_v^*(\boldsymbol{k}_m, \boldsymbol{x})(\boldsymbol{x} - \boldsymbol{R}) \varphi_c(\boldsymbol{k}_\ell, \boldsymbol{x}) d^3\boldsymbol{x} = \int \varphi_v^*(\boldsymbol{k}_m, \boldsymbol{x}) \boldsymbol{x} \varphi_c(\boldsymbol{k}_\ell, \boldsymbol{x}) d^3\boldsymbol{x} \ . \tag{D.6}$$

Using this we may write

$$m_{ij} = -\frac{e}{N^2} \sum_{\ell m} e^{i(\boldsymbol{k}_\ell \cdot \boldsymbol{R}_j - \boldsymbol{k}_m \cdot \boldsymbol{R}_i)} \times \sum_n \int \varphi_v^*(\boldsymbol{k}_m, \boldsymbol{x})(\boldsymbol{x} - \boldsymbol{R}_n) \varphi_c(\boldsymbol{k}_\ell, \boldsymbol{x}) d^3\boldsymbol{x} \tag{D.7a}$$

$$= -\frac{e}{N^2} \sum_{\ell m} e^{i(\boldsymbol{k}_\ell \cdot \boldsymbol{R}_j - \boldsymbol{k}_m \cdot \boldsymbol{R}_i)} \sum_n e^{i(\boldsymbol{k}_\ell - \boldsymbol{k}_m) \boldsymbol{R}_n} \times \int \varphi_v^*(\boldsymbol{k}_m, \boldsymbol{x}) \boldsymbol{x} \varphi_c(\boldsymbol{k}_\ell, \boldsymbol{x}) d^3\boldsymbol{x} \ . \tag{D.7b}$$

In the last step we have used the translational property of (D.1). As the integral in (D.7b) is independent of $\boldsymbol{R}_n$, the lattice sum gives a Kronecker delta $\delta_{\ell m}$. Consequently

$$m_{ij} = -\frac{e}{N} \sum_{\boldsymbol{k}} e^{i\boldsymbol{k}(\boldsymbol{R}_j - \boldsymbol{R}_i)} \int \varphi_v^*(\boldsymbol{k}, \boldsymbol{x}) \boldsymbol{x} \varphi_c(\boldsymbol{k}, \boldsymbol{x}) d^3\boldsymbol{x} \ . \tag{D.8}$$

Using standard commutator relations between the operators $\boldsymbol{x}$ and $\boldsymbol{p} = (\hbar/i)\nabla$, we may express $m_{ij}$ in terms of the momentum matrix element instead of the $\boldsymbol{x}$ matrix element. According to Davydov (1966)

$$\boldsymbol{p}(\boldsymbol{k}) \equiv \int \varphi_v^*(\boldsymbol{k}, \boldsymbol{x}) \frac{\hbar}{i} \nabla \varphi_c(\boldsymbol{k}, \boldsymbol{x}) d^3\boldsymbol{x} = \frac{m_0}{i\hbar}(E_c(\boldsymbol{k}) - E_v(\boldsymbol{k})) \int \varphi_v^*(\boldsymbol{k}, \boldsymbol{x}) \boldsymbol{x} \varphi_c(\boldsymbol{k}, \boldsymbol{x}) d^3\boldsymbol{x} \tag{D.9}$$

where $m_0$ is the free electron mass. Using (D.7) we get

$$m_{ij} = \frac{e\hbar}{i m_0 N} \sum_{\boldsymbol{k}} \frac{\boldsymbol{p}(\boldsymbol{k}) \exp\left[i\boldsymbol{k}(\boldsymbol{R}_j - \boldsymbol{R}_i)\right]}{E_c(\boldsymbol{k}) - E_v(\boldsymbol{k})} \ . \tag{D.10}$$

In the limit of large crystals we may replace the sum over $\boldsymbol{k}$ by an integration

$$m_{ij} = \frac{e\hbar\Omega}{im_0(2\pi)^3} \int_{BZ} \frac{\boldsymbol{p}(\boldsymbol{k})\exp\left[i\boldsymbol{k}(\boldsymbol{R}_j - \boldsymbol{R}_i)\right]}{E_c(\boldsymbol{k}) - E_v(\boldsymbol{k})} d^3\boldsymbol{k} \ . \tag{D.11}$$

## E. Exciton and Polariton Green's Functions

When solving the interband edge equations or the polariton equations under various circumstances it is convenient to derive the relevant Green's functions. A very complete derivation given by Frank (1985) and Dörpelkus (1985) will be summarized here.

We first search for Green's functions $G(\boldsymbol{R}, \boldsymbol{R}', \boldsymbol{r}, \boldsymbol{r}', t, t)$ satisfying

$$\left(\Omega_{eh} - i\frac{\partial}{\partial t} - i\gamma\right)G = \delta(\boldsymbol{R} - \boldsymbol{R}')\delta(\boldsymbol{r} - \boldsymbol{r}')\delta(t - t') \ . \tag{E.1}$$

$G$ is the relevant material Green's function when the electric field is assigned to an external source. As (E.1) is translationally invariant in $\boldsymbol{R}$ and $t$ we may write

$$G(\boldsymbol{R}, \boldsymbol{R}', \boldsymbol{r}, \boldsymbol{r}', t, t') = (2\pi)^{-4} \int \tilde{G}(\boldsymbol{q}, \omega, \boldsymbol{r}, \boldsymbol{r}') \times \exp\left[i\boldsymbol{q}(\boldsymbol{R} - \boldsymbol{R}') - i\omega(t - t')\right] d^3\boldsymbol{q}\, d\omega \tag{E.2}$$

where

$$(\Omega_{eh}(\boldsymbol{q}) - \omega - i\gamma)\tilde{G}(\boldsymbol{q}, \omega, \boldsymbol{r}, \boldsymbol{r}') = \delta(\boldsymbol{r} - \boldsymbol{r}') \tag{E.3}$$

and

$$\Omega_{eh}(\boldsymbol{q}) = \omega_g - \frac{\hbar}{2\mu}\nabla_r^2 + \frac{\hbar}{2m}q^2 + \frac{V_{eh}}{\hbar} \ . \tag{E.4}$$

An expansion of $\tilde{G}(\boldsymbol{q}, \omega, \boldsymbol{r}, \boldsymbol{r}')$ in spherical coordinates is convenient:

$$\tilde{G}(\boldsymbol{q}, \omega, \boldsymbol{r}, \boldsymbol{r}') = \sum_{\ell=0}^{\infty} \sum_{m=-\ell}^{\ell} g_{\ell m}(\boldsymbol{q}, \omega, r, r') Y_{\ell m}^*(\theta', \phi') Y_{\ell m}(\theta, \phi) \tag{E.5}$$

where $g_{\ell m}$ satisfies

$$(\omega_g - \omega - i\gamma) + \frac{\hbar}{2m}q^2 - \left\{\frac{\hbar}{2\mu}\left[\frac{\partial^2}{\partial r^2} + \frac{2}{r}\frac{\partial}{\partial r} - \frac{\ell(\ell+1)}{r^2}\right] + \frac{V_{eh}}{\hbar}\right\} g_{\ell m} = \frac{\delta(r - r')}{4\pi r^2} \ . \tag{E.6}$$

It is usually required that $g_{\ell m}(q, \omega, r, r')$ has no singularity at $r = 0$ and $r = \infty$. Any allowed solutions to the homogeneous version of (E.6) must be matched at $r = r'$ so that $g_{\ell m}$ is continuous at $r'$ and $dg_{\ell m}/dr$ fulfills a suitable jump condition at $r'$. In the simple case of vanishing electron-hole potential $V_{eh}$, (E.6) defines the solutions

$$g_{\ell m} = \begin{cases} -\dfrac{2\mu}{4\pi\hbar}(rr')^{1/2} I_{\ell+1/2}(\kappa r) K_{\ell+1/2}(\kappa r') \ ; & r<r' \\ -\dfrac{2\mu}{4\pi\hbar}(rr')^{1/2} I_{\ell+1/2}(\kappa r') K_{\ell+1/2}(\kappa r) \ ; & r>r' \end{cases} \tag{E.7}$$

where $\sqrt{\pi/2z} I_{\ell+1/2}(z)$ and $\sqrt{\pi/2z} K_{\ell+1/2}(z)$ are modified spherical Bessel functions of first and third kind, respectively. The quantity $\kappa$ in (E.7) is given by

$$\kappa = \left[ (\omega_g - \omega - i\gamma) \frac{2\mu}{\hbar} + \frac{\mu}{m} q^2 \right]^{1/2} . \tag{E.8}$$

The solution $g_{00}$ relevant for spherical symmetry is

$$g_{00} = \begin{cases} \dfrac{2\mu}{4\pi\hbar\kappa rr'} \sinh(\kappa r) \mathrm{e}^{-\kappa r'} & r<r' \\ \dfrac{2\mu}{4\pi\hbar\kappa rr'} \sinh(\kappa r') \mathrm{e}^{-\kappa r} & r>r' \end{cases} . \tag{E.9}$$

When $V_{eh}$ is the Coulomb potential $V_{eh} = -\hbar^2/(\mu r a_B)$, then (E.6) defines the excitonic Green's functions which are analogous to the Coulomb Green's functions in atomic physics (Hostler, 1969, 1975) and can be expressed by the Kummer functions $U(a,b,z)$ and $M(a,b,z)$ (Stahl and Balslev, 1982; Frank, 1985; Dörpelkus 1985):

$$g_{\ell m} = \begin{cases} C(4\pi^2 rr')^{\ell} \mathrm{e}^{-\kappa(r+r')} M(a_\ell, b_\ell, 2\kappa r) U(a_\ell, b_\ell, 2\kappa r') \ ; & r<r' \\ C(4\pi^2 rr')^{\ell} \mathrm{e}^{-\kappa(r+r')} M(a_\ell, b_\ell, 2\kappa r') U(a_\ell, b_\ell, 2\kappa r) \ ; & r>r' \end{cases} \tag{E.10a}$$

$$C = \frac{\mu\kappa}{\pi\hbar} \frac{\Gamma(a_\ell)}{\Gamma(b_\ell)} , \tag{E.10b}$$

$$a_\ell = \ell + 1 - (\kappa a_B)^{-1} , \tag{E.10c}$$

$$b_\ell = 2(\ell + 1) , \tag{E.10d}$$

where the function symbols of Abramowitz and Stegun (1965) are used.

We finally turn to the situation when the edge equation and the field equation describe a coupled system driven by an external source. If the source is in the edge equation then it is convenient to work with polariton Green's functions $G_{YY}(\boldsymbol{R}, \boldsymbol{R}', \boldsymbol{r}, \boldsymbol{r}', t, t')$ and $G_{EY}(\boldsymbol{R}, \boldsymbol{R}', \boldsymbol{r}, \boldsymbol{r}', t, t')$ satisfying

$$\left( \Omega_{eh} - i\frac{\partial}{\partial t} - i\gamma \right) G_{YY} = M(\boldsymbol{r}) G_{EY} + \delta(\boldsymbol{R} - \boldsymbol{R}')\delta(\boldsymbol{r} - \boldsymbol{r}')\delta(t - t') \tag{E.11a}$$

$$\left(c^2\nabla_R^2 - \tilde{\varepsilon}\frac{\partial^2}{\partial t^2}\right)\varepsilon_0 G_{EY} = \frac{\partial^2}{\partial t^2}\int M(\boldsymbol{r})(G_{YY} + G_{YY}^*)d^3\boldsymbol{r} \; . \tag{E.11b}$$

The solutions $G_{YY}$, $G_{EY}$ are conveniently expressed by the partial Fourier transform $\tilde{G}(\boldsymbol{q},\omega,\boldsymbol{r},\boldsymbol{r}')$ of the material Green's function [see (E.3)]. After straightforward algebra one finds

$$\tilde{G}_{YY}(\boldsymbol{q},\omega,\boldsymbol{r},\boldsymbol{r}') = \tilde{G}(\boldsymbol{q},\omega,\boldsymbol{r},\boldsymbol{r}') + \frac{\int M(\boldsymbol{x})\tilde{G}(\boldsymbol{q},\omega,\boldsymbol{r},\boldsymbol{x})d^3\boldsymbol{x}\int M(\boldsymbol{x})(\tilde{G}(\boldsymbol{q},\omega,\boldsymbol{x},\boldsymbol{r}') + \tilde{G}(-\boldsymbol{q},-\omega,\boldsymbol{x},\boldsymbol{r}'))d^3\boldsymbol{x}}{\varepsilon_0(c^2q^2/\omega^2 - \tilde{\varepsilon} - \chi_+(\omega,\boldsymbol{q}) - \chi_-(\omega,\boldsymbol{q}))} \tag{E.12}$$

$$G_{EY}(\boldsymbol{q},\omega,\boldsymbol{r}') = \frac{\int M(\boldsymbol{x})(\tilde{G}(\boldsymbol{q},\omega,\boldsymbol{x},\boldsymbol{r}') + \tilde{G}(-\boldsymbol{q},-\omega,\boldsymbol{x},\boldsymbol{r}'))d^3\boldsymbol{x}}{\varepsilon_0(c^2q^2/\omega^2 - \tilde{\varepsilon} - \chi_+(\omega,\boldsymbol{q}) - \chi_-(\omega,\boldsymbol{q}))} \; . \tag{E.13}$$

Here, the quantities $\chi_-$ and $\chi_+$ given by

$$\chi_\mp(\omega,\boldsymbol{q}) = \frac{1}{\varepsilon_0\hbar}\int M(\boldsymbol{r})G(\pm\boldsymbol{q},\pm\omega,\boldsymbol{r},\boldsymbol{r}')M(\boldsymbol{r}')d^3\boldsymbol{r}\,d^3\boldsymbol{r}' \tag{E.14}$$

are the resonant and antiresonant susceptibilities discussed in Sect. 4.2. The relation between the functions $G_{YY}$, $G_{EY}$ and the partial Fourier transforms $\tilde{G}_{YY}$, $\tilde{G}_{EY}$ is analogous to (E.2). In the shell source approximation

$$M(r) = M_0\frac{\delta(r - r_0)}{4\pi r_0^2} \tag{E.15}$$

the Green's functions are simplified as follows

$$\tilde{G}(\boldsymbol{q},\omega,\boldsymbol{r},\boldsymbol{r}') = g_0(\boldsymbol{q},\omega,r,r') \; , \tag{E.16}$$

$$\chi_\mp(\omega,\boldsymbol{q}) = \frac{1}{\varepsilon_0\hbar}M_0^2 g_0(\pm\boldsymbol{q},\pm\omega,r_0,r_0) \; , \tag{E.17}$$

$$G_{YY}(\boldsymbol{q},\omega,\boldsymbol{r},\boldsymbol{r}') = g_0(\boldsymbol{q},\omega,r,r') + \frac{g_0(\boldsymbol{q},\omega,r,r_0)(g_0(\boldsymbol{q},\omega,r_0,r') + g_0(-\boldsymbol{q},-\omega,r_0,r'))}{\varepsilon_0(c^2q^2/\omega^2 - \tilde{\varepsilon} - \chi_+(\omega,\boldsymbol{q}) - \chi_-(\omega,\boldsymbol{q}))} \; , \tag{E.18}$$

$$G_{EY}(\boldsymbol{q},\omega,\boldsymbol{r}') = \frac{g_0(\boldsymbol{q},\omega,r_0,r') + g_0(-\boldsymbol{q},\omega,r_0,r')}{\varepsilon_0(c^2q^2/\omega^2 - \tilde{\varepsilon} - \chi_+(\omega,\boldsymbol{q}) - \chi_-(\omega,\boldsymbol{q}))} \tag{E.19}$$

where $g_0$ is the $(\ell,m) = (0,0)$ partial wave Green's function, see (E.9) and (E.10a).

## F. Operator Algebra

We compile important commutation relations used in the text.
Fermion-operators:

$c_j, c_j^\dagger$ electron operators of the conduction band

$d_j, d_j^\dagger$ hole operators of the valence band

$$[c_i, c_j^\dagger]_+ = \delta_{ij} \qquad [d_i, d_j^\dagger]_+ = \delta_{ij}$$

$$[c_i, c_j]_+ = [c_i, d_j]_+ = [c_i d_j^\dagger]_+ = 0 \ . \tag{F.1}$$

Pair operators:

$$\hat{Y}_{ij} = d_i c_j \qquad \hat{C}_{ij} = c_i^\dagger c_j \ ;$$

$$\hat{Y}_{ij}^\dagger = c_j^\dagger d_i^\dagger \qquad \hat{D}_{ij} = d_i^\dagger d_j \ . \tag{F.2}$$

From (F.1) and the definitions (F.2) the following commutation relations are derived:

$$[\hat{Y}_{ij}, \hat{Y}_{k\ell}] = \delta_{ki}\delta_{\ell j} - \delta_{ki}\hat{C}_{\ell j} - \delta_{\ell j}\hat{D}_{ki} \tag{F.3}$$

$$[\hat{Y}_{ij}, \hat{C}_{k\ell}] = \delta_{jk}\hat{Y}_{i\ell} \tag{F.4}$$

$$[\hat{Y}_{ij}, \hat{D}_{k\ell}] = \delta_{ik}\hat{Y}_{\ell j} \tag{F.5}$$

$$[\hat{C}_{ij}, Y_{k\ell}^\dagger] = \delta_{\ell j}\hat{Y}_{ki}^\dagger \tag{F.6}$$

$$[\hat{D}_{ij}, Y_{k\ell}^\dagger] = \delta_{jk}\hat{Y}_{i\ell}^\dagger \tag{F.7}$$

$$[\hat{C}_{ij}, \hat{C}_{k\ell}] = \delta_{kj}\hat{C}_{i\ell} - \delta_{i\ell}\hat{C}_{kj} \tag{F.8}$$

$$[\hat{D}_{ij}, \hat{D}_{k\ell}] = \delta_{kj}\hat{D}_{i\ell} - \delta_{i\ell}\hat{D}_{kj} \tag{F.9}$$

$$[\hat{Y}_{ij}, \hat{Y}_{k\ell}] = 0 \tag{F.10}$$

$$[\hat{C}_{ij}, \hat{D}_{k\ell}] = 0 \ . \tag{F.11}$$

## G. Scattering Intensity

The general structure of a source for scattered light is that it is composed of two factors, one describing an incident polariton while the other represents an excitation $\xi$ of the crystal. The scattered field then is represented as a convolution of the source with an appropriate Green's function, as e.g. in (7.8)

$$E_s = G * Y_i \xi \ . \tag{G.1}$$

In general the spontaneous fluctuations of $\xi$ are the thermal noise and quantum noise (zero-point fluctuation). Hence it is appropriate to consider $\xi$ as a quantized amplitude

$$\xi = a w \mathrm{e}^{-i\omega_0 t} + a^\dagger w^* \mathrm{e}^{i\omega_0 t} \tag{G.2}$$

$w$ being a mode function. In the simplest case $w$ is a plane wave

$$w = w_0 \mathrm{e}^{i\boldsymbol{q}\boldsymbol{r}} \ . \tag{G.3}$$

$a^\dagger$ and $a$ are operators creating or annihilating the quanta of the mode. The mode is treated as a Boson oscillator (compare Sect. 2.6). Assuming $Y_i$ to be of the form

$$Y_i = y_i(\boldsymbol{r}) \mathrm{e}^{i\omega_i t} \tag{G.4}$$

(G.1) is found to consist of two terms representing Stokes and Antistokes radiation

$$E_s = a^\dagger G * y_i w^* \exp\left[-i(\omega_i - \omega_0)t\right] + aG * y_i w \exp\left[-i(\omega_i + \omega_0)t\right] \tag{G.5}$$

abbreviated as

$$E_s = a^\dagger E_{St} + a E_{AS}$$

The scattered intensity is (apart from a factor)

$$I_s = \langle E_s^\dagger E_s \rangle \tag{G.6}$$

where the average refers to the equilibrium ensemble of $\xi$. Inserting (6.5) one finds

$$I_s \propto |E_{St}|^2 \langle a a^\dagger \rangle + |E_{AS}|^2 \langle a^\dagger a \rangle \ . \tag{G.7}$$

Introducing the average equilibrium occupation number $\overline{n}$ as

$$\overline{n} = \langle a^\dagger a \rangle = \langle a a^\dagger \rangle - 1 \tag{G.8}$$

(G.7) becomes

$$I_s \propto |E_{St}|^2 (\overline{n} + 1) + |E_{AS}|^2 \overline{n} \ . \tag{G.9}$$

In the low temperature limit ($kT \ll \hbar\omega_0$) we have $\overline{n} \approx 0$ and only the Stokes signal due to zero point quantum noise survives

$$I_s(T \to 0) \propto |E_{St}|^2 \ . \tag{G.10}$$

The same result would have been obtained by simply taking the ground-state expectation of (G.7). In this way it turns out that the picture attributing Raman scattering to zero point fluctuations is equivalent to the usual golden rule treatment (Loudon, 1963; Bechstedt and Henneberger, 1977).

# H. Screening of the Electron-Hole Interaction and Band Renormalization

Consider two point charges brought as a probe into the medium. The probe charge density is

$$\varrho^F = e(\delta(\boldsymbol{r} - \boldsymbol{r}_h) - \delta(\boldsymbol{r} - \boldsymbol{r}_e)) \tag{H.1}$$

with a Fourier transform

$$\tilde{\varrho}^F = \tilde{\varrho}_h + \tilde{\varrho}_e = e[\exp(i\boldsymbol{q}\cdot\boldsymbol{r}_h) - \exp(i\boldsymbol{q}\cdot\boldsymbol{r}_e)] \ . \tag{H.2}$$

The potential produced by $\varrho^F$ is

$$\phi^F = \frac{e}{4\pi\varepsilon_0}(|\boldsymbol{r} - \boldsymbol{r}_h|^{-1} - |\boldsymbol{r} - \boldsymbol{r}_e|^{-1}) \ . \tag{H.3}$$

Let $\varrho^{in}$ be the induced charge density due to the polarization of the medium around the charge $\varrho^F$. Then the energy content of the polarization cloud is derived from the standard textbook formula (Jackson, 1975)

$$W = \tfrac{1}{2}\int \varrho^{in}\phi^F d^3\boldsymbol{r} \ . \tag{H.4}$$

Let us apply (H.4) to the screening by a dense *e-h* plasma. Then it is convenient to do calculations in Fourier representation, symbolized by a tilde. The induced charge is determined from Lindhard's theory (Sect. 6.4)

$$\tilde{\varrho} = \frac{R}{\varepsilon_0 q^2 - R}\tilde{\varrho}^F \quad \text{where} \tag{H.5}$$

$$R = \frac{e^2}{(2\pi)^3}\int\left[\frac{f_e(\boldsymbol{k}) - f_e(\boldsymbol{k}-\boldsymbol{q})}{E_e(\boldsymbol{k}) - f_e(\boldsymbol{k}-\boldsymbol{q})} + \frac{f_h(\boldsymbol{k}) - f_h(\boldsymbol{k}-\boldsymbol{q})}{E_h(\boldsymbol{k}) - E_h(\boldsymbol{k}-\boldsymbol{q})}\right]d^3\boldsymbol{k} \tag{H.6}$$

is the instantaneous charge response in a bipolar plasma (6.72).

In $\boldsymbol{q}$-space the polarization energy (H.4) becomes

$$W = \frac{1}{2}\frac{1}{(2\pi)^3}\int \tilde{\varrho}^{in}(\tilde{\Phi}^F)^* d^3\boldsymbol{q} \ . \tag{H.7}$$

Inserting (H.5) and the Fourier transform of (H.3) one finds

$$W = \frac{e^2}{\varepsilon_0(2\pi)^3}\int q^{-2}\frac{R}{\varepsilon_0 q^2 - R}[1 - \cos(\boldsymbol{q}(\boldsymbol{r}_e - \boldsymbol{r}_h))]d^3\boldsymbol{q} \ . \tag{H.8}$$

The part of $W$ obtained in the limit $|\boldsymbol{r}_e - \boldsymbol{r}_h| \to \infty$ is interpreted as a renormalization of the gap energy

$$\delta E_g = \frac{e^2}{\varepsilon_0 (2\pi)^3} \int q^{-2} \frac{R}{\varepsilon_0 q^2 - R} d^3\boldsymbol{q} \,. \tag{H.9}$$

The remainder contributes to a screening of the electron-hole interaction. The screened potential is

$$\begin{aligned} \overline{V}_{eh} = & -\frac{e^2}{4\pi\varepsilon_0 |\boldsymbol{r}_e - \boldsymbol{r}_h|} \\ & - \frac{e^2}{\varepsilon_0 (2\pi)^3} \int q^{-2} \frac{R}{\varepsilon_0 q^2 - R} \cos\left(\boldsymbol{q}(\boldsymbol{r}_e - \boldsymbol{r}_h)\right) d^3\boldsymbol{q} \,. \end{aligned} \tag{H.10}$$

If, as a crude approximation, it is assumed that the response function (H.6) is independent of $\boldsymbol{q}$

$$R = -\varepsilon_0 \kappa_0^2 \tag{H.11}$$

then it is easy to evaluate the integrals in (H.9,10), and it is found that

$$\delta E_g = -\frac{e^2}{4\pi\varepsilon_0} \kappa_0 \tag{H.12}$$

$$\overline{V}_{eh} = -\frac{e^2}{4\pi\varepsilon_0} \frac{\exp\left(-\kappa_0 |\boldsymbol{r}_e - \boldsymbol{r}_h|\right)}{|\boldsymbol{r}_e - \boldsymbol{r}_h|} \,. \tag{H.13}$$

Above we have neglected the dielectric screening caused by the interband part and the ionic part of $\varrho^{in}$. Such contributions can be included if one replaces $\varepsilon_0$ in the above formulas by $\varepsilon_0 \varepsilon_s$ where $\varepsilon_s$ is the static dielectric constant in the absence of the plasma.

## J. Proof Concerning Coherent Pair States

In this appendix we show that the model Hamiltonian of Sect. 2.4 generates a development through coherent pair states of type (2.82) if the motion starts from the ground state.

The model Hamiltonian of Sect. 2.4 was

$$H = H_0 + H_1 \tag{J.1a}$$

$$H_0 = \frac{1}{2}\hbar\omega_g \sum_j (c_j^\dagger c_j + d_j^\dagger d_j) + \sum_{ij} (T_{ij}^c c_i^\dagger c_j + T^v d_i^\dagger d_j) \tag{J.1b}$$

$$H_1 = -M_0 \sum_j E_j(t) (d_j c_j + c_j^\dagger d_j^\dagger) \,. \tag{J.1c}$$

In a first step we transform from the atomic states to a new Bloch-type basis such that $H_0$ is diagonalized. In the new basis the operators $H_0$ and $H_1$ are written as follows

$$H_0 = \sum_k (E_k^c c_k^\dagger c_k + E_k^v d_k^\dagger d_k) \ , \tag{J.2a}$$

$$H_1 = \sum_{pk} (A_{pk}(t) d_p c_k + A_{pk}^*(t) c_k^\dagger d_p^\dagger) \ . \tag{J.2b}$$

In order to study the evolution of the system under the influence of the model Hamiltonian $H$ we apply the interaction representation. In this representation the state vector $|\psi\rangle$ obeys the equation

$$\frac{\partial}{\partial t}|\psi\rangle = (i\hbar)^{-1}\tilde{H}_1|\psi\rangle \ . \tag{J.3}$$

The interaction representation operator $\tilde{H}_1$ differs from $H_1$ in (J.2b) only by a phase transformation of the coefficients $A_{pk}(t)$ into which the field has been absorbed

$$\tilde{H}_1 = \sum_{pk} (B_{pk} d_p c_k + B_{pk}^* c_k^\dagger d_p^\dagger) \ ; \quad B_{pk} = A_{pk} \exp\left[(E_p^v + E_k^c)t/i\hbar\right] \ . \tag{J.4}$$

For the proof that integration of (J.3) generates a coherent pair state we apply the inductive method.

The initial state at $t = 0$ is assumed to be the ground state $|0\rangle$ with neither electrons nor holes present. This state is evidently a coherent pair state with all $v_q = 0$. Next we consider the infinitesimal transformation equivalent to the Schrödinger equation (J.3)

$$|\psi(t+\tau)\rangle = \left[1 - \frac{i\tau}{\hbar}\sum_{pk}(B_{pk} d_p c_k + B_{pk}^* c_k^\dagger d_p^\dagger)\right]|\psi(t)\rangle \ . \tag{J.5}$$

To first order in $\tau$ it is replaced by

$$|\psi(t+\tau)\rangle = \prod_{pk}\left(1 - \frac{i\tau}{\hbar} B_{pk} d_p c_k\right)\prod_{pk}\left(1 - \frac{i\tau}{\hbar} B_{pk}^* c_k^\dagger d_p^\dagger\right)|\psi(t)\rangle \ . \tag{J.6}$$

Suppose that $|\psi(t)\rangle$ is already a coherent pair state. Then the product

$$\prod_{pk}\left(1 - \frac{i\tau}{\hbar} B_{pk}^* c_k^\dagger d_p^\dagger\right) \tag{J.7}$$

acting on $|\psi(t)\rangle$ will preserve this property as can be seen almost immediately. Thus there remains the problem whether a product of the form

$$\prod_{pk}\left(1 - \frac{i\tau}{\hbar} B_{pk} d_p c_k\right) \tag{J.8}$$

when applied to a coherent pair state transforms it to a state of the same class.

For convenience we shall work out this problem in a representation where the original state $|\psi\rangle$ is given in the diagonal representation (2.84). To make the writing easier we use the abbreviation

$$b_{pk} = -\frac{i\tau}{\hbar} B_{pk} \ .$$

Then we are facing the following problem: Can it be shown that

$$|\tilde{\psi}\rangle = \prod_{pk}(1 + b_{pk} d_p c_k) N \prod_q (1 + \eta_q c_q^\dagger d_q^\dagger)|0\rangle \tag{J.9}$$

is a coherent pair state? It is clear that if the statement is true for one arbitrary factor out of the product (J.8), then it is true for the whole product. Let us therefore consider two representative cases; one where a factor with $p = k$ is applied and one with $p \neq k$.

A) $p = k$ :

$$\begin{aligned} |\tilde{\psi}\rangle &= (1 + b d_k c_k)|\psi\rangle \\ &= N \prod_{q \neq k} (1 + \eta_q c_q^\dagger d_q^\dagger)(1 + b d_k c_k)(1 + \eta_k c_k^\dagger d_k^\dagger)|0\rangle \\ &= N \prod_{q \neq k} (1 + \eta_q c_q^\dagger d_q^\dagger)(1 + b\eta_k + \eta_k c_k^\dagger d_k^\dagger)|0\rangle \ . \end{aligned} \tag{J.10}$$

Renormalizing the last bracket the standard form of a coherent pair state is reestablished:

$$|\tilde{\psi}\rangle = \tilde{N} \prod_{p \neq k} (1 + \eta_q c_q^\dagger d_q^\dagger)(1 + \tilde{\eta}_k c_k^\dagger d_k^\dagger)|0\rangle \tag{J.11}$$

$$\tilde{\eta}_k = \eta_k (1 + b\eta_k)^{-1} \ . \tag{J.12}$$

If it happens that $(1 + b\eta_k) = 0$, then one should switch over to the form (2.87) with $u_k = 0$. Then the result is seen to be again a coherent pair state.

B) $p \neq k$ :

$$\begin{aligned} |\tilde{\psi}\rangle &= (1 + b d_p c_k)|\psi\rangle \ , \\ &= N \prod_{q \neq p,k} (1 + \eta_q c_q^\dagger d_q^\dagger)(1 + b d_p c_k)(1 + \eta_k c_k^\dagger d_k^\dagger)(1 + \eta_p c_p^\dagger d_p^\dagger)|0\rangle \\ &= N \prod_{q \neq p,k} (1 + \eta_q c_q^\dagger d_q^\dagger)(1 + \eta_k c_k^\dagger d_k^\dagger + \eta_p c_p^\dagger d_p^\dagger \\ &\quad + \eta_k \eta_p c_k^\dagger d_k^\dagger c_p^\dagger d_p^\dagger - b\eta_k \eta_p c_p^\dagger d_k^\dagger)|0\rangle \end{aligned}$$

$$= N \prod_{q \neq p,k} (1 + \eta_q c_q^\dagger d_q^\dagger)(1 + \eta_k c_k^\dagger d_k^\dagger - \tfrac{1}{2} b \eta_k \eta_p c_p^\dagger d_k^\dagger)$$

$$\times (1 + \eta_p c_p^\dagger d_p^\dagger - \tfrac{1}{2} b \eta_p \eta_k c_p^\dagger d_k^\dagger)|0\rangle \ . \tag{J.13}$$

In the form given in the last line of (J.13) $|\tilde{\psi}\rangle$ is easily recognized as a coherent pair state. One only has to go back the step leading from (2.81) to (2.82) to find out that $|\tilde{\psi}\rangle$ is of the form

$$|\tilde{\psi}\rangle = N \exp\left[\sum_{kp} \tilde{\eta}_{kp} c_p^\dagger d_k^\dagger\right] \tag{J.14}$$

and therefore a coherent pair state.

This completes the inductive proof of our assertion that the time evolution in the band edge model of Sect. 2.4 goes through coherent pair states of the BCS type.

# K. Quantum Mechanical Wiener-Khintchin Theorem

A suitable definition of intensity in a quantized electromagnetic field is provided by the expectation value (Glauber, 1965)

$$I(\boldsymbol{r},t) = \langle \hat{E}^\dagger(\boldsymbol{r},t)\hat{E}(\boldsymbol{r},t)\rangle \tag{K.1}$$

$\hat{E}$ is the annihilation part of the electric field operator, $\hat{E}^\dagger$ the creation part; the average $\langle \ldots \rangle$ in general is understood in the sense of a quantum ensemble represented by a statistical operator. Let us assume that we are dealing with a stationary ensemble. This means that the correlation function of the field has the property

$$\langle \hat{E}^\dagger(t+\tau)\hat{E}(t)\rangle = \langle \hat{E}^\dagger(\tau)\hat{E}(0)\rangle \ . \tag{K.2}$$

(For simplicity the dependence on $\boldsymbol{r}$ has been omitted.) From (K.2) it follows that the correlation of the Fourier transformed operators becomes

$$\langle \hat{E}_\omega^\dagger \hat{E}_{\omega'}\rangle = 2\pi\delta(\omega - \omega') \int \mathrm{e}^{-i\omega\tau} \langle \hat{E}^\dagger(\tau)\hat{E}(0)\rangle d\tau \ . \tag{K.3}$$

Let us define the filtered signal in a frequency interval $\Delta\omega$ around the mid frequency $\omega$ by

$$\hat{E}(t;\omega,\Delta\omega) = \frac{1}{2\pi} \int_{\omega-\Delta\omega/2}^{\omega+\Delta\omega/2} \hat{E}_{\omega'} \mathrm{e}^{-i\omega' t} d\omega' \ . \tag{K.4}$$

The intensity of the filtered signal formed according to (K.1) becomes

$$I(\omega,\Delta\omega) = \langle \hat{E}^{\dagger}(t;\omega,\Delta\omega)\hat{E}(t;\omega,\Delta\omega)\rangle$$

$$= \frac{1}{(2\pi)^2} \iint\limits_{\omega-\Delta\omega/2}^{\omega+\Delta\omega/2} \langle \hat{E}^{\dagger}_{\omega'}\hat{E}_{\omega''}\rangle e^{i(\omega'-\omega'')t} d\omega' d\omega'' \ . \qquad \text{(K.5)}$$

Using (K.3) one finds

$$I(\omega,\Delta\omega) = \frac{1}{2\pi} \iint \langle \hat{E}^{\dagger}(\tau)\hat{E}(0)\rangle e^{-i\omega'\tau} d\omega' d\tau \ . \qquad \text{(K.6)}$$

Defining the spectral density by the limit

$$\frac{dI}{d\omega} = \lim_{\Delta\omega\to 0} \frac{1}{\Delta\omega} I(\omega,\Delta\omega) \qquad \text{(K.7)}$$

one derives from (K.6) a formula which is the quantum-mechanical version of the well known Wiener-Khintchine Theorem

$$\frac{dI}{d\omega} = \frac{1}{2\pi} \int \langle \hat{E}^{\dagger}(\tau)\hat{E}(0)\rangle e^{i\omega\tau} d\tau \ . \qquad \text{(K.8)}$$

This theorem relates the spectral density to the Fourier transform of the autocorrelation function of $\hat{E}$.

# References


Abella, I.D. (1969): *Progress in Optics*, ed. by E. Wolf, Vol. VII, 139 (North-Holland Publishing Comp., Amsterdam-London) (Sects. 3.7, 7.1)

Abramowitz, M., J. Stegun, eds. (1965): *Handbook of Mathematical Functions* (Dover Publications, New York) (Sects. 4.3, 4.5)

Abstreiter, G. (1984): *Inelastic Light Scattering in Semiconductor Heterostructures*, in "Festkörperprobleme" XXIV, ed. by P. Grosse (Vieweg, Braunschweig) (Sect. 7.2)

Agarwal, G.S., D.N. Pattanayak, E. Wolf (1974): Phys. Rev. B**10**, 1447 (Sect. 5.2)

Agranovich, V.M. (1984): in *Surface Excitations* ed. by V.M. Agranovich, R. Loudon (Elsevier Science Publishers B.V., Amsterdam) p. 513 (Sect. 5.6)

Agranovich, V.M., V.L. Ginzburg (1966): *Spatial Dispersion in Crystal Optics and the Theory of Excitons* (Wiley, New York) (Sect. 4.7)

Agranovich, V.M., V.L. Ginzburg (1984): *Crystal Optics with Spatial Dispersion and Excitons* (Springer, Berlin, Heidelberg, New York) (Sects. 2.2, 2.3, 5.2)

Akhmediev, N.N., V.V. Yatsishen (1978): Sol. State Commun. **27**, 357 (Sect. 5.2)

Altarelli, M., N. Lipari (1976): Phys. Rev. Lett. **36**, 619 (Sects. 4.1, 4.9)

Ando, T., A.B. Fowler, F. Stern (1982): Rev. Mod. Phys. **54**, 437 (Sects. 5.8, 6.2)

Arecchi, F.T., E. Courtens, R. Gilmore, H. Thomas (1972): Phys. Rev. A**6**, 2211 (Sect. 2.7)

Arya, K., W. Hanke (1980): Sol. State Commun. **33**, 739 (Sect. 7.7)

Askary, F., P.Y. Yu (1985): Phys. Rev. B**31**, 6643 (Sect. 5.1)

Aspnes, D. (1974): Phys. Rev. B**10**, 4228 (Sect. 4.1)

Balslev, I. (1972): *Semiconductors and Semimetals*, Vol. 9, ed. by R.K. Willardson, A.C. Beer (Academic Press, New York) p. 403 (Sect.4.1)

Balslev, I. (1978): Phys. Stat. Sol. (b) **88**, 155 (Sect. 5.3)

Balslev, I. (1981): Phys. Rev. B**23**, 3977 (Sect. 5.3)

Balslev, I. (1983): Sol. State Commun. **45**, 661; Phys. Rev. B**28**, 5665 (Sect. 5.3)

Balslev, I. (1984): Sol. State Commun. **52**, 351 (Sect. 3.7, 4.10)

Balslev, I.(1985): J. Lumin. **30**, 162 (Sects. 4.3, 7.7)

Balslev, I., A. Stahl (1982): Phys. Stat. Sol. (b) **111**, 531 (Sect. 5.4)

Balslev, I. (1987): to be published in Comments on Condensed Matter Physics (Sect. 5.4)

Baldereschi, A., N.O. Lipari (1973): Phys. Rev. B**8**, 2697 (Sect. 4.9)

Bányai, L., S.W. Koch (1986): Z. Phys. B**63**, 283 (Sect. 7.7)

Baraff, G.A., A.J. Appelbaum (1972): Phys. Rev. B**5**, 475 (Sect. 6.2)

Bardeen, J., L.N. Cooper, J.R. Schriefer (1957): Phys. Rev. **108**, 1175 (Sect. 2.7)

Barker, A.S., R. Loudon (1972): Rev. Mod. Phys. **44**, 18 (Sect. 7.2)

Bassani, F., G. Pastori Parravici (1975): *Electronic States and Optical Transitions in Solids* (Pergamon Press, Oxford) (Sect. 3.5)

Batke, E., D. Heitmann, J.P. Kotthaus, K. Ploog (1985): Phys. Rev. Lett. **54**, 2367 (Sect. 5.8)

Bechstedt, F., F. Henneberger (1977): Phys. Stat. Sol. (b) **81**, 211 (Sect. 7.6)

Becker, R., F. Sauter (1968): *Theorie der Elektrizität*, Bd. III (B.G. Teubner, Stuttgart) (Sects. 6.1, 6.5)

Bendow, B. (1970): Phys. Rev. B**2**, 5051 (Sect. 7.2)

Bigot, J.Y., B. Hönerlage (1984): Phys. Stat. Sol. (b) **121**, 649 (Sect. 7.6)

Bimberg, D. (1977): in *Festkörperprobleme* (Advances in Solid State Physics) Vol. XVII, ed. by H. Queisser (Vieweg, Braunschweig) p. 195 (Sect. 4.1)

Bir, G.L., G.E. Pikus (1972): *Symmetry and Deformation Effects in Semiconductors* (Nauka, Moscow) (Sect. 3.3)
Birman, J.L., J.J. Sein (1972): Phys. Rev. B**6**, 2582 (Sect. 5.2)
Bishop, M.F., A.A. Maradudin (1976): Phys. Rev. B**14**, 3384 (Sects. 4.8, 5.2)
Bivas, A., Vu Duy Phach, B. Hönerlage, J.B. Grun (1977): Phys. Stat. Sol. (b) **84**, 235 (Sect. 7.6)
Bloch, F. (1946): Phys. Rev. **70**, 460 (Sect. 2.1)
Bloembergen, N. (1967): Am. J. Phys. **35**, 989 (Sect. 7.3)
Blossey, D.F. (1971): Phys. Rev. B**3**, 1383 (Sect. 5.7)
Bohm, D., D. Pines (1953): Phys. Rev. **92**, 609 (Sect. 3.7)
Bohnert, K., G. Schneider, S. El-Dessouki, C. Klingshirn (1978): Sol. State Commun. **27**, 295 (Sect. 7.7)
Brenig, W., R. Zeyher, J. Birman (1972): Phys. Rev. B**6**, 4617 (Sects. 5.1, 7.2)
Brewer, R.G. (1977): *Coherent Optical Spectroscopy*, Proceedings of the International School of Physics "Enrico Fermi", Course LXIV, ed. by N. Bloembergen, Bologna (Sect. 7.1)
Broser, I., et al. (1978): Phys. Stat. Sol. (b) **90**, 77 (Sects. 4.1, 5.6)
Broser, I., M. Rosenzweig (1979): Phys. Stat. Sol. (b) **95**, 141 (Sect. 5.6)
Broser, I., et al. (1980): in *Proc. 15th Int. Conf. Phys. Semicond.* ed. by S. Tanaka, Y. Toyozawa (Phys. Soc. Japan, Tokyo) p. 401 (Sect. 4.1)

Cardona, M. (1969): *Modulation Spectroscopy*, Solid State Phys., Suppl. 11 (Sect. 4.3)
Cardona, M. (1982): In *Light Scattering in Solids II*, ed. by M. Cardona and G. Güntherodt, Topics in Applied Physics, Vol. 50 (Springer, Berlin, Heidelberg, New York) (Sects. 3.3, 7.2)
Chaplik, A.V. (1972): Zh. Exp. Teor. Fiz. **62**, 746 (Sov. Phys. JETP **35**, 395) (Sect. 5.8)
Chemla, D.S., A. Maruani and E. Batifol (1979): Phys. Rev. Lett. **42**, 1075 (Sect. 7.6)
Chemla, D.S., D.A.B. Miller (1985): J. Opt. Soc. America B**2**, 1155 (Sect. 5.8)
Chemla, D.S., D.A.B. Miller, P.W. Smith (1985): Optical Engineering **24**, 556 (Sect. 5.8)
Cho, K. (1976): Phys. Rev. B**14**, 4463 (Sect. 4.9)
Cho, K. (1979): in *Excitons*, Topics in Current Physics, Vol. 14 (Springer, Berlin, Heidelberg, New York) p. 15 (Sect. 4.9)
Comte, C., P. Nozières (1982): J. Physique **43**, 1069 (Sect. 2.7)
Conwell, E. (1982): *Transport: The Boltzmann Equation*, in Handbook on Semiconductors, ed. by T.S. Moss (North-Holland Publ. Comp., Amsterdam) (Sect. 6.3)
Czajkowski, G., I. Balslev (1985): Phys. Stat. Sol. (b) **130**, 655 (Sect. 3.6)

D'Andrea, A., R. Del Sole (1979): Sol. State Commun. **30**, 145 (Sect. 5.3)
D'Andrea, A., R. Del Sole (1984): Phys. Rev. B**29**, 4782 (Sects. 5.3, 5.5)
D'Andrea, A., R. Del Sole (1985): Phys. Rev. B**32**, 2337 (Sect. 5.3)
Das Sarma S., J.J. Quinn (1979): Phys. Rev. B**20**, 4872 (Sect. 6.8)
Davydov, A.S. (1966): *Quantum Mechanics* (NEO Press, Michigan) p. 300 (Appendix D)
Dean, P.J., D.C. Herbert (1979): in *Excitons*, Topics in Current Physics, Vol. 14, ed. by K. Cho (Springer, Berlin, Heidelberg, New York) p. 55 (Sect. 4.1)
De Gennes, P. (1966): *Superconductivity of Metals and Alloys*, Chap. 5 (Benjamin, New York, Amsterdam) (Sect. 2.7)
Deutsche, C.W., C.A. Mead (1965): Phys. Rev. **138A** , 65 (Sect. 5.2)
Dingle, R., W.Wiegmann, C. Henry (1974): Phys. Rev. Lett. **33**, 827 (Sect. 5.8)
Di Rienzo, A., D. Rogovin, M. Scully, R. Bonifacio, L. Lugiato, M. Milani (1978): "Superconductivity and Quantum Optics", in *Coherence in Spectroscopy and Modern Physics*, ed. by F.T. Arecchi, R. Bonifacio and M. Scully (Plenum Press) (Sect. 2.7)
Döhler, G.H. (1983): *Festkörperprobleme* (Advances in Solid State Physics) Vol. XXIII, ed. by P. Grosse (Vieweg, Braunschweig) p. 207 (Sect. 7.8)
Dow, J.D. (1974): in *Proc. 12th Int. Conf. Phys. Semicond.*, Stuttgart, ed. by M.H. Pilkuhn (Teubner, Stuttgart) p. 957 (Sect. 7.7)
Dörpelkus, K.P. (1987): to be published (App. E)
Dresselhaus, G., A.F. Kip, C. Kittel (1955): Phys. Rev. **98**, 368 (Sects. 3.5, 4.9)
Drude, P. (1900): Phys. Zs. **1**, 161 (Sect. 6.5)

Ehlers, D.H., D.L. Mills (1986): Phys. Rev. B**34**, 3939 (Sect. 6.8)
Einstein, A. (1910): Ann. d. Phys. **33**, 1275 (Sect. 7.2)

Egri, I. (1985): Physics Reports **119**, 363 (Sect. 3.2)
Ehrenreich, H., M. Cohen (1959): Phys. Rev. **115**, 786 (Sect. 3.2)
Ekardt, W., M.I. Sheboul (1976): Phys. Stat. Sol. (b) **74**, 523 (Sect. 4.1)
Ekardt, W., K. Lösch, D. Bimberg (1979): Phys. Rev. B**20**, 3303 (Sect. 5.2)
Elliott, R.J. (1957): Phys. Rev. **108**, 1384 (Sect. 4.1, 4.4)
Evangelisti, F., A. Frova, F. Patella (1974): Phys. Rev. B**10**, 4253 (Sects. 4.1, 5.1, 5.2)
Evangelisti, F., J.U. Fischbach, A. Frova (1974A): Phys. Rev. B**9**, 1516 (Sects. 5.2, 5.7)
Evrard, R. (1972): in *Polarons in Ionic Crystals and Polar Semiconductors*, ed. by J.T. Devreese (North Holland Publ. Comp., Amsterdam, London) (Sect. 6.7)

Feibelman, P.J. (1980): Phys. Rev. B**22**, 3654 (Sect. 6.8)
Feynman, R.P., F.L. Vernon, R.W. Hellwarth (1957): J. Appl. Phys. **28**, 49 (Sect. 2.1)
Fischbach, J.U., W. Rühle, D. Bimberg, E. Bauser (1976): Sol. State Commun. **18**, 1255 (Sect.5.7)
Forchel, A., H. Schweitzer, G. Mahler (1983): Phys. Rev. Lett. **51**, 501 (1983) (Sect. 7.5)
Forstmann, E., R. Gerhardts (1982): in *Advances in Solid State Physics*, Vol. XXII, 293, ed. by J. Treusch (Vieweg, Braunschweig) (Sects. 6.5, 6.8)
Frank, D. (1985): Diplomarbeit, Rheinisch Westfälische Technische Hochschule, Aachen (Sect. 3.7, Appendix E)
Frank, D., A. Stahl (1984): Sol. Stat. Comm. **52**, 861 (Sect.7.4)
Frenkel, J.I. (1931): Phys. Rev. **37**, 17 and 1276 (Sect. 4.1)
Frisch, U. (1968): in *Probabilistic Methods in Applied Mathematics*, ed. by A.T. Bharucha-Reid (Academic Press, New York) (Sect. 3.7)
Fröhlich, H., H. Pelzer, S. Zienau (1950): Phil. Mag. **41**, 221 (Sects. 3.3, 6.7)
Fröhlich, D., E. Mohler, P. Weisner (1971): Phys. Rev. Lett. **26**, 554 (Sect. 4.1)
Frova, A., F. Evangelisti, M. Zanini (1974): Phys. Stat. Sol. (a) **24**, 315 (Sect. 5.7)
Fuchs, R., K.L. Kliewer (1969): Phys. Rev. **185**, 905 (Sect. 5.2)

Gan Zi-zhao, Yang Guo-zhen (1982): Phys. Rev. B**26**, 6826 (Sect. 2.1)
Ginzburg, V.L., L.D. Landau (1950): JETP **20**, 1064 (Sect. 1.1)
Glauber, R.J. (1963): Phys. Rev. **131**, 2766 (Sect. 2.6)
Glauber, R.J. (1965): *Optical Coherence and Photon Statistics*, Proceedings of the Les Houches Summer School 1964, ed. by C. de Witt, A. Blandin and C. Cohen Tannoudji (Gordon and Breach New York) (Appendix K)
Göbel, E. (1974): Appl. Phys. Lett. **24**, 429 (Sect. 7.7)
Gogolin, A.A., E.I. Rashba (1973): Zh. Eksp. Teor. Fiz. Pisma **17**, 478 (Sect. 7.6)
Goldmann, A. (1977): Phys. Stat. Sol. (b) **81**, 9 (Sect. 4.1)
Goll, J., H. Haken (1978): Phys. Rev. A**18**, 2241 (Sects. 2.1, 7.5)
Gotthard, L. (1984): Sol. State Commun. **51**, 975 (Sect. 5.6)
Gotthard, L. (1985): Dissertation, Rheinisch-Westfälische Technische Hochschule, Aachen (Sect. 5.5)
Gotthard, L., A. Stahl, G. Czajkowski (1984): J. Phys. C**17**, 4865 (Sect. 5.4)
Gourdon, C., P. Lavallard (1985): Phys. Rev. B**31**, 6654 (Sect. 5.1)
Grosmann, M., J. Bielmann, S. Nikitine (1975): in *Excitons at High Densities*, Springer Tracts in Modern Physics, Vol. 73, ed. by H. Haken, S. Nikitine (Springer, Berlin, Heidelberg, New York) p. 242 (Sect. 5.2)
Gross, E.F. (1956): Nuovo Cimento, Suppl. **3**, 672 (Sect. 4.1)
Grosse, P. (1979): *Freie Electronen in Festkörpern* (Springer Verlag, Berlin, Heidelberg, New York) (Sect. 1.1)
Grun, J.B. (1962): Rev. Opt. **41**, 439 (Sect. 4.5)
Grun, J.B., B. Hönerlage, R. Levy (1982): in *Excitons*, Modern Problems in Condensed Matter Sciences, Vol. 2, ed. by E.I. Rashba, M.D. Sturge (North Holland, Amsterdam) p. 459 (Sect. 4.1)

Haken, H. (1973): *Quantenfeldtheorie des Festkörpers* (B.G. Teubner, Stuttgart) (Sects. 3.2, 6.7)
Haken, H. (1970): *Handbuch der Physik* Vol. XXV, ed. by S. Flügge (Springer Verlag, Berlin, Heidelberg, New York) (Sect. 2.1)
Haken, H., A. Schenzle (1973): Z. Phys. **258**, 231 (Sect. 7.5)

Halevi, T., R. Fuchs (1978): in *Proc. 14th Int. Conf. Phys. Semicond.*, ed. by B.L.H. Wilson (Institute of Physics, Bristol) p. 863 (Sect. 5.6)
Halevi, T., R. Fuchs (1984): J. Phys. C**17**, 3869 (Sect. 5.2)
Hanke, W., L.J. Sham (1980): Phys. Rev. B**21**, 4656 (Sect. 4.10)

Harris, J. (1972): J. Phys. C**5**, 1757 (Sect. 6.8)
Haug, H., S. Schmitt-Rink (1984): Progr. Quant. Electr. **9**, 3 (Sects. 3.4, 7.6., 7.7)
Henry, C.H., K. Nassau, J.W. Shiever (1971): Phys. Rev. B**4**, 2453 (Sect. 4.1)
Hildebrand, O., E. Göbel (1976): in *Proc. 13th Int. Conf. Phys. Semicond.*, ed. by F.G. Fumi (Tipografia Marves) p. 942 (Sect. 7.7)
Hitmair, O., G. Adam (1971): *Wärmetheorie* (Vieweg & Sohn, Braunschweig) (Sect. 6.2)
Hönerlage, B., A. Bivas, V.D. Phach (1978): Phys. Rev. Lett. **41**, 49 (Sect. 4.1)
Hönerlage, B., J.Y. Bigot (1984): Phys. Stat. Sol. (b) **123**, 201 (Sect. 7.6)
Hönerlage, B., R. Lévy, J.B. Grun, C. Klingshirn, K. Bohnert (1985): Physics Reports **124**, 161 (Sects. 4.1, 4.9, 7.6)
Hopfield, J.J. (1958): Phys. Rev. **112**, 1555 (Sects. 2.3, 2.6, 3.4, 4.1)
Hopfield, J.J. (1960): J. Phys. Chem. Solids **15**, 97 (Sect. 4.9)
Hopfield, J.J., D.G. Thomas (1963): Phys. Rev. **132**, 563 (Sects. 1.1, 3.2, 4.1, 4.8, 5.2)
Hopfield, J.J. (1969): Phys. Rev. **182**, 945 (Sects. 2.3, 7.2)
Hornung, T., R.G. Ulbrich (1982): in *Proc. 16th Int. Conf. Phys. Semicond.*, ed. by M. Averous (North Holland, Amsterdam) p. 263 (Sect. 4.1)
Hostler, L. (1969): Phys. Rev. **198**, 126 (Appendix E)
Hostler, L. (1975): J. Math. Phys. **16**, 1585 (Appendix E)
Hümmer, K., (1973): Phys. Stat. Sol. **56**, 249 (Sect. 4.1)
Hümmer, K., P. Gebhardt (1978): Phys. Stat. Sol. (b) **85**, 271 (Sects. 4.9, 5.2)
Huang, K. (1951): Proc. Roy. Soc. A**208**, 352 (Sect. 3.3)
Huhn, W., A. Stahl (1984): Phys. Stat. Sol. (b) **124**, 167 (Sect. 1.1, 3.2)
Huhn, W. (1986): Optics Communications, **57**, 221 (Sect. 7.5)
Hvam, J.M., G. Blattner, M. Reuscher, C. Klingshirn (1983): Phys. Stat. Sol. (b) **118**, 179 (Sect. 4.1)

Ivchenko, E.L. (1982): in *Excitons*, Problems in Condensed Matter Sciences, Vol. 2, ed. by I.E. Rashba, M.D. Sturge (North Holland, Amsterdam) p. 141 (Sects. 4.1, 5.2)

Jackson, J.D. (1975): *Classical Electrodynamics*, Second Edition, (John Wiley and Sons, New York) (Appendix B)
Johnson, E.J. (1967): Phys. Rev. Lett. **19**, 352 (Sect. 4.1)
Johnson, E.J. (1968): in *Proc. 9th Int. Conf. Phys. Semicond.* (Nauka, Leningrad) p. 278 (Sect. 4.1)

Kadanoff, L.P., G. Baym (1962): *Quantum Statistical Mechanics*, in Frontiers in Physics, ed. by D. Pines (Benjamin, Menlo Park, Ca.) (Sect. 3.7)
Kane, E.O. (1956): J. Phys. Chem. Sol. **1**, 83 (Sects. 3.5, 4.3, 4.9)
Kane, E.O. (1969): Phys. Rev. **180**, 852 (Sect. 4.10)
Kane, E.O. (1975): Phys. Rev. B**11**, 3850 (Sect. 4.9)
Kiselev, V.A., B.S. Razbirin, I.N. Ural'tsev (1973): JETP Lett. **18**, 296 (Sect. 4.1)
Kiselev, V.A., B.S. Razbirin, I.N. Ural'tsev (1974): in *Proc. 12th Int. Conf. Phys. Semicond.*, ed. by M. Pilkuhn (Teubner, Stuttgart) p. 996 (Sect. 4.1)
Kittel, C. (1963): *Quantum Theory of Solids* (J. Wiley and Sons, New York, London) (Sects. 3.3, 6.4)
Kliewer, K.L., R. Fuchs (1968): Phys. Rev. **172**, 602 (Sect. 5.6)
Klingshirn, C., H. Haug (1981): Physics Reports **70**, 315 (Sects. 4.1, 7.1, 7.7)
Klitzing v., K., G. Dorda, M. Pepper (1980): Phys. Rev. Lett. **45**, 494 (Sects. 5.8, 6.3)
Knox, R.S. (1963): *Theory of Excitons* (Academic Press, New York) (Sects. 3.2, 3.6, 4.1, 4.4, Appendix D)
Knox, W.H. et al. (1985): Phys. Rev. Lett. **54**, 1306 (Sects. 7.1, 7.5)
Koteles, E.S. (1982): in *Excitons*, Modern Problems in Condensed Matter Sciences, Vol. 2, ed. by E.I. Rashba, M.D. Sturge (North Holland, Amsterdam) p. 83 (Sect. 4.1)
Koteles, E.S., G. Winterling (1980): Phys. Rev. Lett. **44**, 948 (Sect. 4.1)

Koteles, E.S., J.P. Salerno, W. Bloss, E.M. Brody (1984): *Proc. 17th Int. Conf. Phys. Semicond.*, San Francisco, ed. by J.D. Chadi, W.A. Harrison (Springer, New York) p. 1247 (Sect. 5.1)
Koteles, E.S., J. Lee, J.P. Salerno, M.O. Vassell (1985): Phys. Rev. Lett. **55**, 867 (Sect. 4.1)

Lagois, J. (1981): Phys. Rev. B**23**, 5511 (Sects. 4.1, 5.1, 5.6, 5.7)
Lagois, J., B. Fischer (1982): in *Surface Polaritons*, Modern Problems in Condensed Matter Sciences, ed. by V.M. Agranovich, D.L. Mills (North Holland, Amsterdam) p. 69 (Sects. 5.1, 5.6)
Lagois, J., B. Fischer (1979): in *Excitons*, Topics in Current Physics, Vol. 14, ed. by K. Cho (Springer, Berlin, Heidelberg, New York) p. 183 (Sects. 5.1, 5.6)
Lagois, J., K. Hümmer (1975): Phys. Stat. Sol. (b) **72**, 392 (Sect. 4.1)
Lagois, J., E. Wagner, W. Bludau, K. Lösch (1978): Phys. Rev. B**18**, 4325 (Sect. 5.7)
Lamb, W.R. (1971): Rev. Mod. Phys. **43**, 99 (Sects. 2.1, 7.1)
Landau, L.D., E.M. Lifschitz (1960): *Electrodynamics of Continuous Media*, Vol. 8 of Course of Theoretical Physics (Pergamon Press, Oxford) (Sect. 2.5)
Landau, L.D., E.M. Lifschitz (1981): *Physical Kinetics*, Vol. 10 of Course of Theoretical Physics (Pergamon Press, Oxford) (Sects. 2.5, 6.1, 6.4)
Langer, D.W., R.N. Eumena, K. Era, T. Koda (1970): Phys. Rev. B**2**, 4005 (Sect. 4.1)
Lax, M., D.F. Nelson (1971): Phys. Rev. B**4**, 3694 (Sect. 3.3)
Lebedew, N.N. (1973): *Spezielle Funktionen und ihre Anwendungen* (Bibliographisches Institut, Mannheim) (Sect. 5.4)
Lederhofer, R. (1985): Diplomarbeit der RWTH Aachen (Sect. 6.8)
Lee, T.D., D. Pines (1953): Phys. Rev. **92**, 883 (Sect. 6.7)
Leheny, R.F., J. Shah (1976): Phys. Rev. Lett. **37**, 871 (Sect. 7.7)
Lemmens, F., J.T. Devreese (1974): Sol. State Commun. **14**, 1339 (Sect. 6.6)
Lindhard, J. (1954): Kgl. Danske Vidensk. Selskab, Mat. fys. Medd. **28**, 8 (Sects. 3.4, 6.4)
Lorentz, H.A. (1909): *The Theory of Electrons* (Teubner, Leipzig). Reprint: Dover Publications, New York 1952 (Sect. 6.5)
Loudon, R. (1963): Proc. Roy. Soc. A**275**, 218 (Sect. 7.2)
Loudon, R. (1973): *The Quantum Theory of Light* (Clarendon Press, Oxford) (Appendix B)
Low, F.E., D. Pines (1955): Phys. Rev. **98**, 414 (Sect. 6.7)
Lüth, H. (1981): *Festkörperprobleme* (Advances in Solid State Physics), Vol. XXI, ed. by J. Treusch (Vieweg, Braunschweig) (Sect. 6.8)

Macfarlane, G.G., T.P. McLean, J.E. Quarrington, V. Roberts (1957): Phys. Rev. **108**, 1377 (Sect. 4.1)
McCall, S.L., E.L. Hahn (1969): Phys. Rev. **183**, 457 (Sects. 2.1, 7.5)
McLean, T.P. (1960): in *Progress in Semiconductors*, Vol. 5, ed. by A.F. Gibson (Heywood and Company, London) p. 53 (Sect. 4.5)
Madelung, O. (1982): in *Landolt – Börnstein* ed. by O. Madelung (Springer, Berlin, Heidelberg, New York) (Sect. 4.1)
Mahan, G.D., J.J. Hopfield (1964): Phys. Rev. **135**, 428 (Sect. 4.9)
Mahan, G.D. (1981): *Many-Particle Physics* (Plenum Press, New York, London) (Sect. 3.3)
Majunder, F.A., H.-W. Swoboda, K. Kemp, C. Klingshirn (1985): Phys. Rev. B**32**, 2419 (Sect. 7.7)
Maradudin, A.A., D.L. Mills (1973): Phys. Rev. B**7**, 2787 (Sects. 4.8, 5.2, 5.6)
Martin, T.P., H. Schaberg (1979): Z. Physik B**35**, 61 (Sect. 4.1)
Maruani, A., J.L. Oudar, E. Batifol, D.S. Chemla (1978): Phys. Rev. Lett. **39**, 1372 (Sect. 7.6)
Masumoto, Y., Y. Unuma, Y. Tanaka, S. Shionoya (1979): J. Phys. Soc. Japan **47**, 1844 (Sect. 4.1)
Mattis, D.C., G. Beni (1978): Phys. Rev. B**18**, 3816 (Sect. 5.3)
Merzbacher, E. (1970): *Quantum Mechanics*, 2nd ed. (John Wiley & Sons, New York, London, Sydney, Toronto) (Sect. 2.6)
Miller, D.A.B., C.T. Seaton, M.E. Prise, S.D. Smith (1981): Phys. Rev. Lett. **47**, 197 (Sect. 7.4)
Miller, D.A.B. et al. (1984): Phys. Rev. Lett. **53**, 2173 (Sect. 5.8)
Miller, A., D.A.B. Miller, S.D. Smith (1981): Advances in Physics **30**, 697 (1981) (Sect. 7.4)

Mills, D.L., E. Burstein (1969): Phys. Rev. **188**, 1465 (Sect. 7.2)
Mita, T., K. Sotome, M. Ueta (1980): Solid State Comm. **33**, 1135 (Sect. 4.1)
Mohler, E. (1970): Phys. Stat. Sol. **38**, 81 (Sect. 4.1)
Mooradian, A. (1969): *Advances in Solid State Physics*, ed. by O. Madelung (Pergamon, Vieweg, Braunschweig) (Sect. 6.6)
Moss, T.S. (1982): Handbook on Semiconductors, Vol. 1: *Band Theory and Transport Processes*, ed. by W. Paul (North Holland Publ. Comp., Amsterdam) (Sect. 6.3)
Mott, N.F. (1974): *Metal-Insulator Transitions* (Taylor and Francis, Ltd, London) p. 124 (Sect. 7.7)
Mysyrowicz, A., J.B. Grun, R. Levy, A. Bivas, S. Nikitine (1968): Phys. Lett. A**26**, 615 (Sect. 7.6)

Nikitine, S. (1959): Phil. Mag. **4**, 1 (Sect. 4.1)
Nikitine, S. (1975): in *Springer Tracts in Modern Physics*, Vol. 73, ed. by H. Haken, S. Nikitine (Springer, Berlin, Heidelberg, New York) p. 5 (Sect. 4.1)
Nikitine, S., J.B. Grun, M. Sieskind (1961): J. Phys. Chem. Solids **17**, 292 (Sect. 4.5)
Nowak, U., W. Richter, G. Sachs (1981): Phys. Stat. Sol. (b) **108**, 1331 (Sect. 6.6)
Nozières, P., C. Comte (1982): J. Physique **43**, 1083 (Sect. 2.7)
Nozières, P., D. Pines (1958): Phys. Rev. **111**, 442 (Sect. 3.2)
Nye, J.F. (1972): *Physical Properties of Crystals* (Oxford Clarendon Press, 4th ed.) (Sect. 3.3)

Otto, A. (1974): in *Festkörperprobleme* (Advances in Solid State Physics), Vol. XIV, ed. by H.J. Queisser (Vieweg, Braunschweig) p. 1 (Sect. 5.6)

Peierls, R.E. (1932): Ann. Physik (Leipzig) **13**, 905 (Sect. 4.4)
Pekar, S.I. (1957): Zh. exper. teor. Fiz. **33**, 1022 (Eng. transl.: Sov. Phys. - JETP **6**, 785) (Sects. 5.2, 5.6)
Pekar, S.I. (1984): *Crystal Optics and Additional Light Waves*, Frontiers in Physics Vol. 57 (Addison-Wesley, Reading Mass.) (Sect. 5.2)
Penrose, O., L. Onsager (1956): Phys. Rev. **104**, 576 (Sect. 1.1)
Pevtsov, A.B., S.A. Permogorov, Sh.R. Saifullaev, A.V. Selkin (1980): Fiz. Tverd. Tela **22**, 2400 (Eng. tansl.: Sov. Phys. Solid State **22**, 1396) (Sects. 5.1, 5.6)
Pevtsov, A.B., A.V. Selkin (1983): Fiz. Tverd. Tela **25**, 157 (Eng. transl.: Sov. Phys. Solid State **25**, 85) (Sects. 4.1, 5.6)
Pinczuk, A., E. Burstein (1975): in *Light Scattering in Solids*, Topics in Applied Physics, Vol. 8, ed. by M. Cardona (Springer, Berlin, Heidelberg, New York) (Sect. 7.2)
Pines. D. (1964): *Elementary Excitations in Solids* (W.A. Benjamin, New York, Amsterdam) (Sect. 3.2)
Placzek, G. (1934): in *Marx' Handbuch der Radiologie*, Bd. VI (Akademische Verlagsgesellschaft, Leipzig) (Sect. 7.2)
Poluektov, I.A., Y. Popov, V. Roitberg (1975): Sov.Phys. Usp. **17**, 673 (Sects. 7.1, 7.5)
Power, E.A. *Introductory Quantum Electrodynamics* (Longmans, Green and Co., London) (Appendix B, C)

Quattropani, A., J.J. Forney (1977): Il Nuovo Cimento, **39B**, 569 (Sect. 7.6)

Redfield, A.G. (1955): Phys. Rev. **98**, 787 (Sect. 3.7)
Rice, T.M. (1974): Il Nuovo Cim. **23B**, 226 (Sect. 3.4)
Richter, W. (1976): *Resonant Raman Scattering in Semiconductors*, Springer Tracts in Modern Physics, Vol. 78 (Springer, Berlin, Heidelberg, New York) (Sect. 7.2)
Ritz, A., H. Lüth, (1984): Phys. Rev. Letters **52**, 1242 (Sect. 6.8)
Rösler, M., R. Zimmermann, W. Richert (1984): Phys. Stat. Sol. (b) **121**, 609 (Sect. 7.7)
Rössler, U. (1979): in *Festkörperprobleme* (Advances in Solid State Physics). Vol. XIX, ed. by J. Treusch (Vieweg, Braunschweig) p. 77 (Sect. 4.9)

Sakoda, S. (1976): J. Phys. Soc. Japan **40**, 152 (Sect. 5.3)
Satpathy, S. (1983): Phys. Rev. B**28**, 4584 (Sect. 5.3)
Schaich, W.L. 81982): Surf. Sci. **122**, 175 (Sect. 6.8)
Schmitt-Rink, S., D.B.Tran Thoai, H. Haug (1980): Z. Physik B**39**, 25 (Sect. 7.7)

Schöll, E. (1982): Z. Phys. B**46**, 23; **48**, 153; **52**, 321 (Sect. 3.7)
Schultheis, L., I. Balslev (1983): Phys. Rev. B**28**, 2292 (Sects. 4.1, 5.7)
Schweizer, H., A. Forchel. A. Hangleiter, S. Schmitt-Rink, J. Löwenau, H. Haug (1983): Phys. Rev. Lett. **51**, 698 (Sect. 7.7)
Segawa, Y., Y. Aoyagi, K. Azuma, S. Namba (1978): Sol. State Commun. **28**, 858 (Sect. 4.1)
Selkin, A. (1977): Phys. Stat. Sol. (b) **83**, 47 (Sect. 4.8)
Sell, D.D., S.E. Stokowski, R. Dingle, J.V. DiLorenzo (1973): Phys. Rev. B**7**, 4568 (Sects. 4.1, 5.2, 5.7)
Sermage, B., M. Voos (1977): Phys. Rev. B**15**, 3935 (Sect. 4.1)
Sermage, B., M. Voos, C. Schwab (1979): Phys. Rev. B**20**, 3245 (Sect. 4.1)
Shaklee, K.L., R.E. Nahory (1970): Phys. Rev. Lett. **24**, 942 (Sect. 4.1)
Skettrup, T., I. Balslev (1971): Phys. Rev. B**3**, 1457 (Sect. 5.2)
Slichter, C.P. (1980): *Principles of Magnetic Resonance*, Springer Series in Solid-State Sciences, Vol. 1 (Springer, Berlin, Heidelberg, New York) p. 33 (Sect. 3.7)
Stahl, A. (1979): Phys. Stat. Sol. (b) **94**, 221 (Sect. 4.3)
Stahl, A. (1981): Phys. Stat. Sol. (b) **106**, 575 (Sect. 5.4)
Stahl, A. (1983): Surface Science **134**, 297 (Sect. 6.8)
Stahl, A. (1984): Sol. Stat. Comm. **49**, 91 (Sect. 1.1)
Stahl, A. (1985): in *Advances in Solid State Physics*, vol. XXV, ed. by P. Grosse (Vieweg Braunschweig) p. 287 (Sect. 5.5)
Stahl, A., I. Balslev (1982): Phys. Stat. Sol. (b) **113**, 583 (Sect. 4.5)
Stahl, A., C. Uihlein (1979): in *Festkörperprobleme* (Advances in Solid State Physics) Vol. XIX, ed. by J. Treusch (Vieweg, Braunschweig) p. 159 (Sect. 4.2)
Stern, F. (1963): Sol. State Phys. **15**, 300 (Sect. 4.3)
Stern, F. (1967): Phys. Rev. Lett. **18**, 546 (Sect. 5.8)
Suhl, H., N.R. Werthammer (1961): Phys. Rev. **122**, 359 (Sect. 3.2)

Thewalt, M.L.W. (1982): in *Excitons*, Modern Problems in Condensed Matter Sciences, Vol. 2, ed. by E.I. Rashba, M.D. Sturge (North Holland, Amsterdam) p. 393 (Sect. 4.1)
Thomas, D.G., J.J. Hopfield (1959): Phys. Rev. **116**, 573 (Sect. 4.1)
Tokura, Y., T. Koda, I. Hirabayashi, S. Nakada (1981): J. Phys. Soc. Japan **50**, 145 (Sect. 5.6)
Toyozawa, Y. et al. (1967): J. Phys. Soc. Japan, Suppl. **21**, 133 (Sect. 4.10)

Ulbrich, R.G. (1985): in *Festkörperprobleme* XXV, ed. by P. Grosse (Vieweg, Braunschweig) (Sects. 7.1, 7.5)
Ulbrich, R.G., C. Weisbuch (1978): in *Festkörperprobleme* (Advances in Solid State Physics) Vol. XVIII, ed. by J. Treusch (Vieweg, Braunschweig) p. 217 (Sects. 4.1, 5.1)

Van Kampen, N.G. (1985): Physics Reports **124**, 69 (Sect. 3.7)
Varfolomeev, A.V., A.A. Ryskin, R.P. Siesyan (1976): Fiz. Tekh. Poluprov. **10**, 2072 (Eng. Transl.: Sov.Phys. Semicond. **10**, 1234) (Sect. 4.1)
Veliki, B., J. Sak (1966): Phys. Stat. Sol. **16**, 147 (Sect. 4.10)
Venghaus, H. (1979): Phys. Rev. B**19**, 3071 (Sect. 4.1)
Vina, L., S. Logothetides, M. Cardona (1984): Phys. Rev. B**30**, 1979 (Sect. 4.10)
Vogt, H. (1982): *Coherent and Hyper-Raman Techniques* in "Light Scattering in Solids", Topics in Applied Physics, Vol. 50, ed. by M. Cardona and G. Güntherodt (Springer, Berlin, Heidelberg, New York) (Sect. 7.3)
Voigt, J., F. Springelberg, M. Senoner (1979): Phys. Stat. Sol. (b) **91**, 189 (Sect. 4.1)
Vu Duy Phach, A. Bivas, B. Hönerlage, J.B. Grun (1977): Phys. Stat. Sol. (b) **84**, 731 (Sect. 7.6)

Wagner, D. (1966): Zs. Naturf. **21a**, 634 (Sect. 6.8)
Wannier, G.H. (1937): Phys. Rev. **52**, 191 (Sect. 4.4)
Weisbuch, C., R.G. Ulbrich (1982): *Resonant Light Scattering Mediated by Excitonic Polaritons in Semiconductors*, in "Light Scattering in Solids", Topics in Applied Physics, Vol. 51, ed. by M. Cardona and G. Güntherodt (Springer, Berlin, Heidelberg, New York) (Sect. 7.2)
Wherrett, B.S., N.A. Higgins (1982): Proc. R. Soc. London A**379**, 67 (Sect. 7.4)

Wherrett, B.S. (1983): Proc. Royal Soc. London A**390**, 373 (Sect. 7.4)
Wigner, E.P. (1932): Phys. Rev. **40**, 749 (Sect. 6.1)

Yang, C.N. (1962): Rev. Mod. Phys. **34**, 694 (Sect. 1.1)
Yu, P.Y. (1979): in *Excitons*, Topics in Current Physics, Vol. 14, ed. by K. Cho (Springer, Berlin, Heidelberg, New York) p. 211 (Sect. 4.1)
Yuan, Y.R., M.A.A. Pudensi, G.A. Vawter, J.L. Merz (1985): J. Appl. Phys. **58**, 397 (Sect. 7.8)

Zawadzki, W. (1982): *Mechanisms of Electron Scattering in Semiconductors*, in "Handbook on Semiconductors", ed. by T.S. Moss (North-Holland Publ. Comp., Amsterdam) (Sect. 3.3)
Zeyher, R. (1981): *Condensed Matter Physics*, Vol. 1, ed. by J.T. Devreese (Plenum, New York) (Sects. 5.3, 5.5)
Zeyher, R., J.L. Birman, W. Brenig (1972): Phys. Rev. B**6**, 4613 (Sect. 5.2)
Ziman, J.H. (1964): *Principles of the Theory of Solids*, (Cambridge University Press) (Sect. 6.4)
Zimmermann, R. (1976): Phys. Stat. Sol. (b) **76**, 191 (Sect. 3.4)
Zimmermann, R. et al. (1978): Phys. Stat. Sol. (b) **90**, 175 (Sect. 7.7)
Zimmermann, R. (1985): Private communication (Sect. 4.6)

# Subject Index

# Classified Index

## Springer Tracts in Modern Physics, Volumes 36–110

This cumulative index is based upon the Physics and Astronomy Classification Scheme (PACS) developed by the American Institute of Physics. First authors are listed alphabetically under each of the main PACS headings.

**GENERAL**
**04.00 Relativity and Gravitation**
**05.00 Statistical Physics**

Grabert H.:
Projection Operator Techniques in Nonequilibrium Statistical Mechanics.
STMP 95,1-164 (1982) PACS:05.00

Heintzmann H., Mittelstaedt P.:
Physikalische Gesetze in beschleunigten Bezugssystemen.
STMP 47,185-225 (1968) PACS:04.00

Kenkre V.M.:
The Master Equation Approach: Coherence, Energy Transfer, Annihilation, and Relaxation.
STMP 94,1-109 (1982) PACS:05.00 71.00

Reineker P.:
Stochastic Liouville Equation Approach: Coupled Coherent and Incoherent Motion, Optical Line Shapes, Magnetic Resonance Phenomena.
STMP 94,111-226 (1982) PACS:05.00 71.00 75.00

**THE PHYSICS OF ELEMENTARY PARTICLES AND FIELDS**
**11.00 General Theory of Fields and Particles**
**11.30 Symmetry and Conservation Laws**
**11.40 Currents and Their Properties**

Barut A.O.:
Dynamical Groups and their Currents. A Model for Strong Interactions.
STMP 50,1-28 (1969) PACS:11.30

Brandt R.A.:
Physics on the Light Cone.
STMP 57,237-247 (1971) PACS:11.00

Ekstein H.:
Rigorous Symmetries of Elementary Particles.
STMP 37,150-180 (1965) PACS:11.30

Ferrara S., Gatto R., Grillo A.F.:
Conformal Algebra in Space-Time and Operator Product Expansion.
STMP 67,1-64 (1973) PACS:11.00

Genz H.:
Local Properties of sigma-Terms: A Review.
STMP 61,130-136 (1972) PACS:11.40

Jackiw R.:
Canonical Light-Cone Commutators and their Applications.
STMP 62,1-36 (1972) PACS:11.00

Kleiner H.:
Baryon Current Solving SU(3) Charge-Current Algebra.
STMP 49,90-146 (1969) PACS:11.40

Kundt W.:
Canonical Quantization of Gauge Invariant Field Theories.
STMP 40,108-168 (1966) PACS:11.00

Leutwyler H.:
Current Algebra and Lightlike Charges.
STMP 50,29-41 (1969) PACS:11.40

Lopuszanski J.T.:
Physical Symmetries in the Framework of Quantum Field Theory.
STMP 52,201-214 (1970) PACS:11.30

Mendes R.V., Ne'eman Y.:
Representations of the Local Current Algebra.
STMP 60,18-31 (1971) PACS:11.40

Müller V.F.:
Introduction to the Lagragian Method.
STMP 50,42-52 (1969) PACS:11.40

Pauli W.:
Continuous Groups in Quantum Mechanics.
STMP 37,85-104 (1965) PACS:11.30

Pietschmann H.:
Introduction to the Method of Current Algebra.
STMP 50,53-64 (1969) PACS:11.40

Pilkuhn H.:
Coupling Constants from PCAC.
STMP 55,168-173 (1970) PACS:11.40

Pilkuhn H.:
S-Matrix Formulation of Current Algebra.
STMP 50,65-70 (1969) PACS:11.40

Renner B.:
On the Problem of the Sigma Terms in Meson-Baryon Scattering. Comments on Recent Literature.
STMP 61,120-129 (1972) PACS:11.40

Renner B.:
Current Algebra and Weak Interactions.
STMP 52,60-78 (1970) PACS:11.40 12.30

Soloviev L.D.:
Symmetries and Current Algebras for Electromagnetic Interactions.
STMP 46,53-66 (1968) PACS:11.40

Stech B.:
Nonleptonic Decays and Mass Differences of Hadrons
STMP 50,84-99 (1969) PACS:11.40

Stichel P.:
Introduction to Current Algebra.
STMP 50,120-131 (1969) PACS:11.40

Stichel P.:
Current Algebra in the Framework of General Quantum Field Theory.
STMP 50,100-109 (1969) PACS:11.40

Stichel P.:
Current Algebra and Renormalizable Field Theories.
STMP 50,110-119 (1969) PACS:11.40

Symanzik K.:
Small-Distance Behaviour in Field Theory.
STMP 57,222-236 (1971) PACS:11.00

Verzegnassi C.:
Low Energy Photo and Electroproduction, Multipole Analysis by Current Algebra Commutators.
STMP 59,154-163 (1971) PACS:11.40

Weinstein M.:
Chiral Symmetry. An Approach to the Study of the Strong Interactions.
STMP 60,32-73 (1971) PACS:11.40

Wess J.:
Realisations of a Compact, Connected, Semisimple Lie Group.
STMP 50,132-142 (1969) PACS:11.30

Wess J.:
Conformal Invariance and the Energy-Momentum Tensor.
STMP 60,1-17 (1971) PACS:11.30

Zimmermann W.:
Problems in Vector Meson Theories.
STMP 50,143-156 (1969) PACS:11.00

**12.00 Specific Theories and Interaction Models**
**12.20 Quantum Electrodynamics**
**12.30 Models of Weak Interactions**
**12.35 Composite Models of Particles**
**12.40 Models of Strong Interactions**

Ademollo M.:
Current Amplitudes in Dual Resonance Models.
STMP 59,135-153 (1971) PACS:12.40

Amaldi E., Fubini S., Furlan G.:
Pion-Electroproduction. Electroproduction at Low Energy and Hadron Form Factors.
STMP 83,1-162 (1979) PACS:12.00
Barut A.O.:
On the S-Matrix Theory of Weak Interactions.
STMP 53,1-5 (1970) PACS:12.30
Brodsky S.J.:
Quantum Chromodynamics at Nuclear Dimensions.
STMP 100,81-144 (1982) PACS:12.35
Close F.E.:
The Pauli Principle and QCD for Quarks and Nucleons in Hadrons and Nuclei.
STMP 100,57-80 (1982) PACS:12.35 21.40
Collins P.D.B., Gault F.D.:
The Eikonal Model for Regge Cuts in Pion-Nucleon Scattering.
STMP 63,163-189 (1972) PACS:12.40
Collins P.D.B., Squires E.J.:
Regge Poles in Particle Physics.
STMP 45,1-292 (1968) PACS:12.40
Collins P.D.B.:
How Important are Regge Cuts?
STMP 60,204-233 (1971) PACS:12.40
Contogouris A.P.
Regge Analysis and Dual Absorptive Model.
STMP 63,145-162 (1972) PACS:12.40
Contogouris A.P.
Certain Problems of Two-Body Reactions with Spin.
STMP 57,92-118 (1971) PACS:12.40
Dietz K.:
Dual Quark Models.
STMP 60,74-90 (1971) PACS:12.40
Dosch H.G.:
The Decay of the Ko - Ko System.
STMP 52,79-90 (1970) PACS:12.30
Gasiorowicz S.:
A Survey of the Weak Interactions.
STMP 52,1-33 (1970) PACS:12.30 23.00
Gatto R.:
Cabibbo Angle and SU2 x SU2 Breaking.
STMP 53,45-106 (1970) PACS:12.30 11.40
Gehlen G.von:
Weak Interactions at High Energies.
STMP 53,29-44 (1970) PACS:12.30
Heller L.:
Bag Models and Nuclear Forces.
STMP 100,145-185 (1982) PACS:12.35 21.40
Hofmann W.:
Jets of Hadrons.
STMP 90,1-210 (1981) PACS:12.00
Hove L.van:
Theory of Strong Interactions of Elementary Particles in the GeV Region.
STMP 39,1-19 (1965) PACS:12.40
Huang K.:
Deep Inelastic Hadronic Scattering in Dual-Resonance Model.
STMP 62,107-117 (1972) PACS:12.40
Kabir P.K.:
Questions Raised by CP-Nonconservation.
STMP 52,91-112 (1970) PACS:12.30
Källen G.:
Radiative Corrections in Elementary Particle Physics.
STMP 46,67-132 (1968) PACS:12.20
Kummer W.:
Relations for Semileptonic Weak Interactions Involving Photons.
STMP 52,113-125 (1970) PACS:12.30
Michael C.:
Regge Residues.
STMP 55,174-191 (1970) PACS:12.40
Müller V.F.:
Semileptonic Decays.
STMP 52,34-49 (1970) PACS:12.30
Oehme R.:
Duality and Regge Theory.
STMP 57,119-131 (1971) PACS:12.40
Oehme R.:
Complex Angular Momentum.
STMP 57,132-157 (1971) PACS:12.40
Oehme R.:
Rising Cross-Sections.
STMP 61,109-119 (1972) PACS:12.40
Olsen H.A.:
Applications of Quantum Electrodynamics.
STMP 44,84-201 (1968) PACS:12.20

Paul E.:
Status of Interference Experiments with Neutral Kaons.
STMP 79,53-145 (1976) PACS:12.30
Pietschmann H.:
Weak Interactions at Small Distances.
STMP 52,193-200 (1970) PACS:12.30
Primakoff H.:
Weak Interactions in Nuclear Physics.
STMP 53,6-28 (1970) PACS:12.30 23.00
Riazzuddin :
Radiative Corrections to Weak Decays Involving Leptons.
STMP 52,126-160 (1970) PACS:12.30
Rothleitner J.:
Radiative Corrections to Weak Interactions.
STMP 52,161-170 (1970) PACS:12.30
Rubinstein H.R.:
Physical N-Pion Functions.
STMP 57,193-201 (1971) PACS:12.40
Rubinstein H.R.:
Duality for Real and Virtual Photons.
STMP 62,72-91 (1972) PACS:12.40 13.60
Satz H.:
An Introduction to Dual Resonance Models in Multiparticle Physics.
STMP 57,158-190 (1971) PACS:12.40
Schilling K.:
Some Aspects of Vector Meson Photoproduction on Protons.
STMP 63,31-56 (1972) PACS:12.60
Schrempp-Otto B., Schrempp F.:
Are Regge Cuts still Worthwhile?
STMP 61,68-108 (1972) PACS:12.40
Segre G.:
Unconventional Models of Weak Interactions.
STMP 52,171-192 (1970) PACS:12.30
Squires E.J.:
Regge-Pole Phenomenology.
STMP 57,71-91 (1971) PACS:12.40
Stech B.:
Non-Leptonic Decays.
STMP 52,50-59 (1970) PACS:12.30
Tan C.-I.:
High Energy Inclusive Processes.
STMP 60,91-106 (1971) PACS:12.40
Wiik B.H., Wolf G.:
Electron-Positron Interactions.
STMP 86,1-262 (1979) PACS:12.00

**13.40 Electromagnetic Processes and Properties of Hadrons**
**13.60 Photon and Lepton Interactions with Hadrons**
**13.75 Hadron-Induced Reactions**
**14.00 Properties of Spec. Particles and Resonances**
**14.80 Other and Hypothetical Particles**

Atkinson D.:
Some Consequences of Unitary and Crossing Existence and Asymptotic Theorems.
STMP 57,1-21 (1971) PACS:13.75
Basdevant J.L.:
PiPi Theories.
STMP 61,1-24 (1972) PACS:13.75
Brinckmann P.:
Polarization of Recoil Nucleons from Single Pion Photoproduction. Experimental Methods and Results.
STMP 61,135-166 (1972) PACS:13.60
Buchanan C.D., Collard H., Crannell C., Frosch R., Hofstadter R., Ravenhall D.G.:
Recent High Energy Electron Investigations at Stanford University.
STMP 39,20-42 (1965) PACS:13.40
Donnachie A.:
The Nucleon Resonances.
STMP 61,25-48 (1972) PACS:14.20
Donnachie A.:
Exotic Electromagnetic Currents.
STMP 63,121-144 (1972) PACS:13.60
Drees J.:
Deep Inelastic Electron-Nucleon Scattering.
STMP 60,107-137 (1971) PACS:13.60
Drell S.D.:
Special Models and Predictions for Photoproduction above 1 GeV.
STMP 39,71-90 (1965) PACS:13.60
Ebel G., Julius D., Müllensiefen A., Pilkuhn H.,

Schmidt W., Swart J.J.de:
Compilation of Coupling Constants at Low-Energy Parameters.
STMP 55,239-290 (1970) PACS:13.75

Fischer H.:
Experimental Data on Photoproduction of Pseudoscalar Mesons at Intermediate Energies.
STMP 59,188-222 (1971) PACS:13.60

Flügge G.:
Experimental Evidence of Quarks and Gluons.
STMP 100,1-55 (1982) PACS:13.60 14.80

Foa L.:
Meson Photoproduction on Nuclei.
STMP 59,114-134 (1971) PACS:13.60

Froyland J.,
High Energy Photoproduction of Pseudoscalar Mesons
STMP 63,1-30 (1972) PACS:13.60

Furlan G., Paver N., Verzegnassi C.:
Low Energy Theorems and Photo- and Electroproduction Near Threshold by Current Algebra.
STMP 62,118-147 (1972) PACS:13.60 11.40

Gatto R.:
Theoretical Aspects of Colliding Beam Experiments.
STMP 39,106-137 (1965) PACS:13.40

Gehlen G.von:
Pion Electroproduction in the Low-Energy Region.
STMP 59,164-187 (1971) PACS:13.60

Gourdin M.:
Vector Mesons in Electromagnetic Interactions.
STMP 55,192-212 (1970) PACS:13.40

Gustafson G., Hamilton J.:
The Dynamics of Some pi-N Resonances.
STMP 61,49-67 (1972) PACS:13.75

Hamilton J.:
New Methods in the Analysis of pi-N Scattering.
STMP 57,41-70 (1971) PACS:13.75

Heinloth K.:
Experiments on Electroproduction in High Energy Physics.
STMP 65,92-145 (1972) PACS:13.60

Höhler G.:
Special Models and Predictions for Photoproduction (Low Energies).
STMP 39,55-70 (1965) PACS:13.60

Holtey G.von:
Pion Photoproduction on Nucleons in the First Resonance Region (Experimental Situation).
STMP 59,3-26 (1971) PACS:13.60

Huang K.:
Duality and the Pion Electromagnetic Form Factor.
STMP 62,98-106 (1972) PACS:13.40

Kolanoski H.:
Two-Photon Physics at e+e- Storage Rings.
STMP 105,1-187 (1984) PACS:13.65

Kramer G.:
Theory of Jets in Electron-Positron Annihilation.
STMP 102,1-140 (1984) PACS:13.65

Kramer G.:
Nucleon-Nucleon Interactions below 1GeV/c.
STMP 55,152-167 (1970) PACS:13.75 21.40

Landshoff P.V.:
Duality in Deep Inelastic Electroproduction.
STMP 62,37-50 (1972) PACS:13.60 12.40

Lüke D., Söding P.:
Multiple Pion Photoproduction in the s Channel Resonance Region.
STMP 59,39-76 (1971) PACS:13.60

Martin A.D.:
The Landa-KN Coupling and Extrapolation below the KN Threshold.
STMP 55,142-151 (1970) PACS:13.75

Martin B.R.:
Kaon-Nucleon Interactions below 1 GeV/c.
STMP 55,74-141 (1970) PACS:13.75

Morgan D., Pisut J.:
Low Energy Pion-Pion Scattering.
STMP 55,1-42 (1970) PACS:13.75

Oades G.C.:
Coulomb Corrections in the Analysis of pi-N Experimental Scattering Data.
STMP 55,61-72 (1970) PACS:13.75

Osborne L.S.:
Photoproduction of Mesons in the GeV Range.
STMP 39,91-105 (1965) PACS:13.60

Pfeil W., Schwela D.:
Coupling Parameters of Pseudoscalar Meson Photoproduction on Nucleons.
STMP 55,213-237 (1970) PACS:13.60

Pisut J.:
Analytic Extrapolations and the Determination of Pion-Pion Phase Shifts.
STMP 55,43-60 (1970) PACS:13.75

Renard F.M.:
Rho-Omega Mixing.
STMP 63,98-120 (1972) PACS:13.60

Rittenberg V.:
Scaling in Deep Inelastic Scattering with Fixed Final States.
STMP 62,92-97 (1972) PACS:13.60

Rollnik H., Stichel P.:
Compton Scattering.
STMP 79,1-52 (1976) PACS:13.60

Rothleitner J.:
Electromagnetic Mass Differences.
STMP 50,71-83 (1969) PACS:14.20

Rühl W.:
Application of Harmonic Analysis to Inelastic Electron-Proton Scattering.
STMP 57,202-221 (1971) PACS:13.60 11.00 11.30

Schildknecht D.:
Vector Meson Dominance, Photo- and Electroproduction from Nucleons.
STMP 63,57-97 (1972) PACS:13.60

Schwela D.:
Pion Photoproduction in the Region of the delta (1230) Resonance.
STMP 59,27-38 (1971) PACS:13.60

Smith C.H.L.:
Parton Models of Inelastic Lepton Scattering.
STMP 62,51-71 (1972) PACS:13.60

Swart J.J.de, Nagels M.M., Rijken T.A., Verhoeven P.A.:
Hyperon-Nucleon Interaction.
STMP 60,138-203 (1971) PACS:13.75 21.40

Wanders G.:
Analyticity, Unitarity and Crossing-Symmetry Constraints for Pion-Pion Partial Wave Amplitudes.
STMP 57,22-40 (1971) PACS:13.75

Wilson R.:
Review of Nucleon Form Factors.
STMP 39,43-54 (1965) PACS:13.40

Wolf G.:
Photoproduction of Vector Mesons.
STMP 59,77-113 (1971) PACS:13.60

Zinn-Justin J.:
Course on Padê Approximations.
STMP 57,248-270 (1971) PACS:13.75

## NUCLEAR PHYSICS

**21.00 Nuclear Structure**
**21.10 General Properties of Nuclei**
**21.30 Nuclear Forces**
**21.40 Few Nucleon Systems**
**21.65 Nuclear Matter**
**23.00 Weak Interactions**
**25.30 Lepton-Induced Reactions and Scattering**
**28.20 Neutron Physics**
**28.60 Isotope Separation and Enrichment**
**29.00 Experimental Methods**

Arenhövel H., Weber H.J.:
Nuclear Isobar Configurations.
STMP 65,58-91 (1972) PACS:21.00

Baltay C.:
Limits on Like-Sign Dilepton Production in Neutrino Interactions.
STMP 108,7-15 (1986) PACS:29.00

Baym G.:
Quark Matter and Nuclei.
STMP 100,186-213 (1982) PACS:21.65 12.35

Bjorken J.D.:
Electroproduction at Very Small Values of the Scaling Variable.
STMP 108,17-30 (1986) PACS:29.00

Blum W., May J., Rolandi L.:
Space Resolution of a Large Time Projection Chamber.
STMP 108,31-40 (1986) PACS:29.00

Cannata F., Oberall H.:
Giant Resonance Phenomena in Intermediate-Energy Nuclear Reactions.
STMP 89,1-112 (1980) PACS:21.00

Charpak G.:
Forecast on the Future of Gaseous Detectors.
STMP 108,41-48 (1986) PACS:29.00

Drell S.D.:
The Involvement of Scientists in National Security Issues.
STMP 108,49-69 (1986) PACS:29.00

Dydak F., Geweniger C.:
Neutral Currents.
STMP 108,71-93 (1986) PACS:29.00

Eisele F., Turlay R.:
How Well do we Know the Parton Distributions in the Nucleon from Deep Inelastic Scattering Experiments?
STMP 108,95-111 (1986) PACS:29.00

Ellis J.:
Physics at LEP.
STMP 108,113-122 (1986) PACS:29.00

Faessler A.:
How Should or Will QCD Influence Nuclear Physics
STMP 100,214-223 (1982) PACS:21.00 12.35

Fries D., Zeitnitz B. (eds.):
Quarks and Nuclear Forces.
STMP 100,1-223 (1982) PACS:21.00 12.35

Gaillard M.K.:
Future Prospects for Elementary Particle Physics: High and Low Energy Probes.
STMP 108,123-138 (1986) PACS:29.00

Henley E.M., Su R., Li M., Kuyucak S.:
Meson Masses in a Quark Model with Self-Consistent Pion Couplings.
STMP 108,139-147 (1986) PACS:29.00

Kleinknecht K., Lee T.D. (eds.):
Particles and Detectors. Festschrift for Jack Steinberger.
STMP 108,1-291 (1986) PACS:29.00

Kleinknecht K.:
Quark and Lepton Mixing in Weak Interactions.
STMP 108,149-164 (1986) PACS:29.00

Koester L.:
Neutron Scattering Lengths and Fundamental Neutron Interactions.
STMP 80,1-55 (1977) PACS:28.20

Lederman L.M.:
Physics with the Tevatron.
STMP 108,165-184 (1986) PACS:29.00

Lee T.D.:
Quantum Numbers of the Gluon.
STMP 108,185-189 (1986) PACS:29.00

Levinger J.S.:
The Two and Three Body Problem.
STMP 71,88-240 (1974) PACS:21.40

Nauenberg U.:
The Physics in the Z0 Energy Regime.
STMP 108,191-223 (1986) PACS:29.00

Panofsky W.K.H.:
Experimental Techniques.
STMP 39,138-154 (1965) PACS:29.00

Panofsky W.K.H.:
Technical Limits on Colliders for High Energy Physics.
STMP 108,225-235 (1986) PACS:29.00

Putlitz G.zu:
Determination of Nuclear Moments with Optical Double Resonance.
STMP 37,106-149 (1965) PACS:21.10 32.00

Racah G.:
Group Theory and Spectroscopy.
STMP 37,28-84 (1965) PACS:21.00 11.30 32.00

Rubbia C.:
Experimental Observation of Some of the Properties of the Intermediate Vector Bosons.
STMP 108,237-266 (1986) PACS:29.00

Samios N.P.:
Reminiscences of Columbia.
STMP 108,267-268 (1986) PACS:29.00

Schwartz M.:
Jack Steinberger - The Best Teacher I have Known.
STMP 108,269-274 (1986) PACS:29.00

Singer P.:
Emission of Particles Following Muon Capture in Intermediate and Heavy Nuclei.
STMP 71,39-87 (1974) PACS:21.00

Springer T.:
Quasielastic Neutron Scattering for the Investigation of Diffusive Motions in Solids and Liquids.
STMP 64,1-100 (1972) PACS:28.20 64.00

Steyerl A.:
Very Low Energy Neutrons.
STMP 80,57-130 (1977) PACS:28.20

Strauch K.:
The Use of Bubble Chambers and Spark Chambers at Electron Accelerators.
STMP 39,155-168 (1965) PACS:29.00

Theissen H.:
Spectroscopy of Light Nuclei by Low Energy (<70 MeV) Inelastic Electron Scattering.
STMP 65,1-57 (1972) PACS:25.30

Oberall H.:
Study of Nuclear Structure by Muon Capture.
STMP 71,1-38 (1974) PACS:21.00

Oberall H.:
Electron Scattering, Photoexcitation and Nuclear Models.
STMP 49,1-89 (1969) PACS:25.30

Wick G.C.:
J. Steinberger and the Early Days of Pion Physics.
STMP 108,275-278 (1986) PACS:29.00

Wildermuth K., McClure W.:
Cluster Representations of Nuclei.
STMP 41,1-172 (1966) PACS:21.00

Yang C.N.:
Is KNO Scaling Valid for Two Jet Events in e+e- Collisions?
STMP 108,279-283 (1986) PACS:29.00

**ATOMIC AND MOLECULAR PHYSICS**
**31.00 Electronic Structure of Atoms and Molecules, Theory**
**32.00 Atomic Spectra and Interactions with Photons**
**34.00 Atomic and Molecular Collisions with Photons**

Büttgenbach S.:
Hyperfine Structure in 4d- and 5d-Shell Atoms.
STMP 96,1-97 (1982) PACS:32.00

Dettmann K.:
High Energy Treatment of Atomic Collisions.
STMP 58,119-206 (1971) PACS:34.00

Donner W., Süßmann G.:
Paramagnetische Felder am Kernort.
STMP 37,1-27 (1965) PACS:31.00 21.10

Langbein D.:
Theory of van der Waals Attraction.
STMP 72,1-139 (1974) PACS:34.00 61.00

Seiwert R.:
Unelastische Stöße zwischen engeregten und unangeregten Atomen.
STMP 47,143-184 (1968) PACS:34.00

**CLASSICAL FIELDS OF PHENOMENA**
**41.70 Particles in Electromagnetic Fields**
**41.80 Particle Optics**
**42.00 Optics**
**42.50 Quantum Optics**
**42.72 Optical Sources**

Agarwal G.S.:
Quantum Statistical Theories of Spontaneous Emission and their Relation to Other Approaches.
STMP 70,1-129 (1974) PACS:42.50 05.00

Godwin R.P.:
Synchrotron Radiation as a Light Source.
STMP 51,1-73 (1969) PACS:42.72 07.85 78.70

Graham R.:
Statistical Theory of Instabilities in Stationary Nonequilibrium Systems with Applications to Lasers and Nonlinear Optics.
STMP 66,1-97 (1973) PACS:42.50 05.00

Haake F.:
Statistical Treatment of Open Systems by Generalized Master Equations.
STMP 66,98-168 (1973) PACS:42.50 05.00

Hawkes P.W.:
Quadrupole Optics. The Electron Optical Properties of Rectilinear Orthogonal Systems.
STMP 42,1-126 (1966) PACS:41.80

Olson C.L.:
Collective Ion Acceleration with Linear Electron Beams.
STMP 84,1-144 (1979) PACS:41.70

Schumacher U.:
Collective Ion Acceleration with Electron Rings.
STMP 84,145-231 (1979) PACS:41.70

**FLUIDS, PLASMA**

**47.00 Fluid Dynamics**

**51.00 Kinetics and Transport Theory of Fluids; Physical Properties of Gases**

Ehrfeld W.:
Elements of Flow and Diffusion Processes in Separation Nozzles.
STMP 97,1-140 (1983) PACS:47.00

Geiger W., Hornberg H., Schramm K.H.:
Zustand der Materie unter sehr hohen Drücken und Temperaturen.
STMP 46,1-52 (1968) PACS:51.00

Hess S.:
Depolarisierte Rayleigh-Streuung und Strömungs-Doppelbrechung in Gasen.
STMP 54,136-176 (1970) PACS:51.00

**CONDENSED MATTER, MECHANICAL AND THERMAL PROPERTIES**

**61.00 Structure of Liquids and Solids**

**61.12 Neutron Determination of Structures**

**62.00 Mechanical and Acoustical Properties**

**63.00 Lattice Dynamics**

**64.00 Equation of State, Phase Transitions**

**68.00 Surfaces and Interfaces**

Behringer J.:
Factor Group Analysis Revisited and Unified.
STMP 68,161-199 (1973) PACS:61.00

Dederichs P.H., Zeller R.:
Dynamical Properties of Point Defects in Metals.
STMP 87,1-170 (1980) PACS:61.00

Dorner B.:
Coherent Inelastic Neutron Scattering in Lattice Dynamics.
STMP 93,1-96 (1982) PACS:61.12 63.00

Engel T., Rieder K.-H.:
Structural Studies of Surfaces with Atomic and Molecular Beam Diffraction.
STMP 91,55-180 (1982) PACS:68.00

Heinz K., Müller K.:
LEED Intensities - Experimental Progress and New Possibilities of Surface Structure Determination.
STMP 91,1-53 (1982) PACS:68.00

Lacmann R.:
Die Gleichgewichtsform von Kristallen und die Keim bildungsarbeit bei der Kristallisation.
STMP 44,1-81 (1968) PACS:61.00

Lechner R.E., Riekel C.:
Applications of Neutron Scattering in Chemistry.
STMP 101,1-84 (1983) PACS:61.12 82.00

Leibfried G., Breuer N.:
Point Defects in Metals I, Introduction to the Theory.
STMP 81,1-342 (1978) PACS:61.00

Ludwig W.:
Recent Developments in Lattice Theory.
STMP 43,1-299 (1966) PACS:63.00

Press W.:
Single-Particle Rotations in Molecular Crystals.
STMP 92,1-129 (1981) PACS:61.12

Richter D.:
Transport Mechanisms of Light Interstitials in Metals.
STMP 101,85-222 (1983) PACS:66.30

Schramm K.H.:
Dynamisches Verhalten von Metallen unter Stoßwellenbelastung.
STMP 58,207-265 (1971) PACS:63.00

Schroeder K.:
Theory of Diffusion Controlled Reactions of Point Defects in Metals.
STMP 87,171-262 (1980) PACS:61.00

Steeb S.:
Evaluation of Atomic Distribution in Liquid Metals and Alloys by Means of X-Ray, Neutron, and Electron Diffraction.
STMP 47,2-66 (1968) PACS:61.00

**MAGNETIC AND OPTICAL PROPERTIES**

**71.00 Electron States**

Bagaev V.S.:
Properties of Electron-Hole Drops in Germanium Crystals.
STMP 73,72-90 (1975) PACS:71.00

Bendow B.:
Polariton Theory of Resonance Raman Scattering in Solids.
STMP 82,69-114 (1978) PACS:71.00

Bennemann K.H.:
A New Self-Consistent Treatment of Electrons in Crystals.
STMP 38,158-188 (1965) PACS:71.00

Daniels J.,Festenberg C.v.,Raether H.,Zeppenfeld K.:
Optical Constants of Solids by Electron Spectroscopy.
STMP 54,77-135 (1970) PACS:71.00 78.00

Dornhaus R., Nimtz G.:
The Properties and Applications of the Hg(1-x)Cd(x)Te Alloy System.
STMP 98,119-304 (1983) PACS:71.00 72.00 78.00

Dornhaus R., Nimtz G.:
The Properties and Applications of the Hg(1-x)Cd(x)Te Alloy System.
STMP 78,1-120 (1976) PACS:71.00 72.00 78.00

Forstmann F., Gerhardts R.R.:
Metal Optics Near the Plasma Frequency.
STMP 109,1-132 (1986) PACS:71.45 73.40 78.65

Grosse P.:
Die Festkörpereigenschaften von Tellur.
STMP 48,1-208 (1969) PACS:71.00 72.00

H. Haken, S. Nikitine (eds.): Excitons at High Densities.
STMP 73,1-298 (1975) PACS:71.00 78.00

Haken H.,J., Schenzle A.:
Polaritons at High Light Intensities and in Bose Condensed Exciton Systems.
STMP 73,285-295 (1975) PACS:71.00

Haken H., Nikitine S.:
Theory of Stimulated Emission by Excitons.
STMP 73,192-210 (1975) PACS:71.00 78.00

Hanamura E.:
Biexcitons - Bose Condensation and Optical Response.
STMP 73,43-69 (1975) PACS:71.00

Levy R., Bivas A., Grun J.B., Nikitine S.:
Interaction between Excitons at High Concentration
STMP 73,171-190 (1975) PACS:71.00 78.00

Mahr H.:
Medium and High Polariton Densities.
STMP 73,265-284 (1975) PACS:71.00

Nikitine S.:
Properties of Biexcitons.
STMP 73,18-42 (1975) PACS:71.00

Nikitine S.:
Introduction to Exciton Spectroscopy.
STMP 73,5-16 (1975) PACS:71.00

Nimtz G., Schlicht B.:
Narrow-Gap Lead Salts.
STMP 98,1-117 (1983) PACS:71.00 72.00 78.00

Novikov B.V.:
Spectroscopic Study of Exciton-Exciton Interaction (Biexcitons, Drops) in Semiconducting Crystals.
STMP 73,106-126 (1975) PACS:71.00

Pick H.:
Struktur von Störstellen in Alkalihalogenidkristallen.
STMP 38,2-83 (1965) PACS:71.00

Raether H.:
Solid State Excitations by Electrons. Plasma Oscillations and Single Electron Transitions.
STMP 38,85-157 (1965) PACS:71.00 78.00

Rashba E.I.:
Gigantic Oscillator Strengths Inherent in Exciton Complexes.
STMP 73,150-170 (1975) PACS:71.00

Rice T.M.:
Theory of Electron-Hole Drops in Germanium and Silicon.
STMP 73,91-104 (1975) PACS:71.00

Rogachev A.A.:
Exciton Condensation in Germanium.
STMP 73,129-148 (1975) PACS:71.00

Schwentner N., Koch E.-E., Jortner J.:
Electronic Excitations in Condensed Rare Gases.
STMP 107,1-239 (1985) PACS:71.00 71.25 71.35

Shaklee K.L.:
Experimental Studies of Excitons at High Densities
STMP 73,222-240 (1975) PACS:71.00

Stahl A., Balslev I.:
Electrodynamics of the Semiconductor Band Edge.
STMP 110,1-205 (1987) PACS:71.35 72.20 78.20

**72.00 Electronic Transport in Condensed Matter**
**73.00 Properties of Surfaces and Thin Films**
**74.00 Superconductivity**
**75.00 Magnetic Properties**
**76.00 Magnetic Resonances and Relaxation**
**77.00 Dielectric Properties and Materials**

Bauer G.:
Determination of Electron Temperatures and of Hot Electron Distribution Functions in Semiconductors.
STMP 74,1-106 (1974) PACS:72.00

Bussmann-Holder, Bilz H., Vogt P.:
Electronic and Dynamical Properties of IV - VI Compounds.
STMP 99,51-98 (1983) PACS:77.00 64.00 78.00

Coufal H., Lüscher E., Micklitz H.:
Electron Spin and Nuclear Gamma Resonance Studies of Rare Gas Matrix-Isolated Atoms and Ions.
STMP 103,1-57 (1984) PACS:76.80 76.30

Feitknecht J.:
Silicon Carbide as a Semiconductor.
STMP 58,48-118 (1971) PACS:72.00

Fischer K.:
Magnetic Impurities in Metals: The s-d Exchange Model.
STMP 54,1-76 (1970) PACS:75.00

Hölzl J., Schulte F.K.:
Work Function of Metals.
STMP 85,1-150 (1979) PACS:73.00

Jantsch W.:
Dielectric Properties and Soft Modes in Semiconducting (Pb, Sn, Ge)Te.
STMP 99,1-50 (1983) PACS:77.00 64.00 78.00

Lüders G., Usadel K.-D.:
The Method of the Correlation Function in Superconductivity Theory.
STMP 56,1-215 (1971) PACS:74.00

Müller K.:
How Much can Auger Electrons Tell us about Solid Surfaces?
STMP 77,97-125 (1975) PACS:73.00

Norberg R.E.:
Nuclear Magnetic Resonance in Condensed Rare Gases
STMP 103,59-95 (1984) PACS:76.60

Raether H.:
Excitation of Plasmons and Interband Transitions by Electrons.
STMP 88,1-196 (1980) PACS:73.00

Schmid D.:
Nuclear Magnetic Double-Resonance - Principles and Applications in Solid-State Physics.
STMP 68,1-75 (1973) PACS:76.70 21.10 71.00

Schnakenberg J.:
Electron-Phonon Interaction and the Boltzmann Equation in Narrow-Band Semiconductors.
STMP 51,74-120 (1969) PACS:72.00

Stierstadt K.:
Der Magnetische Barkhausen-Effekt.
STMP 40,3-106 (1966) PACS:75.00

Ullmaier H.:
Irreversible Properties of Type II Superconductors
STMP 76,1-165 (1975) PACS:74.00

Wagner H.:
Physical and Chemical Properties of Stepped Surfaces.
STMP 85,151-221 (1979) PACS:73.00

Wissmann P.:
The Electrical Resistivity of Pure and Gas Covered Metal Films.
STMP 77,1-96 (1975) PACS:73.00

**78.00 Optical Properties**

Bäuerle D.:
Vibrational Spectra of Electron and Hydrogen Centers in Ionic Crystals.
STMP 68,76-160 (1973) PACS:78.00

Borstel G., Falge H.J., Otto A.:
Surface and Bulk Phonon-Polaritons Observed by Attenuated Total Refelction.
STMP 74,107-148 (1974) PACS:78.00 71.00

Claus R.,Merten L., Brandmüller J.:
Light Scattering by Phonon-Polaritons.
STMP 75,1-237 (1975) PACS:78.00 71.00

Grossmann M., Bielmann J., Nikitine S.:
Tests of Validity of Spatial Dispersion Theories on Lead Iodide Crystal Spectra.
STMP 73,243-264 (1975) PACS:78.20

Lengeler B.:
De Haas - van Alphen Studies of the Electronic Structure of the Noble Metals and Their Dilute Alloys.
STMP 82,1-67 (1978) PACS:78.00

Levy R., Grun J.B., Nikitine S.:
Experimental Investigation of the Competition of Stimulated Emission Involving Exiton.
STMP 73,211-219 (1975) PACS:78.00 71.00

Pockrand I.:
Surface Enhanced Raman Vibrational Studies at Solid/Gas Interfaces.
STMP 104,1-164 (1984) PACS:78.30 68.10

Richter W.:
Resonant Raman Scattering in Semiconductors.
STMP 78,121-272 (1976) PACSd:78.00

**79.00 Electron and Ion Emission by Liquids and Solids; Impact Phenomena**

Kirschner J.:
Polarized Electrons at Surfaces.
STMP 106,1-158 (1985) PACS:79.20

**RELATED AREAS OF SCIENCE AND TECHNOLOGY**
**85.70 Magnetic Devices**

Lehner G.:
Ober die Grenzen der Erzeugung sehr hoher Magnetfelder.
STMP 47,67-110 (1968) PACS:85.70

**GEOPHYSICS, ASTRONOMY, AND ASTROPHYSICS**
**95.00 Theoretical Astrophysics**
**97.00 Stars**

Börner G.:
On the Properties of Matter in Neutron Stars.
STMP 69,1-67 (1973) PACS:97.00

Kundt W.:
Survey of Cosmology. Is "Our World" Implied by Thermal Equilibrium in the Hadron Era?
STMP 58,1-47 (1971) PACS:95.00

Kundt W.:
Recent Progress in Cosmology. Isotropy of the 3 deg Background Radiation, and the Occurrence of Spacetime Singularities.
STMP 47,111-142 (1968) PACS:95.00

Stewart J., Walker M.:
Black Holes: the Outside Story.
STMP 69,69-115 (1973) PACS:97.60 04.00

# Author Index

## Springer Tracts in Modern Physics, Volume 36–110

Alphabetical list of all contributors together with volume number, page number and year of publication (bracketed) 19XX) in Springer Tracts Mod. Phys. (Sp. Tr. Mod. Phys.).

| Name | First Author | Sp.Tr.Mod.Phys. | | PACS |
|---|---|---|---|---|
| Jackiw R. | Jackiw R. | 62,1-36 | (72) | 11.00 |
| Jantsch W. | Jantsch W. | 99,1-50 | (83) | 77.00 |
| Jortner J. | Schwentner N. | 107,1-239 | (85) | 71.00 |
| Julius D. | Ebel G. | 55,239-290 | (70) | 13.75 |
| Kabir P.K. | Kabir P.K. | 52,91-112 | (70) | 12.30 |
| Källen G. | Källen G. | 46,67-132 | (68) | 12.20 |
| Kenkre V.M. | Kenkre V.M. | 94,1-109 | (82) | 05.00 |
| Kirschner J. | Kirschner J. | 106,1-158 | (85) | 79.20 |
| Kleiner H. | Kleiner H. | 49,90-146 | (69) | 11.40 |
| Kleinknecht K. | Kleinknecht K. | 108,1-291 | (86) | 29.00 |
| Kleinknecht K. | Kleinknecht K. | 108,149-164 | (86) | 29.00 |
| Koch E.-E. | Schwentner N. | 107,1-239 | (85) | 71.00 |
| Koester L. | Koester L. | 80,1-55 | (77) | 28.20 |
| Kolanoski H. | Kolanoski H. | 105,1-187 | (84) | 13.65 |
| Kramer G. | Kramer G. | 55,152-167 | (70) | 13.75 |
| Kramer G. | Kramer G. | 102,1-140 | (84) | 13.65 |
| Kummer W. | Kummer W. | 52,113-125 | (70) | 12.30 |
| Kundt W. | Kundt W. | 40,108-168 | (66) | 11.00 |
| Kundt W. | Kundt W. | 47,111-142 | (68) | 95.00 |
| Kundt W. | Kundt W. | 58,1-47 | (71) | 95.00 |
| Kuyucak S. | Henley E.M. | 108,139-147 | (86) | 29.00 |
| Lacmann R. | Lacmann R. | 44,1-81 | (68) | 61.00 |
| Landshoff P.V. | Landshoff P.V. | 62,37-50 | (72) | 13.60 |
| Langbein D. | Langbein D. | 72,1-139 | (74) | 34.00 |
| Lechner R.E. | Lechner R.E. | 101,1-84 | (83) | 61.12 |
| Lederman L.M. | Lederman L.M. | 108,165-184 | (86) | 29.00 |
| Lee T.D. | Kleinknecht K. | 108,1-291 | (86) | 29.00 |
| Lee T.D. | Lee T.D. | 108,185-189 | (86) | 29.00 |
| Lehner G. | Lehner G. | 47,67-110 | (68) | 85.70 |
| Leibfried G. | Leibfried G. | 81,1-342 | (78) | 61.00 |
| Lengeler B. | Lengeler B. | 82,1-67 | (78) | 78.00 |
| Leutwyler H. | Leutwyler H. | 50,29-41 | (69) | 11.40 |
| Levinger J.S. | Levinger J.S. | 71,88-240 | (74) | 21.40 |
| Levy R. | Levy R. | 73,171-190 | (75) | 71.00 |
| Levy R. | Levy R. | 73,211-219 | (75) | 78.00 |
| Li M. | Henley E.M. | 108,139-147 | (86) | 29.00 |
| Lopuszanski J.T. | Lopuszanski J.T. | 52,201-214 | (70) | 11.30 |
| Lüders G. | Lüders G. | 56,1-215 | (71) | 74.00 |
| Ludwig W. | Ludwig W. | 43,1-299 | (66) | 63.00 |
| Lüke D. | Lüke D. | 59,39-76 | (71) | 13.60 |
| Lüscher E. | Coufal H. | 103,1-57 | (84) | 76.80 |
| Mahr H. | Mahr H. | 73,265-284 | (75) | 71.00 |
| Martin A.D. | Martin A.D. | 55,142-151 | (70) | 13.75 |
| Martin B.R. | Martin B.R. | 55,74-141 | (70) | 13.75 |
| May J. | Blum W. | 108,31-40 | (86) | 29.00 |
| McClure W. | Wildermuth K. | 41,1-172 | (66) | 21.00 |
| Mendes R.V. | Mendes R.V. | 60,18-31 | (71) | 11.40 |
| Merten L. | Claus R. | 75,1-237 | (75) | 78.00 |
| Michael C. | Michael C. | 55,174-191 | (70) | 12.40 |
| Micklitz H. | Coufal H. | 103,1-57 | (84) | 76.80 |
| Mittelstaedt P. | Heintzmann H. | 47,185-225 | (68) | 04.00 |
| Morgan D. | Morgan D. | 55,1-42 | (70) | 13.75 |
| Müllensiefen A. | Ebel G. | 55,239-290 | (70) | 13.75 |
| Müller K. | Müller K. | 77,97-125 | (75) | 73.00 |
| Müller K. | Heinz K. | 91,1-53 | (82) | 68.00 |
| Müller V.F. | Müller V.F. | 50,42-52 | (69) | 11.40 |
| Müller V.F. | Müller V.F. | 52,34-49 | (70) | 12.30 |
| Nagels M.M. | Swart J.J.de | 60,138-203 | (71) | 13.75 |
| Nauenberg U. | Nauenberg U. | 108,191-223 | (86) | 29.00 |
| Ne'eman Y. | Mendes R.V. | 60,18-31 | (71) | 11.40 |
| Nikitine S. | Nikitine S. | 73,5-16 | (75) | 71.00 |
| Nikitine S. | Nikitine S. | 73,18-42 | (75) | 71.00 |
| Nikitine S. | Levy R. | 73,171-190 | (75) | 71.00 |
| Nikitine S. | Haken H. | 73,192-210 | (75) | 71.00 |
| Nikitine S. | Levy R. | 73,211-219 | (75) | 78.00 |
| Nikitine S. | Grossmann M. | 73,243-264 | (75) | 78.20 |
| Nimtz G. | Dornhaus R. | 78,1-120 | (76) | 71.00 |
| Nimtz G. | Nimtz G. | 98,1-117 | (83) | 71.00 |
| Nimtz G. | Dornhaus R. | 98,119-304 | (83) | 71.00 |
| Norberg R.E. | Norberg R.E. | 103,59-95 | (84) | 76.60 |
| Novikov B.V. | Novikov B.V. | 73,106-126 | (75) | 71.00 |
| Oades G.C. | Oades G.C. | 55,61-72 | (70) | 13.75 |
| Oehme R. | Oehme R. | 57,119-131 | (71) | 12.40 |
| Oehme R. | Oehme R. | 57,132-157 | (71) | 12.40 |
| Oehme R. | Oehme R. | 61,109-119 | (72) | 12.40 |
| Olson C.L. | Olson C.L. | 84,1-144 | (79) | 41.70 |
| Olsen H.A. | Olsen H.A. | 44,84-201 | (68) | 12.20 |
| Osborne L.S. | Osborne L.S. | 39,91-105 | (65) | 13.60 |
| Otto A. | Borstel G. | 74,107-148 | (74) | 78.00 |
| Panofsky W.K.H. | Panofsky W.K.H. | 39,138-154 | (65) | 29.00 |
| Panofsky W.K.H. | Panofsky W.K.H. | 108,225-235 | (86) | 29.00 |

| Name | First Author | Sp.Tr.Mod.Phys. | | PACS |
|---|---|---|---|---|
| Paul E. | Paul E. | 79,53-145 | (76) | 12.30 |
| Pauli W. | Pauli W. | 37,85-104 | (65) | 11.30 |
| Paver N. | Furlan G. | 62,118-147 | (72) | 13.60 |
| Pfeil W. | Pfeil W. | 55,213-237 | (70) | 13.60 |
| Pick H. | Pick H. | 38,2-83 | (65) | 71.00 |
| Pietschmann H. | Pietschmann H. | 50,53-64 | (69) | 11.40 |
| Pietschmann H. | Pietschmann H. | 52,193-200 | (70) | 12.30 |
| Pilkuhn H. | Pilkuhn H. | 50,65-70 | (69) | 11.40 |
| Pilkuhn H. | Pilkuhn H. | 55,168-173 | (70) | 11.40 |
| Pilkuhn H. | Ebel G. | 55,239-290 | (70) | 13.75 |
| Pisut J. | Morgan D. | 55,1-42 | (70) | 13.75 |
| Pisut J. | Pisut J. | 55,43-60 | (70) | 13.75 |
| Pockrand I. | Pockrand I. | 104,1-164 | (84) | 78.30 |
| Press W. | Press W. | 92,1-129 | (81) | 61.12 |
| Primakoff H. | Primakoff H. | 53,6-28 | (70) | 12.30 |
| Putlitz G.zu | Putlitz G.zu | 37,106-149 | (65) | 21.10 |
| Racah G. | Racah G. | 37,28-84 | (65) | 21.00 |
| Raether H. | Raether H. | 38,85-157 | (65) | 71.00 |
| Raether H. | Daniels J. | 54,77-135 | (70) | 71.00 |
| Raether H. | Raether H. | 88,1-196 | (80) | 73.00 |
| Rashba E.I. | Rashba E.I. | 73,150-170 | (75) | 71.00 |
| Ravenhall D.G. | Buchanan C.D. | 39,20-42 | (65) | 13.40 |
| Reineker P. | Reineker P. | 94,111-226 | (82) | 05.00 |
| Renard F.M. | Renard F.M. | 63,98-120 | (72) | 13.60 |
| Renner B. | Renner B. | 52,60-78 | (70) | 11.40 |
| Renner B. | Renner B. | 61,120-129 | (72) | 11.40 |
| Riazzuddin | Riazzuddin | 52,126-160 | (70) | 12.30 |
| Rice T.M. | Rice T.M. | 73,91-104 | (75) | 71.00 |
| Richter D. | Richter D. | 101,85-222 | (83) | 66.30 |
| Richter W. | Richter W. | 78,121-272 | (76) | 78.00 |
| Rieder K.-H. | Engel T. | 91,55-180 | (82) | 68.00 |
| Riekel C. | Lechner R.E. | 101,1-84 | (83) | 61.12 |
| Rijken T.A. | Swart J.J.de | 60,138-203 | (71) | 13.75 |
| Rittenberg V. | Rittenberg V. | 62,92-97 | (72) | 13.60 |
| Rogachev A.A. | Rogachev A.A. | 73,129-148 | (75) | 71.00 |
| Rolandi L. | Blum W. | 108,31-40 | (86) | 29.00 |
| Rollnik H. | Rollnik H. | 79,1-52 | (76) | 13.60 |
| Rothleitner J. | Rothleitner J. | 50,71-83 | (69) | 14.20 |
| Rothleitner J. | Rothleitner J. | 52,161-170 | (70) | 12.30 |
| Rubbia C. | Rubbia C. | 108,237-266 | (86) | 29.00 |
| Rubinstein H.R. | Rubinstein H.R. | 57,193-201 | (71) | 12.40 |
| Rubinstein H.R. | Rubinstein H.R. | 62,72-91 | (72) | 12.40 |
| Rühl W. | Rühl W. | 57,202-221 | (71) | 13.60 |
| Samios N.P. | Samios N.P. | 108,267-268 | (86) | 29.00 |
| Satz H. | Satz H. | 57,158-190 | (71) | 12.40 |
| Schenzle A. | Haken H. | 73,285-295 | (75) | 71.00 |
| Schildknecht D. | Schildknecht D. | 63,57-97 | (72) | 13.60 |
| Schilling K. | Schilling K. | 63,31-56 | (72) | 12.60 |
| Schlicht B. | Nimtz G. | 98,1-117 | (83) | 71.00 |
| Schmid D. | Schmid D. | 68,1-75 | (73) | 76.70 |
| Schmidt W. | Ebel G. | 55,239-290 | (70) | 13.75 |
| Schnakenberg J. | Schnakenberg J. | 51,74-120 | (69) | 72.00 |
| Schramm K.H. | Geiger W. | 46,1-52 | (68) | 51.00 |
| Schramm K.H. | Schramm K.H. | 58,207-265 | (71) | 63.00 |
| Schrempp-Otto B. | Schrempp-Otto B. | 61,68-108 | (72) | 12.40 |
| Schrempp F. | Schrempp-Otto B. | 61,68-108 | (72) | 12.40 |
| Schroeder K. | Schroeder K. | 87,171-262 | (80) | 61.00 |
| Schulte F.K. | Hölzl J. | 85,1-150 | (79) | 73.00 |
| Schumacher U. | Schumacher U. | 84,145-231 | (79) | 41.70 |
| Schwartz M. | Schwartz M. | 108,269-274 | (86) | 29.00 |
| Schwela D. | Pfeil W. | 55,213-237 | (70) | 13.60 |
| Schwela D. | Schwela D. | 59,27-38 | (71) | 13.60 |
| Schwentner N. | Schwentner N. | 107,1-239 | (85) | 71.00 |
| Segre G. | Segre G. | 52,171-192 | (70) | 12.30 |
| Seiwert R. | Seiwert R. | 47,143-184 | (68) | 34.00 |
| Shaklee K.L. | Shaklee K.L. | 73,222-240 | (75) | 71.00 |
| Singer P. | Singer P. | 71,39-87 | (74) | 21.00 |
| Smith C.H.L. | Smith C.H.L. | 62,51-71 | (72) | 13.60 |
| Söding P. | Lüke D. | 59,39-76 | (71) | 13.60 |
| Soloviev L.D. | Soloviev L.D. | 46,53-66 | (68) | 11.40 |
| Springer T. | Springer T. | 64,1-100 | (72) | 28.20 |
| Squires E.J. | Collins P.D.B. | 45,1-292 | (68) | 12.40 |
| Squires E.J. | Squires E.J. | 57,71-91 | (71) | 12.40 |
| Stahl A. | Stahl A. | 110,1-205 | (87) | 71.35 |
| Stech B. | Stech B. | 50,84-99 | (69) | 11.40 |
| Stech B. | Stech B. | 52,50-59 | (70) | 12.30 |
| Steeb S. | Steeb S. | 47,2-66 | (68) | 61.00 |
| Stewart J. | Stewart J. | 69,69-115 | (73) | 97.60 |
| Steyerl A. | Steyerl A. | 80,57-130 | (77) | 28.20 |
| Stichel P. | Stichel P. | 50,100-109 | (69) | 11.40 |
| Stichel P. | Stichel P. | 50,110-119 | (69) | 11.40 |
| Stichel P. | Stichel P. | 50,120-131 | (69) | 11.40 |
| Stichel P. | Rollnik H. | 79,1-52 | (76) | 13.60 |
| Stierstadt K. | Stierstadt K. | 40,3-106 | (66) | 75.00 |
| Strauch K. | Strauch K. | 39,155-168 | (65) | 29.00 |

| Name | First Author | Sp.Tr.Mod.Phys. | PACS |
|---|---|---|---|
| Su R. | Henley E.M. | 108,139-147 (86) | 29.00 |
| Süßmann G. | Donner W. | 37,1-27 (65) | 31.00 |
| Swart J.J.de | Ebel G. | 55,239-290 (70) | 13.75 |
| Swart J.J.de | Swart J.J.de | 60,138-203 (71) | 13.75 |
| Symanzik K. | Symanzik K. | 57,222-236 (71) | 11.00 |
| Tan C.-I. | Tan C.-I. | 60,91-106 (71) | 12.40 |
| Theissen H. | Theissen H. | 65,1-57 (72) | 25.30 |
| Turlay R. | Eisele F. | 108,95-111 (86) | 29.00 |
| Oberall H. | Oberall H. | 49,1-89 (69) | 25.30 |
| Oberall H. | Oberall H. | 71,1-38 (74) | 21.00 |
| Oberall H. | Cannata F. | 89,1-112 (80) | 21.00 |
| Ullmaier H. | Ullmaier H. | 76,1-165 (75) | 74.00 |
| Usadel K.-D. | Lüders G. | 56,1-215 (71) | 74.00 |
| Verhoeven P.A. | Swart J.J.de | 60,138-203 (71) | 13.75 |
| Verzegnassi C. | Verzegnassi C. | 59,154-163 (71) | 11.40 |
| Verzegnassi C. | Furlan G. | 62,118-147 (72) | 13.60 |
| Vogt P. | Bussmann-Holder | 99,51-98 (83) | 77.00 |
| Wagner H. | Wagner H. | 85,151-221 (79) | 73.00 |
| Walker M. | Stewart J. | 69,69-115 (73) | 97.60 |
| Wanders G. | Wanders G. | 57,22-40 (71) | 13.75 |
| Weber H.J. | Arenhövel H. | 65,58-91 (72) | 21.00 |
| Weinstein M. | Weinstein M. | 60,32-73 (71) | 11.40 |
| Wess J. | Wess J. | 50,132-142 (69) | 11.30 |
| Wess J. | Wess J. | 60,1-17 (71) | 11.30 |
| Wick G.C. | Wick G.C. | 108,275-278 (86) | 29.00 |
| Wiik B.H. | Wiik B.H. | 86,1-262 (79) | 12.00 |
| Wildermuth K. | Wildermuth K. | 41,1-172 (66) | 21.00 |
| Wilson R. | Wilson R. | 39,43-54 (65) | 13.40 |
| Wissmann P. | Wissmann P. | 77,1-96 (75) | 73.00 |
| Wolf G. | Wolf G. | 59,77-113 (71) | 13.60 |
| Wolf G. | Wiik B.H. | 86,1-262 (79) | 12.00 |
| Yang C.N. | Yang C.N. | 108,279-283 (86) | 29.00 |
| Zeitnitz B. | Fries D. | 100,1-223 (82) | 21.00 |
| Zeller R. | Dederichs P.H. | 87,1-170 (80) | 61.00 |
| Zeppenfeld K. | Daniels J. | 54,77-135 (70) | 71.00 |
| Zimmermann W. | Zimmermann W. | 50,143-156 (69) | 11.00 |
| Zinn-Justin J. | Zinn-Justin J. | 57,248-270 (71) | 13.75 |